I0797233

FROM
VIRILE
to
Sterile

SCIENCE AND CULTURE IN THE NINETEENTH CENTURY

Bernard Lightman, *Editor*

UNIVERSITY *of* PITTSBURGH PRESS

FROM VIRILE

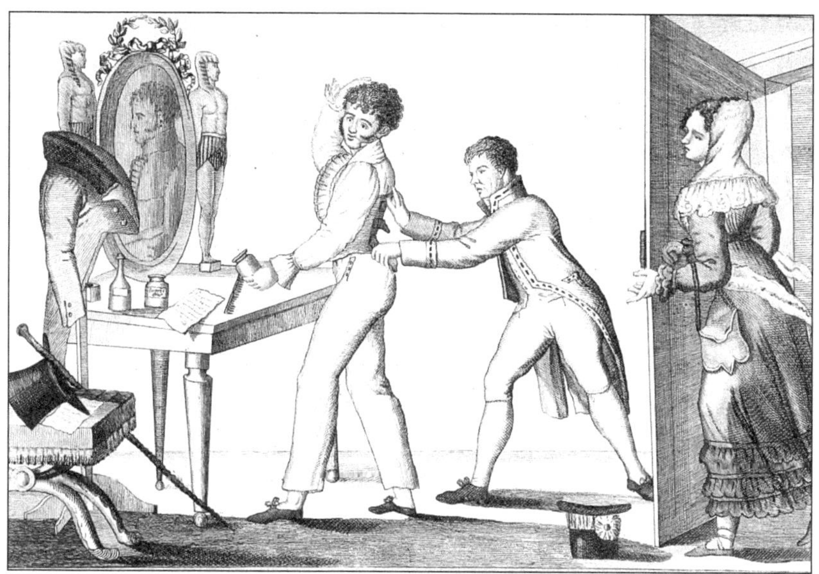

to Sterile

SCIENCE, MASCULINITY, AND MODERNITY IN ARGENTINA, 1776–1852

ADRIANA NOVOA

Published by the University of Pittsburgh Press, Pittsburgh, Pa., 15260

Manufactured in the United States of America
Printed on acid-free paper
10 9 8 7 6 5 4 3 2 1

Cataloging-in-Publication data is available from the Library of Congress

Hardcover: 978-0-8229-4852-0
Paperback: 978-0-8229-6777-4

Cover art: "The Elegant *Petrimetre* or the *Lechugino* in His Dresser," 1770–1800s. Courtesy of Museo de la Historia de Madrid. Inventory no. 4916.

Cover design: Alex Wolfe

Publisher: University of Pittsburgh Press, 7500 Thomas Blvd., 4th floor, Pittsburgh, PA 15260, United States, www.upittpress.org

EU Authorized Representative: Easy Access System Europe, Mustamäe tee 50, 10621 Tallinn, Estonia, gpsr.requests@easproject.com

To Maitis, who makes me believe every day that I have sown the future.

CONTENTS

ACKNOWLEDGMENTS

This book took years to complete, involving many people at various stages of the lengthy research and writing process. My life is enriched by a remarkable network of loving, brilliant, and loyal people, without whom this work would not have been possible. Gratitude feels essential, especially given the profound challenges of being an academic in Florida in recent years. The loss of academic freedom, institutional autonomy, and tenure compelled me to become a plaintiff in two ongoing lawsuits, thanks to the dedicated work of attorneys Greg Greubel and Gary Edinger. Balancing research with the urgent fight against censorship and a rancid dogma that I have not heard this loudly since 1983 has been difficult, making their efforts even more vital.

I am deeply thankful to Bernard Lightman and Abby McAllister of the University of Pittsburgh Press for their patience, encouragement, and support in helping me bring this manuscript to completion. Intellectually, I owe much to extraordinary mentors in Argentina and the US, including Francis Korn, Torcuato di Tella, and Eric Van Young. Eric's superb scholarship on Mexican liberalism often guided my work, and his example gave me the confidence to tackle a project of this scope, for which, regardless of its reception, I hope I will make him proud.

Eileen Kahl's great editorial support enabled me to complete the first manuscript for submission. The Humanities Institute at the University of South Florida (USF) supported my research with a 2022 summer grant, and additional backing came from the Dean of the College of Arts and Sciences and the university's Library Committee. I am grateful to my USF colleagues and friends, including Golfo Alexopoulos, Michael Decker, Davide Tanasi, Denise Cali, Scott Perry, Dan Belgrad, Kees Boterbloem, David Johnson, Brian Connolly, Roger and Susie Ariew, whose encouragement helped me to complete research over the years.

I am especially thankful to Mónica Szurmuk, whose insights shaped many chapters, and to Patricia Chomnalez, whose humor and wisdom lightened the challenges of academia. Mariana Ortega deserves special recognition for encouraging me to pursue the core arguments of this book decades ago.

My Tampa support system—Pablo Brescia, Anat and Max Pollack, Christina Richards, Rosario Hidalgo, Marcela van Olphen, Tamara Nemirosvky, and their families—has been indispensable. Deep appreciation to Beatriz Padilla for inspiring me to meet my walking goals while patiently listening to my venting. To my "sistras" Sandra Jaramillo, Naomi Yavneh, Madeline Cámara, Mabel Cuesta, Neysi Romero, and Sonia Labrador, thank you for your unwavering camaraderie.

Finally, my siblings in Argentina and Spain, Mónica, Susana, Javier, and Laura, along with my mother, María Solla Louro, have supported me during research visits, offering their homes, food, and care. Watching the next generation grow brings me joy, pride, and a sense of possibility; for this, I am grateful to the Novoa-Laclau, Bruzzesse-Novoa, Levine-Kanetani, Levine-Mendieta, Calabria-Chomnalez, Brescia-Labrador, Bergman, Susmel, Ramos, Gottschamer, Padilla Correia, and the Ruiz, Shattenkirk, and Goldberg families.

I met Megan Altman, Andrea Pitts, Elisabeth Paquette, Stephanie Rivera, and Robert Koch when they were all students, and they are now inspiring presences in my life. Andrea provided critical feedback on early drafts that made me reshape my project.

Alex Levine read every draft, offered translations, and provided invaluable guidance throughout the process. Our decades of collaboration remain my greatest intellectual joy. Sharing our daughter Maia, whose presence radiates joy is our most cherished achievement. I hope she will forgive the plans I canceled to finish this book and find the result worth the sacrifice, as nothing brings me greater happiness than spending time with her.

FROM
VIRILE
to
Sterile

INTRODUCTION

What Do Modern Men Want?

This book is the result of many years of research toward understanding the complex gendering of modernity and civilization in Argentina's history.[1] The existence of a tension that resulted from competitive ideas about masculinity in the context of building the modern nation was part of a revolutionary understanding of gender that began in the late colonial period. The new male subject that emerged in the last quarter of the eighteenth century was formed by contradictory views that arose from political, philosophical, and scientific ideas organized within a system that was understood as universal and encompassing all aspects of society. The failure of these systems to achieve political stability and coherence produced constant ideological renewals that are analyzed in the different chapters.

My goal in this book is to analyze how universal characterizations of masculinity shaped the politics of the River Plate Viceroyalty and later the creation of the Argentine Republic based on scientific and philosophical ideas that were considered modern and gendered. I need to clarify, though, that I did not write a history of masculinity, but the history of a relationship that involves masculinity. I am interested in analyzing how and why gender, knowledge, and politics were connected in the formation of Argentina to shape the modern culture that resulted after the end of colonialism. This historical process was characterized by a simultaneity made possible by networks that allowed the spread of knowledge across multiple locations, lending credence to the notion of universality, a concept to which I return throughout the book.

The association between universality and Enlightenment culture is well established, but less understood is how the men who embraced its scientific and philosophical ideas experienced universality during a time of rapid change, including the failure of republicanism in Europe and the transformation of science, which eventually separated from philosophy. Despite these ideological collapses that exposed its limitations, this ideology remained relevant by fostering shared forms of association and sociability that supported political transformation grounded in ideas believed to be as precise as the science and philosophy that inspired them. This experience was tangible, widely communicated, and sustained by transnational networks of philosophers and scientists who also served as political leaders.

This book examines the history of that experience in Argentina, exploring the models followed by enlightened men over time in the Americas and Europe, as well as the failure to establish a stable national model due to the constant emergence and expiration of new ideas. This process explains the title *From Virile to Sterile*, which reflects Argentina's transition from a culture that embraced a feminized modern virility to one that cast the virile man as a sterile force in the formation of the modern nation.

This phenomenon cannot be fully understood by focusing solely on the internal political life of a single country. Instead, it becomes evident when examining the external context, which underscored the power of universality to make men equal through interconnected knowledge. The awareness that their actions mirrored those occurring simultaneously in Europe and the Americas offered the advocates of the Enlightenment compelling proof that humanity had entered a new historical age. This simultaneity profoundly transformed individuals and reaffirmed the existence of a new historical force that brought men up to date through knowledge and modern sociability. Each chapter of this book examines the simultaneous experiences that influenced the development of modern Argentina.

In brief, I do not assert the existence of a universal, uniform man, but demonstrate the concerted effort to forge one through the process of nation building, which gave rise to a new male identity grounded in a shared culture of sociability and association, allegedly emancipating men from archaic societal norms through the power of knowledge. As elucidated in various chapters, this masculinity was predominantly representational, and the associated representations were understood as universal, which over time clashed with the very purpose of building a new nation. The formation of Creoles and Argentines that led the nation-building process was the result of experiencing the Enlightenment as a global culture. Each chapter analyzes the changing international context with which these men were interacting, and the resulting philosophical and scientific ideas that changed or were changed by different models of masculinity.

In the case of Argentina, the tortuous path to change began with the expectation of forming a republic based on a philosophical system that through science would regulate all aspects of society ended up in a civil war won by conservatives that banned the teaching of science and liberal ideas in the name of natural law, setting limits to what ideology could achieve on its own.

Thomas Abercrombie affirms that by 1800, thinking through the "sex/gender system which separates gender as a performative social construct from 'biological' sex, did not exist," and it would be a mistake to apply it to this era.[2] While it is clear that during the period covered the normative social behavior and socialization for individuals of the masculine sex (*sexo masculino*) was believed to be either created by God or natural, it is also evident that a strict association between being a man and doing strictly manly things became more elusive due to the effect of the scientific revolution and the ideological changes that followed. In Latin America, the enlightened elite believed that nature was finally tamed by human will and that men were less natural as a result, because they had achieved an understanding and manipulation of its laws. As a result, in Spanish America, the natural world began to be treated as an instrument rather than a subject, and this included men's bodies. The emphasis on freedom and individualism, the rise of consumerism, new scientific ideas, and the adoption of feminine socialization collectively expanded the meaning of what it was to be a man.

Finally, I need to make some clarifications. While "Illustration" is the preferred term among scholars of Spain and Spanish America, I have chosen to use "Spanish Enlightenment" to align with the shared objectives of this intellectual movement across Europe and the Americas and to avoid narrative confusion. It is also important to explain that the chronological order followed in the book does not imply that the ideological changes described began and ended on precise dates and developed in vertical succession; ideas coexisted over time and, in some cases, emerged or reemerged in a complex, chaotic manner. Presenting them in a strict sequential order is a methodological choice to help readers identify trends, but a simple or linear development should not be assumed—on the contrary, continuities and discontinuities characterized each ideological path addressed.

SCIENTIFIC, POLITICAL, AND GENDER REVOLUTION IN ARGENTINA

The chapters that follow this introduction are part of a contextual and cultural history of the ideas that formed a man aligned with the ideas of the Enlightenment in Argentina from the 1770s to 1850s. While I am a cultural historian and my understanding of masculinity is defined by this field, the definition of culture that guides me comes from the anthropologist Clifford Geertz, whose work I started to read and apply to gender thanks to my friend, and a historian of masculinity, John Pettegrew. After many discussions, I came to agree with him about the advantage that Geertz's ideas presented for historians like us trying to determine the meaning of a very complex idea like masculinity. In John's words, in appropriating "this understanding of culture as humanly woven 'webs of significance,' the historian comes to the meaning of a symbolic act by 'uncover[ing] the conceptual structures' that inform it."[3] For Geertz, semiotics offered a much better vehicle to understand culture, but he was also interested in the dynamism of human action removed from causality. "As interworked systems of construable signs, (what, ignoring provincial usages, I would call symbols), culture is not power, something to which events, behaviors, institutions, or processes can be causally attributed; it is a context, something within which they can be intelligibly—that is, thickly—described."[4] Consequently, this work focuses on discovering, describing, and analyzing the networks developed through philosophy and science to guide the creation of a modern nation and its male citizen in Argentina within international contexts. My main purpose is to convey the simultaneity of events that connected different cultures, traditions, and continents. I believe that since nation-making was experienced at the same time in many places aligned with universal ideas, the communications that resulted needed to be considered to understand the double view that shaped the political environment. At the same time, I am not denying the historical specificity of the local experience, I am just focusing on the side developed through a belief in universality and through international networks.

The models of masculinity that will be analyzed in each chapter explain the different moments in the evolution of liberal and conservative cultures in global and local contexts, which resulted in the emergence of masculinities defined by sensibility and sentiment, revolutionary patriotism, citizenship, eclecticism and Romanticism, race, and sodomy. I am very aware that the masculinist nature of the project can raise eyebrows among feminist historians, a group in which I include myself; but I must clarify that research on masculinity is not necessarily related to the supremacy of men or the erasure of women. Quite to the contrary, the narrative that emerges from the Argentina case exemplifies the struggle caused by defining the meaning of the masculinity of its male citizen. Also, the struggles between liberals and conservatives that tore the country apart during the period analyzed ended with a battle among men split between masculine and feminine roles, which provides a unique national narrative in Argentina, as we will see.

The historiography on masculinity has greatly evolved since R. W. Connell's "hegemonic masculinity" was formulated, and the scholarship published today mostly avoids this kind of analysis. For example, many categories previously defined as sexual are today analyzed as part of cultural intersections.[5] More important, we need to recognize that gender and sexuality were very connected to science in the eighteenth century. For example, as Londa L. Schiebinger points out, plant sexuality "was not incidental to but indeed lies at the heart of the eighteenth-century revolution in the study of the plant kingdom."[6] Linnaeus's new botanical taxonomy recapitulated the sexual hierarchy of Western Europe and stressed the importance of sexual reproduction in plants, considering that male parts had priority in determining status. According to Schiebinger, this illustrates that the "scientific revolution and the revolution in sexuality and gender" were juxtaposed. She complains that historians seldom "brought the revolution in science to bear on another fundamental transformation of European society in this period: the revolution of sexuality and gender—the remaking of relations between men and women that began in the late seventeenth century and culminated in the French Revolution."[7] I agree with this assessment, and this study on masculinity within the context of Argentina considers scientific and gender revolutions as part of revolutionary politics that shaped Argentina during the first half of the nineteenth century.

The uses of feminine allegories to represent science, as Schiebinger explains, were common during early modernity and continued in philosophy and science until the end of the eighteenth century. Denis Diderot and Jean le Rond d'Alembert's *Encyclopédie* presented "an elaborate allegory of feminine hegemony in science."[8] In England, though, a masculinist allegory of science had been associated with the development of the country's scientific community, particularly in the work of Francis Bacon, whose "masculine philosophy" involved being active, virile, and generative—an experimental science drawn from the "light of nature, not from the darkness of antiquity" (677). Thinkers like Bacon saw French intellectual culture as "effeminate, especially in light of the highly visible role played by French noblewomen in Parisian salons." The rejection of a civilization represented by a woman, and civilized socialization as mediated by the creation of a feminine space, split those who were involved in the culture of the Enlightenment. In the case of Bacon, science needed "to be distinctively English (not French), empirical (not speculative), and practical (not rhetorical)" (678). The gendering of science as masculine or feminine according to national traditions continued in the nineteenth century.

Philosophically, the depiction of science as feminine was rooted in neo-Platonic ideas that defined masculinity as reason and body as feminine, while those like Bacon "invoked well-worn Aristotelian categories where masculinity signified hot active spirit and femininity signified cold sluggish matter." Bacon rejected "a passive, speculative, and effeminate philosophy," and called for "an active philosophy, one which would act as a formative principle upon feminine nature" (678). This tension surrounding the gender of science was also inseparably linked to the changes that emerged in the eighteenth century through the culture of sensibility that spread in Europe from both England and France. Syndy McMillen Conger observes that by the 1790s "it was equally possible to believe either that the sentimental ethic could precipitate the decline of established institutions or that it could reinforce the status quo."[9] She understands the eighteenth-century sensibility as the result of a new consciousness in the ideological and political sense.

Marina Benjamin has pointed out the tension that existed between notions of "refined" and "vernacular" in eighteenth-century Great Britain, which she transposed "to accommodate an illustration of the tensions between different modes of scientific discourse."[10] Many scholars have worked on this topic since Benjamin's publication, but there is still an absence of a comprehensive analysis of how philosophy and science influenced the notion of masculinity as a universal experience. More recently, Heather Ellis also noted that while the "scholarship exploring connections between the history of science and the history of gender in nineteenth-century Britain is not new," the resulting publications have been chiefly concerned "not with the formation of masculine identity *per se* as with the construction of narratives of female inferiority through the language, discourse, and practices of male-dominated science."[11] She explains this omission partly "by the apparent success of male scientists" who appeared to enjoy "not merely high socioeconomic status, but also the considerable advantages of intellectual and medical authority. Viewed superficially, the male scientist appears to be one of the most powerful and secure masculine identities" in modern British history.[12]

Ellis also noted that there has been "comparatively little interest shown in exploring the masculine identities of Victorian scientists," though she recognizes the contributions of Jan Golinski to the study of "the man of science."[13] Ellis argued that "modes of self-representation within the scientific community were intimately tied to models of masculinity," and called for investigation of how the natural philosopher's or scientist's identities were "formed from a variety of cultural resources, including those used to shape masculine identity in society at large."[14] Like Ellis, I was inspired by Golinski's observations; but while Ellis's project aims to bring together gender history and the mainstream history of science "to apply insights from both in investigating the construction of male scientific identities in nineteenth-century Britain," my project is not focused on scientists but on the creation of citizens who used science to represent the republic's new values.[15]

I am not the first scholar interested in the topic of masculinity in Argentina. Since the publication of Jorge Salessi's foundational book, many other works have been published in the past thirty years.[16] But, as is the case in other countries, the study of masculinity in relationship to science has been less explored, even less so for the period covered in this book. Starting with the Republic of Letters as it was in the eighteenth century, I determine how philosophy and science were intrinsically linked with the creation over time of a new model of man, and how this dynamic worked both globally and locally. Steve Shapin provided me with an excellent guide to understanding the formation of the man of science and his culture; he explained the "characters of the men of science" through the figures of the Godly Naturalist, the Moral Philosopher, the Polite Philosopher, and the Civic Expert. Natural theology and the priest/scientist culture were more influential in Protestant cultures, supporting "a character of the man of science as godly and the doing of science as the acquittal of religious goals."[17] This was the science attacked in the eighteenth century by David Hume, Immanuel Kant, the philosophes, and the encyclopédistes.

The moral philosopher of nature is for Shapin the proponent of "a natural order bearing the sure evidence of divine creation and superintendence was understood to uplift those who dedicated themselves to its study." He was "virtuous beyond the normal run of scholars."[18] But it is the emergence of a "new natural philosophy whose products were socially useful and whose practitioners were suitable for membership in civil society" that are at the center of his book. They introduced "a new civility" that made "the practitioners of natural knowledge fit for the drawing room and the salon."[19] Here the scientist was an example of a new way for men to conduct themselves, which introduced a new sociability, manners, and masculinity.

This interpretation coincides with George Sebastian Rousseau's evidence that by the beginning of the nineteenth century the treatment of man was "less as a fixed and final object of creation, an 'Adam'" and more often analyzed "as the product of time, circumstances, and milieu—the creature of education (as Locke, Condillac, and Helvetius especially stressed), of climate and physical environment (Montesquieu), of physical evolution (Buffon, Erasmus Darwin, and Lamarck), or of history (Vico, Boulanger, Ferguson, Miller, Herder)."[20] This provided a more dynamic notion of being that was also supported since the last quarter of the eighteenth century by the revolutionary culture that worked behind republicanism. "Man was thus a creature less of fixed being than of becoming."[21] This tension between stability and becoming appears clearly by the 1840s in Argentina.

Each chapter in this book introduces a different model of masculinity according to its chronological appearance. However, this does not mean that they emerged this neatly in real life. They coexisted and interacted with each other in a dynamic way that I had to separate clearly to explain the process. I start with the analysis of the Republic of Letters as it worked in the eighteenth century, and how it changed the culture of masculinity, followed by the French response to English Empiricism, and sensationism. The latter implied that the senses could reveal the new political subject's bringing possibilities that enhanced the role of imagination. Modern ideas during the 1800s disassembled the old definitions of man as a fixed creation, often fueled by the sensorial experience and shaped by a postcolonial reality. Education became a catalyst for social transformation, as environmental influences on human beings were increasingly seen as tools for self-improvement, particularly through the study of philosophy and emerging sciences. John C. O'Neal, understands sensationism as a crucial transitional phase between the spiritual rationalism of the seventeenth century and the early stages of modern scientific investigation in the eighteenth century. By combining the mind and body within the concept of the soul, which encompassed both spiritual and physical aspects, this view made the emerging idea of the mind as a physical brain and the nervous system as a material substance seem more acceptable to cautious and devout thinkers.[22] This is the point examined in the first three chapters. To be a man started to be less about acting according to the purpose of his creation and more about representing the civilized citizen of the modern nation that communicated new values through science and philosophy, which meant that the modern man's selfhood was created and perfected over time. These chapters end with the creation of a scientific culture by the 1820s, which failed partly because of the development of a conservatism among those devoted to destroying the spread of philosophical and scientific radicalism associated with feminization and chaos.

This local conservatism expressed concerns like those in Europe that resulted at the same time in the defeat of Napoleon and the restoration of monarchical authority, which, by the 1830s, caused a reassessment of the idea of virility as it had been conceived. The end of Jacobinism's politics favored a philosophy and science that promoted the value of reconciliation, harmony, and a masculinity conceived as more spiritual and linked to aesthetics and a historical science that produced a sense of origin and continuity after revolutionary and republican failures, which led to an adjustment of the humanist model. More conservative positions led to the identification of modern man according to race, a "natural" division that more scientifically separated the different human populations to identify those that better facilitated the development of the modern nation's goals. At the same time, the influence of French Romanticism contradicted the more materialist conceptions of science, also promoting a more universal culture and subjectivity.

As I will discuss, Juan Manuel de Rosas (1793–1877) replaced Manuel Dorrego (1787–1828), a citizen of the Republic of Letters, as leader of the Federalist Party in the Argentine Confederation after Dorrego was killed by Unitarians fighting a united central power to administer the nation. The radicalism of the decision changed the country's politics; Rosas organized a movement that would not fight for the establishment of constitutional ideology to rule the country. His fight consisted in destroying any symbol of liberal ideology, including the modern representation of gender in society through philosophy and science.

This masculinist culture was a radical departure from the scientific liberalism that dominated the 1820s; this led to competitive models of masculinity: the federalists promoted a masculinist culture and rejection of political speculation, while the liberals defended the identification of the culture of civilization associated with the feminine and based on spiritual ideas, both sides clashing about the levels of virility that needed to be expressed in society. This explains the narratives of sodomy that were popular among those liberals who fought Rosas from exile in the 1840s, and the uses of science to defend such an ideology.

From the 1770s to 1852, the period covered in this book, the contact with scientific and philosophical knowledge became one of the signs used to divide or unite the nation. The federalists saw in liberal science an irrational force that limited the capacity for maintaining rigid categories required for good government; they believed in change created over time, the way it happened in nature. Unlike what their enemies claimed, they did not hate science per se, but its uses in politics. In their view, the formation of a new nation could not be based on ideology and speculation alone, it needed to evolve rooted in a place.

The fact that the leaders of this conservatism were cattle ranchers interested in the study of nature as an applied field explains how they saw nation-building as if it were a natural organism that needed to grow in the soil. Ironically, this position would be closer to post-Darwinian notions of civilization, which explains the changes in liberalism in the 1870s in Argentina. In terms of gender, liberals struggled to demonstrate that their project would lead to the emergence of a virile and dominant citizen (a "natural" created man in conservative terms) because of their interest in representation and the notion of becoming, which were coded as feminine. Their supporters' embrace of this culture made the popularization of this kind of politics more difficult, which led to the polarization between civilization and barbarism with a clear gendered component.

HISTORIOGRAPHY AND METHODOLOGY

Since the identity of the modern/civilized man was conceived as universal, the case of Argentina can only be understood when analyzed with the same simultaneity that originated its development, which explains why each chapter locates Argentina in the global context to which it belonged first; it is also examined through the life of the country's main intellectuals of each generation. In the different chapters, I pay attention to how the politics of emancipation were related to evolving philosophies and scientific ideas, and how they are linked to the root of gendered politics in Argentina. My goal is to indicate that science was not a field that secured the supremacy of man; on the contrary, the evolution of philosophy and science in this period weakened the notion of modern masculinity as a stable category that could provide a foundation to the nation.

My analysis of simultaneous historical processes and their impact on Argentina's notions of nationhood and citizenship was made possible by consulting secondary sources authored by experts in the history of the United States and Europe. These sources illuminated parallel developments occurring concurrently. I selected these sources based solely on their utility in reconstructing how individuals in Argentina perceived the international context, rather than on any personal historiographical preference on my part. Finally, I need to make clear that what I have stated does not imply that I see thinkers from Argentina, and Latin America at large, as just imitators of foreign ideas. Quite the contrary, the role of the Americas in the philosophical and scientific thought of Europe is one of the constants in this book. It is often forgotten that in the context of the nineteenth century the place where radicalism triumphed was in the Americas, and Europe's failure to build republics was what slowed down emancipatory politics everywhere. The first Black nation ended slavery and was located in the Caribbean, Haiti. There, an institution that had existed since biblical times ended, which helped to form a much more expansive idea of emancipation in Europe.

The end of colonialism did not happen because European powers decided to dissolve their empires; it was radical emancipatory politics in the Americas that made it possible, making this continent the land of realization against the European failure to turn its more radical philosophy and science into political reality. Therefore, I present a juxtaposition of the national and international dynamics in each chapter, inspired by the scholarship of global historians published in the past twenty years.

Patrick Manning has defined the work of the world historian as needing "to link speculation, logic, and evidence into a coherent analysis with the goal of developing a broad, interpretative and well-documented assessment of past transformations and connections."[23] Manning does not make a clear differentiation between "world" and "globe." Bruce Mazlish does, however; he criticizes those who have used world history as a synonym for global history and have refused to "understand that something different has emerged in the last half of the twentieth century." What matters is "the recognition that something important has happened" and that this requires "a new openness and a new mindset if we are to understand it well enough to grapple with it effectively."[24] Still, "there is no consensus on what global history actually means" and from what point one can write it. But it tends "to deal more with the period after the explorations of the fifteenth century and often refers to the process of globalization" since the last third of the twentieth century.[25]

James Poskett recommends that historians of science "study the relationship between the global as the analytic category and the global as an actor's category." This analytic element means to focus on the existence, or not, of a traffic of knowledge that "affected the content and political uses of science." Poskett conceives the actor's element as the relationship between "the external world" and "the constitution of man." Here, science's conceptualizations of the world are closely related to the circulation and communication of ideas, another important aspect of global science.[26] As a discipline, the history of science can easily be adjusted to global history; science itself was built on the study of knowledge produced in different places in the world. James Secord has described the importance of the idea of circulation for scholars in this field, regardless of their methodology and theoretical sympathies.[27]

In 2009, on the anniversary of the publication of Darwin's *On the Origin of Species*, Secord wrote that "to understand the phenomenon of Darwin and Darwinism it makes sense to start with communication."[28] According to him, Darwinism became "a shared context to work through highly local, specific experiences in relation to transformations across the world," and this understanding of science has deeply influenced my emphasis on the study of communicable knowledge that characterizes the period studied.[29] Kapil Raj's account of "the mobility and spread of the sciences beyond their site of origin" helped me to understand the dynamism that a historical narrative needs to describe the historical events covered by this book. I was aware while writing that "scientific propositions, artifacts, and practices are neither innately universal nor forcibly imposed on others" and that I needed to be careful in how I switch from local to global contexts.[30]

Work produced over five decades also supports the concept that the circulation of scientific knowledge goes "through complex processes of accommodation and negotiation, as contingent as those involved in their production" and this is what I try to convey.[31] Consulting Bernard Lightman's edited collection on Spencerian philosophy in the global context confirmed the importance of paying attention to "modes of communication in different national contexts, as well as the circulation of knowledge worldwide," to understand the different ways in which the work of a single author could be used by thinkers belonging to different cultures.[32]

The development of scholarship examining the science of Iberian Empires and colonialism was also indispensable for this book. It taught me about the importance of science in the colonies and the absence of work explaining how men at the time related to masculinity and gender at large, which stimulated my interest in the Republic of Letters. The ideas of Miguel de Asúa were constant companions as I thought about science in the River Plate area and its international connections, particularly in the context of Jesuit science; his work is the foundation of my interpretation of how the latter influenced the context of the River Plate Viceroyalty's Republic of Letters.

1

The Citizen of the Republic of Letters and Sensibility

The scientific revolution had an impact on Iberian America by the second half of the eighteenth century along with programs for the economic and cultural reorganization of the colonies according to the reforms ordered by Charles III (1716–88) in Spain and the Marquis of Pombal in Portugal. Creoles were not directly involved in the process of creating the reforms, but the changes coincided with a time in which they had started to think about an American identity independent from Spain and based on modern philosophical and scientific ideas. This process was reinforced by the Spanish Crown's attempt to turn the colonies into agro-export producers, which triggered scientific expeditions to identify available resources enhancing the spreading of new ideas among the Creole elites.[1]

According to Mauricio Nieto, the introduction of modern science by the Spaniards was intended to gain better control of the colonies; but, at the same time, since it was also related to political philosophy, it influenced a revision of colonialism.[2] Jorge Cañizares-Esguerra also believes that Linnaean botany took on a life of its own when it arrived in the American colonies. Deployed by local patriot-naturalists it undermined "the very goals that Linnaean natural history had set out to accomplish in Spanish America, namely, to revamp and strengthen the imperial state."[3] Those who made this culture possible were part of the "Republic of Letters" that changed the way in which nature was understood.

This republic was "a network of the scholarly and scientific community of the sixteenth, seventeenth, and eighteenth centuries" that emphasized the communicability of knowledge. As a result, the only requirement to belong to this association was to write letters: "Those scholars who failed or refused to establish sustained lines of communication, could not be reckoned as citizens of this republic."[4] Responsible primarily for the education of boys, the citizens of this republic were modeled by Jesuits, and their science shaped the knowledge that was produced in Spain and its colonies.

Edward Jones Corredera has concluded that transnational imperial experiences "fostered institutional reform in Spain." The resulting changes were part of a "broader European effort to resolve the quarrel of the Ancients and Moderns, reframe and reform the antiquarian tradition, and foster the study of philosophical history and political economy."[5] Educational changes allowed Spanish officers to learn how to socialize in a heterogeneous environment, including studying "mathematics and dancing." By the 1730s they needed to learn trigonometry "and to study and discuss the works of Tycho Brahe (1546–1601) and Copernicus (1473–1543)." Men in the service of the nation had "a great degree of influence over Spanish imperial pursuits" and needed to socialize according to the demands of its global possessions.[6] In 1713 the Royal Spanish Academy (Real Academia Española) was created to prepare a Spanish dictionary that regulated the use of language, following the examples of what had happened with other language academies, such as Italian (1583) and French (1635), and in 1779 the Royal Academy of Sciences was created in Portugal.[7]

In addition, an erudite Spanish soldier, Álvaro Navia-Osorio y Vigil, Marquis de Santa Cruz de Marcenado (1684–1732), proposed in 1722 the creation of an encyclopedic dictionary (*diccionario*) to gather all the old and new knowledge as "a service to the Republic of Letters and the Fatherland." Authors needed to know four languages to translate the books that Marcenado deemed necessary for the project, many of which were not available in Spain and had to be obtained in bookstores located in Turin, Venice, Geneva, Paris, and Amsterdam. Since 1730, the latter has been part of the province of Holland in the Netherlands and an important center for printing books banned in societies dominated by Catholicism.[8]

The Academy of Good Letters (Buenas Letras) of Barcelona was founded in 1729 and put under royal protection in 1756; a celebratory volume of the event was published the same year. The publication included an article by the writer and expert in classical literature Ignacio Luzán (1702–54), an admirer of classicism, who asserted that "the fortune of the Letters was always tied to the fortune of empires." Greece and Italy in antiquity, and, in his lifetime, Spain and France, had "their happiest times when they saw shining with better light in their horizons the good taste in the Sciences, Arts, Poetry, Rhetoric, Criticism, and style," which confirmed that peoples were more content "when the philosophers were Kings, or Kings philosophized," following Plato's philosophy.

Luzán also made clear that the "solitary speculative experiences" that characterized theologians did not make men happy because what they learned was not communicable to others, and happiness consisted in being able to transmit truths to "enlighten" (*ilustrar*), motivate the will, "reduce and moderate customs; and, in general, to make virtues kind and vices atrocious." The purpose of "the Good Letters" was to "make a good citizen" who was a man able and willing to receive all the knowledge brought by Sciences and Arts and not only understood his own happiness but also that of other men. This was a new kind of man that was clearly defined.[9] He was interested "in public affairs [*Repúblico*]," loved and searched "the prosperity of his Fatherland, the goodness of his Nation," and was "a good Vassal [*vasallo*]" who obeyed and respected "the Laws, the precepts, and the glory of his King." In addition, he was "a *good Man*" who wanted everyone to experience "his humanity" to set an example that could be followed by others through imitation; in this new social environment, "good faith, culture, affability, generosity and true human happiness" would spread over all nations the practice of the most sociable virtues.[10]

This new figure, emerging simultaneously from modern culture in Europe and Spain's American colonies, linked the concept of fatherland (patria) with a literature and culture that were both local and universal. Consequently, a distinct form of patriotism arose—one that integrated knowledge with institutions fostering communication, circulation, and intellectual dynamism. In 1769, a volume dedicated to Charles III by the priests Rafael and Pedro Rodríguez Mohedano commended the contributions of Father Feijóo and French literature to the enrichment of the Spanish Republic of Letters. These writings celebrated the "perfection that the sciences" had reached in the eighteenth century. Through modern readings, Ancient Greek and Latin works reemerged after the "flooding of the Oriental Barbarians," which had "disfigured and forgotten the sciences, burying their splendor and beauty."

According to these priests, scientific thought began to thrive under the protection of princes after the fifteenth century, but it took another three centuries for it to reach the strength needed to produce groundbreaking discoveries.

The publication aimed to rectify Spain's exclusion from the ranks of "literary and cultured nations," a slight that tarnished the nation's honor. By then, "the taste of true literature" and the "happy progress in the sciences" were regarded as markers of national glory. The literary history of France, written by the Benedictine fathers of the Congregation of Saint Maur, had emerged as an influential model for promoting cultural pride and national achievements.[11]

THE CITIZEN OF THE REPUBLIC OF LETTERS AND JESUIT SCIENCE

Paola Giuli understands seventeenth century Arcadia as the first modern example of the Republic of Letters; founded in Rome in 1690, the Accademia dell'Arcadia was the "most representative literary institution of eighteenth-century Italy."[12] For all Europe it meant "the continued metaphor of literary Italy," a reference to "a golden age as much as a temple of arts and letters."[13] It is defined as "a meeting place for prominent and influential Italian writers and intellectuals and an entity that can only be understood taking into account the history of philosophy and of science." This understanding of a place of knowledge organized around the sociability of men who were there to communicate ideas about literature, arts, and sciences was the novelty of this institution. Its culture was gendered because it implied a new way for men to relate to each other, which explains why Enlightenment and Risorgimento critics would reject this republic's focus "on the pastoral and supposedly 'effeminate' aspect of the Academy," a tension that would continue over the eighteenth century.[14]

Mordechai Feingold notes that, as heirs to Renaissance Humanism, the Jesuits "proved remarkably successful at modeling their schools on the humanist program," particularly classical education.[15] But by the seventeenth and eighteenth centuries this model was challenged by new ideas that departed from ancient knowledge, though they originated from it.[16] Luigi Antonio Muratori (1672–1750), the Jesuit archivist and librarian of Modena's library, advocated for a reform of the literary world that diverged from the Arcadian model. He suggested one that focused on the pursuit of "truth" and serious critical analysis. This approach included integrating philosophy and science into a new form of male subjectivity. Poetry also played a crucial role in his vision, linking imaginative power with judgment, intellect, and "good taste," which in turn guided and regulated its expression.[17] In 1708 Muratori emphasized the relationship between good government and knowledge of the arts and sciences as the easiest "path to Glory" for the Italian sovereigns. He believed that the man of letters loved public good and honor.[18]

In addition, the religious conflicts that divided European Christianity in the seventeenth century also involved the Jesuits when the theological interpretation of one of its members, Luis de Molina, was challenged by the Augustinian priest and professor of Louvain, Cornelius Jansen (1585–1638). The debate between Molinists and Jansenists received much attention, and it was focused on the problem of grace and predestination. Following Molina, the opinion "championed by many Jesuits was that *attrition* was sufficient for absolution in the confessional—that is, one needed only to seek forgiveness because one feared divine wrath." Jansenists argued that "only *contrition*—that is, love of God and sorrow for having offended him—sufficed for absolution."[19] The Jansenists lost this first round, but they enjoyed renewed admiration over the eighteenth century when they were popular in France and Spain, though by this time their ideas had merged with the new revolutionary politics.

The Jesuit model of education, its science, also defined an understanding of masculinity, which led to the formation a new male identity in sixteenth-century Iberia that was complementary to that of the conquistador. Ulrike Strasser defines this understanding of male culture as the creation of an emotional community that valued a specific kind of feelings, resulting in "a self-assured and novel Catholic masculinity grounded in homosocial bonds."[20] The foundation of the Society of Jesus (the Jesuits) "marked a watershed moment in the European history of masculinity" through its humanistic plan of education, revolutionary in the sixteenth century, that trained men all over Europe and the Americas (47). This early modern missionary masculinity "helped pave the way for the imperializing masculinities of the modern age" (226). It created a "potent link between clerical masculinity, global mobility, and Europe's religious civilizing mission." Jesuits' missionary letters "combined edifying and educational material from all over the world," targeting the broader European Republic of Letters (208).

During the seventeenth century, the natural and hard sciences led to the development of inquiry through physics and mathematics; their discoveries were tied to a repositioning of philosophy, questioning how effective Scholasticism was as the system that could be useful to understand the meaning of scientific phenomena. The Jesuits' humanism was challenged by the criticism of figures such as René Descartes (1596–1650) and his disciples who contributed to the reputation of the Jesuits as being antiscientific. Despite this characterization, the reality is that this image was created by the debates about religious orthodoxy in which this order engaged; ideas about modern philosophy and science were mostly not in dispute.[21] In fact, it is impossible to understand the scope and influence of the Jesuits, "without considering the science some of its members produced."[22] Since philosophy and science were in a process of constant transformation throughout the seventeenth and eighteenth centuries, members of the Society of Jesus were engaged with this change, and modern ideas were part of the instruction delivered by Jesuit institutions even in the colonies. For example, in the Viceroyalty of New Spain, it was announced that for the term 1796–97, one course in the arts at one institution would open with a "Latin prayer to praise Modern Philosophy, making visible the advantages that were achieved by the studious Youth through it" under the protection of the "Archbishop."[23] This same publication also let readers know that "Linnaeus's botanical philosophy" was for sale in a shop.[24]

French Jesuits' ideas throughout the eighteenth century coincided with some espoused by Enlightenment thinkers in France, unlike the common perception of them as enemies. Jeffrey D. Burson notes that Jesuit moral philosophy "often presupposed a generally more optimistic appraisal of human nature that comported well with its rehabilitation as assumed by many radical thinkers." For example, the Jesuit Claude Buffier (1661–1737) had argued that common sense was "warped after the fall because natural reason became inexorably dependent upon sense perception." Inspired by John Locke, Buffier believed that human understanding was "liable to corruption due to the very nature of sense perception," which explained the limits of natural theology. It was for this reason that for Jesuits only the Catholic Church could be a barrier "against the inherently corrosive tendencies of natural reason."[25] This interest in defining human understanding also enhanced interest in the sciences, and, in the case of France, young Jesuits were obsessed with new science, so much so that they were warned not to "let their zeal for science" affect their commitment to the church (52).

Jesuit education "instilled within their students an ease and comfort, not only in the courts, ministries, and institutions of the Bourbon monarchy and its Gallican Church, but also in the salons and academies of high Enlightenment sociability" (53). This kind of education connected masculinity and science with politics, and the men who were promoting it were praised for their ability to teach useful things to improve society and, simultaneously, keep the youth devoted to the Church. In France the members of this order were responsible for the popularization of "experimental physics and Enlightenment science" while they were battling those associated with the *Encyclopédie* (54).

Similar interests existed among the Jesuits in Spain. In 1753 Esteban de Terreros y Pando, a prominent Jesuit mathematician who taught at the elite institution Real Seminario de Nobles, translated a book written by the French priest Abbé Nöel Antoine Pluche (1688–1761) that was popular all over Europe. The intention of the book, originally published in 1732, was to develop in the youth an interest in the study of nature. Pluche, theologically inspired by some Jansenist ideas, defined men as happy and belonging to "a universal society" defined by the experience "of being united in the body of a city, kingdom, or republic." The "fatherland" (*patria*) was "the world" and men should conserve "the impartiality of a Cosmopolitan."[26] But good government of a specific society was crucial to prevent an individual from experiencing deprivation from "the universal society," each well-organized community "commanded all men to be willing to serve each other."[27] The same happened in kingdoms because "all the links that connected individuals needed to be subordinated to the love of the human species [*género humano*]"; this was the first obligation of all members of society.[28]

Portugal also experienced a similar change; Jaime Cortesão indicates that by the 1740s the introduction of Freemasonry was an example of the influence that foreigners began to have in the kingdom and its colonies; those who were part of the intellectual community were for the most part faithful to a "universalist ideal of a nation of people that formed other peoples and assimilated other cultures" with its own. This was an era of "contradictions and inconsistencies," in which fanaticism coexisted with kind tolerance, and an egalitarian cosmopolitanism with rejection of the foreigner. João V (1689–1750), who ruled the Portuguese from 1706 to 1750, was a representative of this culture. He "cultivated and used the exact sciences as a material to increase a universalism imported from outside," which would eventually undermine his own regime in Cortesão's view.[29]

In Portugal the changes meant a conflict with the Jesuits because the monarchy attacked the latter's system of education for not being modern and wanted to replace it with one that was. Directed by the Marquis de Pombal from Portugal, "a gigantic operation" to destroy the Jesuits' reputation in Europe was well planned and sustained over roughly two decades, likely the "widest and most ambitious propaganda in Europe up to the eighteenth century."[30] Francisco Malta Romeiras concludes that economic and political reasons were the real reason for the opposition to the Jesuits, but, importantly for this book, as part of his plan, Pombal adjusted "the Enlightenment perceptions that equated modernity with scientific progress" expanding the previous narrative about the order being the enemy of science. The conflict ended with the expulsion of the Jesuits from Portugal in 1759, from France (1764), and from the Spanish Empire (1767).[31]

Pombal's reforms included a new educational plan created by the Portuguese cleric Luis António Verney (1713–92), an important figure in the Congregation of the Oratory, rivals of the Jesuits. This congregation was created in Rome in 1550 and later introduced in France (1611) and Portugal (1688). Verney's new method was published in 1746 and supported an eclectic approach that promoted freedom to think outside the limits of Scholasticism and its derivations; a Spanish translation appeared in 1760. In Verney's words, the "modern system is not to have a system: only in this way some truth has been discovered." A philosopher was then liberated from passion, and he could "propose the reasons for the things he observes." Those reasons that were clear were embraced, and those doubtful were despised and rejected, and in this way "the body of the doctrine" was formed.[32] This philosophical turn aligned with advancements in physics and natural sciences, but "a physics that cannot be understood" needed to be rejected, and things that were not proved "must not be admitted." Simplicity and clarity were the foundation of the new thought.[33]

The Jesuits who were sent into exile after the expulsion continued being productive. Pedro Aullón de Haro and Davide Mombelli identify the movement as the Universalist Spanish Enlightenment. Characterized by not being "politically oriented," these scholars were mostly interested in science and education. This community advocated "for a convergence of classical humanism and modern empirical science to establish the Universal History of Literature and Science" that developed interest in comparative linguistics. The geographical center of this group was in Italy, where many of them resided in close communication with each other after the expulsion.[34] This undertaking was "a Christian Enlightenment, humanistic and empiricist, historiographical and scientific, methodologically comparative, ambitious yet not groundbreaking, international and worldly, in harmony with a globalized understanding of the universe and the world."[35] It was led by Juan Andrés, Lorenzo Hervás, and Antonio Eximeno, all Spanish Jesuits who were exiled in Italy. They were inspired by the already mentioned Muratori and Antonio Genovesi (1713–69), a student of Giambattista Vico (1668–1744), the author of *Principles of a New Science*.[36]

The Spanish Jesuit Lorenzo Hervás (1735–1809) explained that the difference between a Muslim philosopher and a Christian one was that Christianity required the study of all sciences and understood knowledge as originating in a "divine source" that was "freely communicated" to all those who desired it.[37] Men's happiness on earth was the result of the use and combination of reason and the study of nature.[38] While this scholar rejected the most radical modern thinkers because they were not compatible with his conception of God, he believed that the modern man was by nature a cosmopolitan subject, an individual open to the universe to learn the lessons taught by it. The cosmopolitan man went further and dissolved national allegiances because his fatherland was reason and truth. In this republic of knowledge its citizens' loyalty was in communicating with God, which was possible through reason and knowledge of nature.

Some of these Jesuits were born in the colonies; Francisco Javier Clavijero (1731–87), from Veracruz, wrote a book about the history of Mexico that indicates how an incipient nationalism was connected with science. In a letter he wrote to the University of Mexico he defined this work as a "History of Mexico written by a Mexican," and as a very local enterprise with universal reverberations. He complained about the "passion and prejudice of some authors" and defended the Indigenous population of the Viceroyalty of New Spain. The only difference between the Spaniards and the descendants of the native population was education; the people encountered by the Spaniards had developed a science that was superior to the one known in Europe. The intelligence of the descendants of this native population made them capable of learning any science. They were good in geometry, architecture, and theology.[39]

Other scholars from the American continent were Pedro José Márquez (1741–1820), from Guanajuato, who promoted philosophy, theology, grammar, history, geography, physics, and astronomy; Pedro Cantón Ubianco, from Guadalajara, specialized in grammar and translation; and José Rafael Campoy (1706–90), born in Los Álamos (Sonora), was interested in history, geography, geometry, botany, and natural history. All of them ended up taking residency in Italy, mostly in Bologna. José Lino Fábrega, born in Tegucigalpa (Honduras), studied history and grammar; Pedro Franco Dávila (1711–86), born in Guayaquil, was interested in botany, geography, and anatomy. In the River Plate Viceroyalty, Joaquín Camaño, born in La Rioja, devoted his studies to philosophy, theology, grammar, and geography; and, in the Captaincy of Chile, Juan Ignacio Molina (1740–1829) became an expert in astronomy, geography, natural history, and botany. While Molina and Fábrega ended up in Bologna after the expulsion, the others went to France and Spain.[40] This impressive network was influential in Europe because they circulated knowledge about the American continent that was based on direct contact and experience.

The expulsion of the Jesuits also signaled a diminished emphasis on maleness. For example, in documents related to the transition of Spain and its colonies to a society without Jesuits, published in 1767, it was made clear that "the education of the youth should not be limited to boys." Girls needed to be educated to be ready for motherhood because "good customs depended mainly on primary education." Representatives of the church were already founding schools for girls.[41] Charles III ordered that "in the main towns, where it seemed best, schools be established for girls, with honest and educated teachers that take care of their education." Girls needed to learn "principles and obligations of civilian and Christian life" together with "abilities proper to their sex," which were linked to family life.[42] The government needed to pay for the education of the daughters of peasants and artisans, leaving wealthier parents to pay for theirs.

In this illustrated Iberian culture, knowledge needed to be useful and applied to the economic and social improvement of the kingdom through a government that ruled according to philosophical and scientific principles, which made the study of economics a favorite subject among the Spanish illustrated elite. According to Mónica Ricketts, the Enlightenment's universalism "spread the fundamental concept that the world was interconnected," creating a double view of the local and the external that resulted in a thought that was broader in its conceptualization and required debates and public communications to create "a republic of letters," which was the ideal place of knowledge.[43] This opening was not only intellectual; more people were allowed to visit the Spanish colonies, and by the "late eighteenth century, scientists, military officers, and men of letters from different places began to arrive in Spanish America"—Alexander von Humboldt (1769–1859) being the best known.[44]

In 1747 Juan Bautista Corachán (1661–1741), a priest who was also a professor of mathematics at the Spanish University of Valencia, wrote a book that included a warning from Gregorio Mayans y Siscar (1699–1781), the censor who approved the publication; it was made clear that the activities characterizing those who belonged to "the Republic of Letters" were specific: laziness was forbidden and the erudite could not socialize with idle men. The enlightened could only enjoy "an honest pastime that strengthens the body to avoid the physical damage caused by a sedentary life among the literate community" (*comunidad letrada*).[45] In the Spanish American colonies, this republic was also very important for developing a new hierarchy based on intellectual bonds among those who belonged to it. In a ceremony in 1762 to celebrate the arrival of Don Manuel de Amat y Junient (1707–82), the previous captain general of Chile, as viceroy of Peru, he was portrayed as having been sent by providence "to increase the society of letters of this republic." At this time "the sciences were shining more than in past centuries," but in the region the conditions for their advancement were underdeveloped.[46]

Science was viewed as indispensable for good government, "the Exact Sciences, even the Belle Lettres," were needed "for the person who holds on his shoulders the enormous weight of the republic" because to reign was the "most difficult art of the arts." The text also used Plato's *Republic*, and the absolute power of the monarch was related here to knowledge; monarchies would be happy "when the wise rule them, or the philosophers became kings."[47] The supremacy given to a governing erudite class based on their understanding of philosophical and scientific principles meant that blood lines alone were not enough to rule.

THE REPUBLIC OF LETTERS IN THE SPANISH COLONIES

Jesuit networks explain the access of educated elites born in Spanish colonies to the study of literature, arts, philosophy, and sciences. Two towering members of the Republic of Letters were born in New Spain in the seventeenth century: Sor Juana Inés de la Cruz (1648–95), a nun, poet, and woman of letters, and her friend Carlos Singüenza y Góngora (1645–1700), a priest, mathematician, philosopher, astronomer, and poet. This priest was born in Mexico to Spanish parents and was educated by the Jesuits to whose society he unsuccessfully sought admission, though he did become a priest. He eventually became a professor of mathematics and astrology, a field taught since 1637 at the request of the students at the Royal University of Mexico, North America's first institution of higher learning; the first on the continent was the University of San Marcos, established in Lima, Viceroyalty of Peru, in 1551.

In 1690 Singüenza y Góngora published an analysis of comets, which demonstrated his rejection of Aristotelian science and his familiarity with the works of Copernicus, Galileo, Descartes, Kepler, and Brahe. This study contradicted the ideas of the astronomer and Jesuit missionary Eusebio Kino (1645–1711) with whom he had established a friendship after the Tyrolean arrived in Mexico. Singüenza y Góngora wrote that he supported his own notions "for no other reason than that they are modern," his intention being to dispute the superstitions and predictions associated with the sight of comets that had terrified humans for centuries.[48]

This native of Mexico was certain about the abilities that Creoles had for natural observation in New Spain; he ironically criticized Kino, "who had come from the educated Germany" to teach mathematics to those living "in the ignorant America."[49] His self-confidence was not misplaced; in her study of the 1680 comet, Laura E. Bland concludes that by this time, unlike Harvard, which was collapsing, "the universities of Peru and Mexico had matured into stable and productive institutions in the century and a half since their foundation." This meant that the educated elite could access a wide network of scholars, which explains how the debate between Kino and Singüenza y Góngora "attracted attention from all over the Americas," an indication that a scientific community existed on the continent.[50]

The interest in scientific exploration increased in the eighteenth century because since his coronation in 1759, the Spanish king Charles III (1716–88) supported the organization of expeditions to improve the economy of the kingdom through modern science. The first was directed to the Viceroyalty of Peru and the Captaincy of Chile and was active from 1777 to 1788. Alessandro Malaspina's journey is well known because of its circumnavigation of the world from 1789 to 1794. Another was sent to New Spain traveling from 1797 to 1803; in his Royal Order for this botanical exploration, Charles mentioned that he expected "the examination, drawing and methodical description of the natural productions of [his] fertile domains" on the American continent. His general objective was "to promote the progress of the physical sciences, and the elimination of doubts and adulterations existing in medicine, the dye process, and other useful arts that increase commerce."[51] While it was commanded by the Spanish naturalist Martín de Sessé y Lacasta (1751–1808), Creoles had crucial roles in it; the artists Atanasio Echeverría (1771–1803) and Vicente de la Cedra were employed to create the drawings; both had been educated in Mexico's San Carlos Art Academy. They produced around two thousand drawings following directions based on Linnaeus's classification, which explains their focus on stamens and pistils.

Another Creole, the botanist José Mariano Mociño (1757–1829) was also invited to join; he was trained by Alzate y Ramírez and Clavijero, defended the work of Linneaus, and was ideologically aligned with the Universalist School. In 1791 Mociño led the expedition that gathered botanical collections in Mexico and Guatemala. Later he continued to collect samples in Cuba, Puerto Rico, and the Northern territories of the Viceroyalty of New Spain, arriving at Nutka (Nootka), which today is part of British Columbia. This expedition was able to create an herbarium that ended up in Spain as part of the Botanical Garden's collection in Madrid.

As Daniela Bleichman has noted, the Hispanic world experienced a surge of scientific activity by the last decades of the eighteenth century, when science proved valuable not only as part of Enlightenment ideologies and policies but also as a continuation of endeavors initiated during the peak of imperial expansion in the sixteenth century.[52] This meant that monarchical institutions and authority were regenerated through the scientific enterprise.

The increased circulation of ideas in this century resulted in the publication of newspapers and journals in New Spain. In 1722 the first newspaper appeared. *Gaceta de México* was edited by Juan Ignacio de Castorena Ursúa y Goyeneche (1668–1733), a priest born in Zacatecas who was a friend of Sor Juana Inés de la Cruz and the editor of some of her work. The first journal in the Spanish colonies devoted to science appeared in 1768 in Mexico, edited and written by Jose Antonio Alzate y Ramírez (1737–99). In the first issue, the editor clarified that he was following the newspapers (*periódicos*) and journals being published in Europe for an audience that was a mix of aficionados and scientific experts. Creating *Diario literario de México* would contribute "to the good of the Spanish nation" because agriculture, commerce, and mining needed scientific knowledge to improve.

Following the eclecticism that characterized the Iberian Enlightenment, Alzate y Ramírez's critical approach would avoid dogmatic attacks, and because religious and political entanglements existed, the editor promised to be impartial and avoid "becoming a partisan to simply informing the opinions and doctrines proposed by all parties." He assured that he would only be critical of those ideas that were clearly wrong and not useful, work that was "as needed as [it was] useful to the Republic of Letters," received a different treatment.[53] The second issue was devoted to a thesis defended in Querétaro that with the aid of experimental physics explained how God's creation had happened. Sciences could support with the best evidence the actions of a rational creator who acted according to principles that were analyzed by writers of the classical Greek period, not to mention such "modern physicists" (*físicos modernos*) as Buffon, Galileo, Descartes, and Kepler.[54]

Also in Mexico, the mathematician and physician José Ignacio Bartolache (1739–90), a professor at the Royal University of Mexico, began publishing the medical journal *Mercurio volante* (Flying Mercury) in 1772, continuing publication until February of the following year. The first issue made clear that the "barbarism and ignorance of the Indians" was followed by the "glorious conquest" of America. Since then, the "enlightenment" (*luces*) and "good taste" had made "stupendous progress" including the foundation of the University of Mexico in 1553. The delayed progress in the medical sciences at the time was viewed as due to the dogmatic acceptance of the "writings of Aristotle, Galen, and Avicenna." Furthermore, since the last educational reform in the colony in 1645, "useful physics and the related medical field" had not been advanced. In addition, belles lettres needed more attention according to the desired humanistic plan. In the second issue, women are defined as "endowed with the same potential men had, maybe even better than those graduate students so respected by their reputation" in the viceroyalty.[55]

This characterization of gender was not strange. Women had a central role in the Spanish Enlightenment because gender was a central preoccupation of the movement, giving gender a pivotal role in shaping Spain as an enlightened nation.[56] The article in *Mercurio volante* also criticized the established philosophy, praising the study of "good physics" that was understood as a "science that gives us enough knowledge of the bodies to explain their nature, properties, and the sensible effects that resulted from the combination of one with the other"; a body was all that "existed in the created world," everything that was "perceived through one of our material and external senses," indicating the importance that sensorial perception had by this time among those interested in science because of the replacement of traditional Scholasticism with sensationism.[57]

In the case of the River Plate Viceroyalty, which included today's Argentina, Uruguay, Paraguay, and part of Bolivia, Miguel de Asúa has determined that the reception of Copernicus in the River Plate was "a protracted and erratic process." He concluded that the instruction in Córdoba during the Jesuit period was notably more supportive of the Copernican position compared to later Franciscan times, and the same applied to the secular clergy's teaching in Buenos Aires. De Asúa sees the expulsion of the Jesuits from Spain and its colonies as the result of the evolving political needs of the most powerful Catholic kingdoms that desired a Crown with more autonomy from Rome and more power over the "climate of opinion fostered in the European courts by those aligned with the Enlightenment and other anti-Jesuitical parties" that were powerful in this century.[58]

This culture favored local expression in educational affairs, and the eclecticism present in Spain also appeared in the River Plate area after the expulsion of the Jesuits; the report written in 1771 to request the creation of a university in Buenos Aires specified that "the teachers will not have any obligation of following a particular system, particularly in the teaching of physics, where it will not be required to follow Aristotle." Teachers would be able to "teach the principles of Descartes, Gassendi, Newton, or any of the other systems that explained natural effects through the light of experience based on observations and experiments developed by modern institutions."[59] In a speech in 1778, the philosophy professor Manuel José de Labardén (1754–1809) defended the sciences as aids in the process of knowing God, particularly in the study of nature: "The perfect coordination of the universe, the harmonious correspondence of its parts, the conformity of effects, [and] the perfection of the smallest thing are showing the wise hand of the Supreme creator."[60] The creation of the River Plate Viceroyalty in 1776 provided stronger arguments to improve education, particularly in the service of the merchants who were a powerful group in Buenos Aires.

The Colegio of San Carlos was created in 1783, but the desired public university was not. The creation of the Naval Academy (Escuela Náutica) in 1779 also promoted the study of mathematics in the colony and continued with the scientific impulse of the last part of the eighteenth century. In terms of gender these revolutionary ideas happened at a time when traditional categories were also changing; the importance of science and philosophy was related to the fact that by this time being a man or a woman was defined by education; men needed to learn how to become men, and women learned to use knowledge to grow their natural predispositions to socialization and motherhood. The original request for the creation of a university was supported by local leaders who wanted their children to learn some mathematics, geometry, and navigation because the sciences "gave men rules to become useful in the combat to win through art the resistances of nature."[61] The need for applied knowledge became even more important when the River Plate Viceroyalty's capital, Buenos Aires, required better-trained bureaucrats.

Jesuit education, not as rigid as its detractors claimed, allowed the study of sciences suited to local needs. For example, Aristotelian philosophy claimed that "life was impossible in the torrid zone," but Jesuit scientists knew by experience that this was not true. So, they did what scholars in the Americas would continue to do: they adjusted canonical writings to fit reality outside Europe. For example, José de Acosta (1539–1600) "corrected Aristotle's mistakes using Aristotelian precepts to soften his deviations from the master."[62] Similarly, José Sánchez Labrador (1717–98) of Paraguay and his associates continued to work on the understanding of the natural world according to their observations.[63]

Miguel de Asúa noted the work produced by Spanish American Jesuits "who wrote in defense of their lost homelands against those authors who argued for the inferiority of the New World, such as Buffon, the Dutch philosopher and diplomat de Pauw, the French Jesuit Guillaume Raynal (1713–1796), and the Scottish Episcopalian William Robertson (1721–1793)."[64] Those born in the colonies developed a "proto-nationalistic sentiment of a regional kind" related to local nature and science that made the transition to political patriotism easier by the end of the eighteenth century.[65]

SENSATIONISM AND THE CREATION OF A NEW MAN

Charles II, the last Habsburg king of Spain, was in power from 1665 to 1700 until his death without descendants allowed the House of Bourbon to rule in Spain; the first king of this family, Philip V (1683–1746), was considered "foreign" by many Spaniards. It is not surprising that in the same year as his coronation in 1700, a book written by Pedro Portocarrero y Guzmán (1640–1708) was published and dedicated to the dead king. Its author was a powerful priest, a chaplain of the court, a member of the aristocracy that had been part of the process that ended up choosing the new king. Written during previous years to provide a model of education for future monarchs, this work is described by Alejandro Cañeque as "a moderate compendium of the core political ideas that formed the foundation of the Habsburg monarchy"[66] One of the concerns expressed in the book was about the culture that would be introduced in Spain by a French king, since Philip was born Philippe, Duke of Anjou, and was the grandson of Louis XIV of France.

This cultural concern appears in the book's first chapter, titled "About the Loss of Spain," in which the reader is warned that when "the Spanish Nation kept its old clothes, it was always victorious," but once it admitted foreign fabrics, it was defeated.[67] The entrance of foreign fashions was partly the cause of a national decline, something that had already happened to the Visigoths, who had buried "the old courage because of the freedom of the customs," which made the vassals "effeminate" and eventually victims of tyranny (3). In another chapter, Portocarrero y Guzmán dealt with the problems caused by men who wore too many embellishments; he complained about the recent impossibility of recognizing whether a person was a man or a woman if they were not naked.

Foreign influence in Spain was making men less strong, and it was impossible to deny the "great decay of the Spanish Nation, if not completely in the lack of courage, at least in the strength of the body" because men were extremely devoted "to the cult of their persons," which was a foreign practice

(402). Since men were wearing these adornments themselves, they did not reprimand women for their dangerous use of them, which made them "not men but women, effeminate men who tried everything to deny their sex, including the way they walked, spoke, and laughed, and the clothes and shoes they wore" (404). They assimilated to women in everything because they wore clothing and behaved in ways that were foreign to real men, a denial of their true sex that put the nation in danger.

This problem originated in France, from the introduction of textiles that competed with the Spanish fabrics to the ideology that was replacing the ideas that had shaped the Spanish Empire. The text expressed clear concerns about Spain lagging in Europe, which explained why it was important for the new king to know the principles that would preserve the Spanish nation. Portocarrero's fears were prophetic. According to Richard Herr, the most important Enlightenment books read in this century were written in French. Étienne Bonnot de Condillac (1714–80) was a priest renowned for taking John Locke's epistemology to the extreme, asserting that all knowledge, judgments, and passions are merely different forms of sensation.[68]

Together with Condillac, Jean-Jacques Rousseau (1712–78) made sensationism relevant among those seeking to modernize the nation. The empiricism of John Locke (1632–1704), which had been resisted by the Catholic Church early on, dominated the formation of the Spanish Illustration (Enlightenment). René Descartes granted metaphysics independence from Scholastic theology, and, in turn, Locke established the groundwork for dispelling the notion that God instilled certain basic ideas in the minds of humans at birth.[69]

Philosophically, as explained by John Torrance, Descartes separated his physics from his metaphysics, which led to the development of the metaphysical systems of the seventeenth-century rationalists. However, the ideas of Malebranche, Spinoza, and Leibniz were ultimately discredited and supplanted in France by the empiricism of Locke and Condillac.[70] At the same time, those who only followed Cartesian physics, mostly physicians, became anti-metaphysicists and initiated a crucial shift from philosophy toward science. This was viewed at the time as a mechanistic view of man that inspired the term "ideology," and emerged from the same early French Positivist circle that made the distinction between science and metaphysics.[71] These divisions were essential to understanding the progression of ideas into the nineteenth century.

In 1746 Condillac published *Essai sur l'origine des connaissances humaines*, considered his addition to Locke's *An Essay Concerning Human Understanding*; three years later *Traité des systems* was published, and was followed by his *Treatise on Sensation* on 1754. His impact in Spain begun to be felt within a couple of decades as is reflected by the Spanish translations that were published to spread his ideas, including in Mexico by 1828.[72] His book of logic appeared in 1780, and a translation in Spanish, done by Bernardo María de Calzada, came out four years later.[73] He became known in Europe because in 1758 he was the supervisor of the education of Ferdinand (1751–1802), the future Duke of Parma, and his sister, María Luisa (1751–1819), the future queen of Spain and mother of Ferdinand VII (1784–1833) through her marriage to Charles IV (1748–1819). Condillac remained as a tutor until 1767; the next year he was elected as a member of the French Academy.

The plan of education he developed was printed in 1775, *Cours d'études*, which connected sensationism with a concrete pedagogy that differed from Jesuit education.[74] The growing influence of the French philosopher resulted in the banning of his *Cours* by the Inquisition in 1789, but the enlightened Spaniards read the philosophy of Condillac and Montesquieu because they were unaware that these authors could pose a threat to religious faith.[75] It was difficult to see the new ideas as dangerous at a time in which many priests were advancing the idea that humanity had entered a new era of transformation.

Henry Martin Lloyd points out that the philosophy of the French Enlightenment was not marked by a tendency toward abstract reasoning as was the case with Kant; it was largely characterized by a distrust of such an understanding of reason and a shift toward corporeal sensibility.[76] This meant that rationality was predicated on the senses and "not on a transcendent faculty of reason."[64] Empiricism was a thought associated with Great Britain at this time, but, as Rebecca Haidt establishes, the interest in embodiment that defined Spain's Illustration did not meant that the illustrated man was a philosophe. The Spanish refrained from addressing broader philosophical issues, such as those questioning the power of God.[77] As mentioned earlier, the method that replaced Jesuit education was eclectic and not defined by a particular philosophical dogma.

This attitude is exemplified in *Espiritu de los mejores diarios literarios que se publican en Europa*, a journal edited by the well-known theologian and lawyer Cristóbal Cladera (1760–1816) with other prominent supporters of Charles III's reforms. It was modeled after a popular French publication approved by the king of France and published in Paris by a "society of authors" (Société des gens de lettres), titled *L'Esprit des journaux français et étrangers* (*The Spirit of French and Foreign Newspapers*). This Parisian monthly promoted Enlightenment ideas in Europe, too, and its Spanish imitator did the same in the colonies. The first issue came out in 1787 to advocate for a modern Spanish culture, but its publication ended in 1791 when the ideas it presented became dangerous due to the French Revolution. The intended audience was "curious and literary people," and the material came from articles selected from publications committed to economic and ideological reforms.

In November 1787 the description of *Espíritu*'s mission highlighted that the eighteenth century was "*the most scientific*" of the previous seven thousand years. The "modern" culture was superior to the ancient in "*Philosophy, Moral, Mathematics, Industry, Commerce, and everything that was not sublime poetry*," though it could be imagined that poets would also reach new heights soon. This knowledge was directed to "*the perfection of what man*" needed to know, and these ideas had revealed that man was the "*inhabitant of the World of Stars*," a new position in the universe supported by physics, natural history, and chemistry. In brief, "all nature was deemed to subject itself to the study and inquiry of the rational being that inhabited it."[78] This cosmic positioning was part of the aforementioned universalism.

Since letters were the preferred medium for transmitting knowledge, some were published to introduce complex philosophical ideas in a simpler manner. One of them, written by a future participant in the French Revolution, François-Martin Poultier d'Elmotte (1753–1826), described a young woman interested in philosophy. He explained that she should give preference to the ideas of Condillac, which explained "in an ingenious, clear, and methodical way the operations of the soul."[79] Acknowledging that the ideas of this philosopher could be too difficult for a woman's intellect, he tried to explain this philosophy in a simple form, proceeding to mention how language was formed in her mind to articulate the properties of objects related to her needs, the different qualities of things, and to indicate the intended use of what she named. From this, it was clear that "all these notions were received from the senses," and without them, she would know "absolutely nothing." This revealed that "*sensations are the beginning, and the origin of all our ideas*."[80] As is clear, Condillac could be presented in a way that was not contrary to religious dogma when his philosophy was described as interested in a basic process of cognition.

Condillac's philosophy emphasized the life of the mind deeply intertwined with conscience, rather than the body.[81] Jean-Jacques Rousseau was another philosopher who developed a plan of education for boys and young men. In *Emile* he asserted that there was "no parity between the two sexes as to the consequence of sex, the male is male only at certain times, the female is female all her life, or at least all her youth," which explained why man's identity was more elusive and in need of education to establish it.[82] Men needed to learn how to become men, which was associated with the modern culture that was becoming more prominent and distinctive. This tied scientific and philosophical education to gender and politics.

In the British case, educational plans were also gendered and proposed forming a new kind of man, a more refined and sensible individual who replaced both the feudal lord and the materialistic merchant. In his writings about education, for example, Locke wrote a section on craving that differentiated "natural wants," such as hunger, thirst, cold, "or any other necessity of nature," from those "of fancy and affectation." The former should be relieved by the parents, but the latter "should never, if once declared, be hearkened to, or complied with." The purpose of these recommendations was to teach children to have "mastery over their inclinations and learn the art of stifling their desires as soon as they rise up in them," something that would be "of great use to them in the future course of their lives." This did not mean they should not be pleased, but that they needed to practice "modesty and temperance" to recognize that what was bestowed on them was "a natural consequence of their good behavior," which naturalized the moral values needed in adulthood.[83]

Anthony Fletcher's and G. J. Barker-Benfield's analyses of seventeenth century England indicate that the polarity of sexual difference was defined by men through the concepts of effeminacy and manhood—or, conversely, strength and weakness. Masculinity, in this context, suggested a constant awareness of the risks associated with allowing emotion to dominate men's actions.[84] The ambiguity toward emotion and sentiment that started to dominate in the next century was related to the tension around the role of men's sensibility that coincided with economic expansion, consumerism, and the introduction of new fashions following economic changes and the new importance of beauty and refinement. This consumer culture that emerged from 1650 to 1750 was the result of the increased wealth circulating in the United Kingdom, which explains why men like Adam Smith (1723–90), for instance, regarded the refinement of feeling as the result of histories shaped by manners. He constructed a historical hierarchy, placing very high value on the most "cultivated" and "civilized" societies.[85] This culture of sensibility that developed in the late eighteenth century identified civilization with the capacity to analyze the sensations that emerged in the presence of beauty and harmony, both central to the formation of morality.

In terms of gender, Brian Vickers suggests that *The Man of Feeling*, written by the Scottish lawyer Henry Mackenzie (1745–1831) and published in 1771, implied that "to 'civilize' was also to 'feminize.'" While this was seen, on one hand, as a positive development, it also carried the risk of rendering civilized men effeminate and weak.[86] As a consequence, being a modern man became a difficult task that required an almost impossible balance, as Gillian Williamson has indicated. The "civilizing influence of female conversation in mixed gatherings—such as at the tea table, in assembly rooms, public walks, and gardens—was considered essential." However, excessive engagement in frivolous interactions with women and the worlds of fashion and shopping associated with them could lead to the feminization of a man.[87] Among proponents of the Enlightenment, this would lead to debates about what was the right formula to avoid the dangers brought by too much or too little feminine influence.

Condillac's ideas on masculinity presented a different angle on this dynamic. He gave the imagination a larger role, defining this concept as occurring when "a perception, by the mere force of the connection that attention has established between it and an object, is recalled at the sight of the object." This was different from Locke's understanding because "we may very well remember a perception which we do not have the power to revive."[88] Imagination revived "perceptions themselves," memory recalled "only the signs or the circumstances," and reminiscence reported "those we have already had" (30). Unlike animals, humans' use "of different kind of signs" contributed to the progress of the imagination, contemplation, and memory (40). These three operations were central to the understanding of the human psyche.

This sensibility emphasized the representational and gave activities such as reading a crucial function in education because of how books dealing with arts and sciences contributed to the training of perception. Whereas Locke understood the education of boys as learning how to master desires, Condillac's principles emphasized how to make boys "masters of the exercise of imagination" (41). Arts were crucial because in seeing a painting, for example, we "recall our knowledge of nature and the rules that teach us to imitate it; and we direct our direction from the painting to our knowledge, or in turn to its different parts." An enlightened man was defined by his will, action, and imagination. If not for imagination, "we could not rule ourselves, for it would obey only the action of objects." Reflection was the most important activity, and education should be organized around teaching how to become a reflective person (42).

Language, the most complex system of signs, acquired great importance in the work of Condillac, as he was the first one to pay attention to it. Language had to be taught together with geometry, because "the best means of facilitating our reflection is to place before the senses the very objects of the ideas that are our concern, for then our consciousness of them is livelier, but this is an artifice that cannot be used in all sciences." Meditations should be clear, precise, and ordered; they depend on the individual's talent and could be acquired by "doing violence to our native faculties" or be "the result of a happy disposition and a great facility for their development" (43). The latter was in "greater conformity with nature, more lively, more active," and produced "greatly superior effects," while the former only led to mediocrity. Education should be a happy experience, but Condillac noted that reading romances was dangerous for women, because they had impressionable brains, "too little engaged in education," and they eagerly seized "fictions that flatter the natural passions of their age." But "even women" were able to master other languages, such as music and dance (126). A gendered language had its origins entirely in "the difference of sex" and nouns introduced "two or three genders" to bring "greater order and clarity to language" (168). The emphasis of ideology was on a precise organization of ideas.

Antonio Gregorio Rossell (1748–1829), a mathematician and naturalist who studied philosophy in Valencia, moved to Madrid in 1770 and became a professor in the Seminario de Nobles where he was among those in charge of providing new educational programs for students. He added courses in arithmetic, algebra, and geometry arguing that this kind of knowledge was useful "for all educated persons," which contributed to the popularity of Condillac.[89] The latter's influence was spread through his *Logic* in Rossell's 1785 book on mathematics, because he reduced "Mathematics to Algebra saying: 'The language of mathematics, algebra, is simplest of all languages.'"[90] In turn, Rossell explained that Condillac had affirmed "that Algebra was a language and could not be anything else," which pointed to a crucial aspect of this philosophy—the intersection between language and sign.[91] In this method a sign emerged in the form of a general idea connected to a direct experience, freeing the mind from being bound by sensation and instinct as animals were.[92] This philosophy implied that the process was natural, and in humans had "to be innate."[93] Imagination and language were what separated humans from animals, which made the creation of a culture of sensibility indispensable.

All philosophical systems derived from Locke emphasized sensation, but there were important differences among them. While this English philosopher did not attribute to matter any vitality, and believed it was ruled by mechanistic laws, Condillac understood matter as having vitality and activity, as having sensitivity, which favored positions related to the understanding of the environment and experience.[94] This explains why the teaching of a new sensibility, of different sensations through a transformed environment, became very important in France, Spain, and their colonies.

This brand of sensationism had detractors—many in powerful institutions. One was Simón de Viegas y García Arnáiz (1740–1811), a powerful lawyer and member of the Spanish Royal Academy of Law, who published a book in 1799 that examined Condillac's ideas. He confessed that each of Condillac's explanations was for him "a very dark enigma" that made it difficult to decide between the reputation Condillac had earned and the one he deserved.[95] Because Viegas believed that truth was found through induction and not in sensations, one of his main objections was the role that sensations played in knowledge.[96] The appealing idea that "all men could be equally knowledgeable" was not acceptable to him; Condillac's view of nature was "a purely mechanistic work" in which reason had no role, which was troubling.[97] While aspects of this interpretation were inaccurate, it was a common criticism in Spain.

Those who attacked Condillac understood nature as a force that conserved and preserved an order, while for the followers of the French philosopher, humans did not distinguish different classes by following the nature of things but according to their way of conceiving ideas.[98] There were as many classes as necessary for classification because of all the relationships that humans discern.[99] Thus, there were different perceptions of nature according to the context and the relationships established in each environment, which was not the same as believing in nature as real, universal, and created by God for humans to enjoy.

The development of the sentimental component linked to this kind of philosophy in France was important. Jessica Riskin explains that recognizing these developments helps us understand the relationship between the natural sciences and moral and political thought in the eighteenth century. This led to a close connection between the natural sciences and the emerging moral sciences, which, in turn, brought together epistemology and psychology with questions of proper conduct and governance. The moral sciences were seen as addressing topics such as political economy, civic education, and law—subjects that were examined through scientific perspectives.[100]

THE TROUBLED MASCULINITY OF *MAJOS*, *PETRIMETRES*, AND *CURRUTACOS*

The War of the Spanish Succession (1701–14) ended with the Bourbon dynasty ruling both Spain and France. This resulted in the encouragement of trade and cultural exchanges between the kingdoms "along with an increasing disregard for the sumptuary laws that had governed dress and commodity consumption throughout the seventeenth century, and a mounting concern about the threat of foreign competition to national industries."[101] This was exactly what Portocarrero had feared because of how foreign ideas also influenced gender roles in society. His prediction turned out to be true when a new practice originating in Italy and France was introduced in Spain in the 1720s. From the Italian word "cicisbeo," the "chichisveo" was defined in 1729 as a kind of courtship, "a courtesan service that a man offered a woman," which was not condemned if it did not lead to a sexual relationship.[102] In 1797 it was translated into English as "to pay court to a lady with gallantry and a refined spirit," involving "a whispering court and attendance paid to the sex [women], even to married ladies," and typically performed by young men. This practice broke the strict social separation between men and women that existed in Spain.[103]

Spanish critics often viewed it as a foreign import that threatened the nation's values at a time when Spain did not have the military and naval prowess of the past, leading to perceptions of decline and weakness, as was the case with Portocarrero. In Italy and Spain, the 'cicisbeo' and the general organization of gender were linked to their classification as unmodern at a time when a distinct differentiation was growing between the north and south of the European continent. Italian and Spanish thinkers who believed in the Enlightenment viewed their nations as decadent and embraced reformism as the only way to become morally and economically updated.[104]

In Italy, for example, the analysis of Italian literature included accusations of Spain's backwardness and lack of good taste, which had damaged Italian literature because of Spanish influence during their occupation of Naples and Sicily. In 1756 the successor of Muratori as Modena's archivist and librarian, the Italian Jesuit Francesco Antonio Zaccaria (1714–95), rejected this view as "insulting to the good name of Spanish literature" because it described the nation that in the sixteenth century had given "so many learned and immortal men in every science" as "buried in a horrid dark night."[105] Since Spain was an ally of the Church, this criticism was politically charged.

In Spain, new forms of sociability were criticized as a threat to virile culture. In 1729 Father Joseph Haro de San Clemente (1658–?) published his criticism of "chichisveo" denouncing "men who made themselves more respectful and venerable when using Spanish clothes," who were "so effeminate" because of the adoption of foreign fashions, that were "confusing the sexes" (*confundidos los sexos*). Men exchanged big swords for smaller ones "that looked like toothpicks" demonstrating by this choice "their womanly inclination [*mugeril inclination*], or, better said, their natural effeminacy [*afeminado natural*]," which ended in an emphatic declaration to define modern times: "Truly, there were not men anymore!"[106] In a chapter devoted to answering the question, "What is a woman and which ones are her properties?" men are warned that "by nature women had ambitions to lead, and to be free, and wanted to invert the natural order [. . .] to dominate men."[107] As a consequence, men needed to abandon the new culture that made them inferior to women, who were natural enemies of men's power and supremacy.

This was not a universal criticism; for example, the Benedictine Benito Jerónimo Feijóo y Montenegro (1676–1764) wrote *Defense of Women* in 1726, a work "intended to prove women's moral, physical, and intellectual equality to men."[108] He attributed the idea that women were inferior to men to the latter's control of education: "If women had written books, we [men] would be the ones being put down."[109] By the end of the book, Feijóo mentioned the tension that existed between the old and new ideas, which left "women unhappy and in a depressing state" to watch because before she "only listen[ed] to contemptuous words," while in the present she saw herself "elevated to the sphere of a deity." In the past "she was told that she was dumb, now she hears that she has divine understanding"; according to her husband "she was full of imperfections," among her gallants she was now "full of graces." After marriage, her spouse "ruled over her as a tyrant owner," but in the salons her male admirer was her "confined slave," which explained why she found "the difference between the two like that between an Angel and a brute." She was presented "with slavery on one side and empire on the other."[110] Both views were considered wrong because they were created to satisfy men's needs, and Feijóo wanted women to develop their own knowledge of who they were.

Feijóo explained that his views about the sexes were supported by Jesuit thinkers. In 1769 he noted that when he first began to think about the "the equal intelligence of women," he did not know of other authors who defended this idea. Later, though, he encountered the French Jesuit P. Buffier, who had written a book made of dialogues, including one "destined to prove the egality of intelligence in the two sexes." Other Jesuits had written that "the vulgar opinion that men had more intelligence than women was an unfounded concern."[111] He continued listing other books written about women's abilities over time.

The socioeconomic and cultural changes Spain experienced caused significant shifts in gender roles and behavior during the eighteenth century, particularly among young men. In 1734 one notable character appeared and was defined in the dictionary—the "majo" was characterized by his demonstration of courage (*guapeza*) and bravery in both actions and words. *Majos* typically lived near the court, suggesting a certain social level.[112] By 1744, a French–Spanish dictionary associated this kind of young man with the "compagnon" in France, understood as a ruffian.[113]

In 1787 Isidro Bosarte (1747–1807), an enlightened individual who served as the secretary of the Count of Aguilar and traveled extensively across Europe, wrote an article about "majeza." He defined this term as "low elegance in clothing," explaining that *majos* and *majas* adorned themselves in an exaggerated fashion in the small towns (pueblos) of Spain, as an expression of popular culture. According to Bosarte, the term originated from the word "mayo" (May), the month when various nations celebrated with dances, decorations made from natural flowers, and other cultural expressions common in rural Spain. Over time, the word evolved into "majo," signifying unrefined and overly ornate embellishment.[114] In Spain Bosarte considered Andalucia the capital of *majeza*, which had left the area in a state of decadence that called for the restoration of its former glory. Bosarte claimed that any society characterized by *majeza* lacked "good taste in the arts." For him, good taste stemmed from the arts, just as truth originated from the sciences. Furthermore, he argued that the arts were perfected through "a deep philosophy and sublime enthusiasm." Clearly, the *majo* symbolized the rustic countryside, which, in Bosarte's view, lacked the elevated knowledge derived from philosophy, science, and the arts—the foundations of modern culture. This perspective reflects a clear dichotomy between rural and urban life, a contrast frequently emphasized in Enlightenment thought.[115]

In 1745 a Spanish translation of the French version of a book written by the Swedish aristocratic philosopher Johan Turesson Oxenstierna (1666–1733) was published. It contains a section on the *petrimetres*, describing them as the "whip of society." They were considered a "disease that had infected the youth," with each country having its own version. The worst offenders, however, were among the nobility, whose customs spread to those in the cities, who, in turn, infected rural populations. In a satirical tone, the author explained that defining trait of these men was their thinking and acting differently from others, as portrayed in a comedy presented in London, where the Marquis de Polainville teaches Milord Houzee to distinguish a *petimetre*. The latter asks how to become one and is told it requires being born presumptuous, vain, a buffoon, with "a taste dominated by pleasure, vices, and an extreme love of

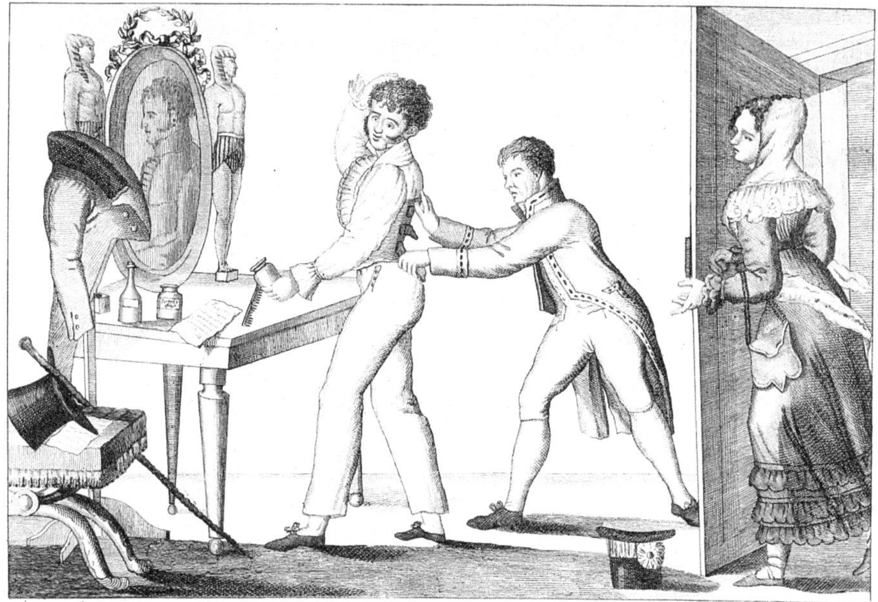

Figure 1.1. “The Elegant *Petrimetre* or the *Lechugino* in His Dresser,” 1770–1800s. Courtesy of Museo de la Historia de Madrid. Inventory no. 4916.

vanity and gallantry.”[116] This reference, from an earlier period around the time of the dictionary definition, highlights the widespread circulation of the figure across Europe, indicating the simultaneous emergence of this type.

In 1792 *Diario de Madrid* reproduced a letter from Thomas Warton (1728–90) about Madrid, presumably written a few years prior, praising Spanish men for their “admirable predisposition for science” and his communications with many who were educated in “mathematics, physics, chemistry, natural history, etc.” They deserved praise because they had the talent to “elevate their knowledge”; they did not have the “lightness and lack of constancy” that made men “effeminate” and concerned only with “the present moment.” Instead, they were “firm” when they made decisions, and it was “impossible to change their minds,” an excellent quality because they were inclined to “the good and beautiful” and were “steadfast in keeping their principles stable.” Warton also praised the Spanish women who wore cheap clothes and “were sweet, pleasant, and circumspect.” Despite the lack of luxury in their clothing, these women possessed great “grace” and had “very natural manners.” More important, they did not “talk much around foreigners and observed with curiosity and admiration" their surroundings. This contrasted with men like the *petimetres,* who were living as if they were leaves in the wind.[117]

Around 1790 the worries about the *petimetres* were replaced by denunciations of a new urban character, the *currutaco*: he was short, unarmed, loud, obsessed with dancing the contra dance (*contradanza*), which represented the "modern taste." This character was not, as in the case of the *petimetre,* a representation of tension between foreign and national. Elizabeth Amann notes a different understanding of French politics in the treatment of the figures of the *petimetre* and the *currutaco,* whose ideological position was not clear-cut. The *currutaco* offered "a space—an alternative society of sorts—for reacting upon, reenacting, and working through the tensions and traumatic scenarios that were being played out across the Pyrenees."[118] By the 1790s "currutaco" and "currutaca" became the terms used to describe young men who understood being modern in a problematic way both in Spain and its American colonies.[119] Indeed, the traumatic events in France after 1789 changed attitudes about the culture of the Enlightenment because they demonstrated the destructive side of the era that made those in government very concerned about change, especially change in the culture of men.

One example of the literature about the *currutaco* was a satire written by the Augustinian priest Juan Fernández de Rojas (1750?–1819) about the gender problems caused by foreign ideas in Spain.[120] This author was a man of the Enlightenment who had a good relationship with intellectuals like Gaspar Melchor de Jovellanos. One of his satires pretended to be a philosophical and scientific study to determine the meaning of the new manhood. He began by criticizing the hyperbolic language with which men spoke to women, and men's adoration for the feminine. This exaggerated praise of the female was according to new philosophical principles claiming that "all is linked by immutable laws." As the character of the *currutaco* lover explains, "The ideal pleasure alienates me. . . . My spirit gives in to the sum of sensations that are sublimely delicious. . . . I am in ecstasy. . . . Ah! I am the happiest of all beings with feelings," which is his creed (v–vi).

This character was the result of obsessive interest in communication, which dominated the century, when "all the globe forms a single nation, a single people, a single family." In ancient times "there was more difference between a habitant of Byzantium and one of Greece than now between an Englishman and an Iroquois, because the nations communicate and know each other more." Those in Scandinavia dressed like Parisians, and those in Asia lived like those in the center of Europe; there was "no difference in customs." This explains

Figure 1.2. Manuel Albuerne and Antonio Rodríguez, "*Currutaco* in Frock Coat," 1801. Text: "I am going to see her before she leaves." Courtesy of Museo del Prado, Madrid, Spain. Catalog no. G005710/007.

why there were *currutacos* everywhere and they could not be differentiated; this was a universal male identity that resulted from all the forms of communication that were established at the time in ideas and economic trade (18–19). A young man was "regarded as a philosophical prodigy" if he understood that philosophy no longer had any plan: "There were few or no plans, and the ones that existed were part of Mathematics. There was no method: on the contrary, I told you that a beautiful mess is the best order" (9–10). There was plenty of talk "about philosophy and humanity," together with "pompous and poetic descriptions." Satire and mordacity were common, and the *currutaco* "let himself be taken by the imagination," inventing comfortable dreams that were mixed with "sublime reflexions" condensed by these activities: "paint, rave, declaim, apostrophize, satirize, hurt," the pure forms of *currutaquería* (10). In another satire written by Fernández de Rojas under the pen name Licenciado Francisco Agustín Florencio, we find the same concerns and doubts about the changes brought by the new philosophy and science. Written in the first person, Florencio, "a *currutaco* philosopher" represented an erudite man who was tired because the abstract sciences had "molded him in a way" that he "only [knew] how to think."[121] This character wanted to be an inventor, and for this reason had decided to "invent a happy, fun, festive science" that reflected "the genius of my compatriots," which resulted in "Crotalogy [*Crotalogía*] or the Science of the Castanets"; important knowledge because it would be used for dancing the bolero, a Spanish dance, unlike the contra dance, the fashion dance of the elite youth.[122]

In another satire, written circa 1804 by Juan Jacinto Rodríguez Calderón (1770–1835?), the main character, Roque, or Don Líquido (Sir Liquid), is a young man who had become a *currutaco* through the study of the "science of dressing according to fashion," a quite complicated feat that required the mastery of several fields. Originating in the threat that this kind of man presented to society, the reactions to men like the protagonist were bad: "Many hypocrites call us half-men, fags, and dolls." In the end, though, Roque realizes that the clothes had been introduced in Spain by "pride and fanatism" and were the source of all the bad things that had happened to him and to society since they had imposed a new idea of "honor, blood, and lineage." Fashion made him "hallucinate," and from being "a good Spaniard" he had become a "doll," "fainthearted, clumsy, and effeminate," who was distracted from "knowing the fatherland [*patria*] that was his passion." The play ends with his promise to change after "the Supreme God" has opened his eyes.[123] The relationship between philosophy, science, and a liquid, soft culture that was against the nature of the fatherland was popularized with references to sodomy as a cure to restore the natural order.

In the "*Currutaco* de Sevilla," the old, natural order dominated in the end. Written by an anonymous author around the 1800s, the poem tells the story of a heterosexual couple of *currutacos* planning to sodomize a young student who had tried to seduce the woman, using a syringe to perform an enema. At the end, the process did not work because the one receiving the treatment is the *currutaco*, who ended up sodomized by his male rival. Finally, the author warned young fashionable men that this could happen to all of them, noting that he "did not know science because he was a peasant," a fact that had saved him. This characterization elucidated the opposition between the urban and rural populations, and the view of the inferiority of science in the face of natural authenticity. Also, this narrative follows Rebecca Haidt's characterization of effeminacy as "the staining by men of male opponents based on the scrutiny of signs of the feminine."[124] Mehl Allan Penrose agrees with this characterization, adding that one way "of marking another man as non-man or inferior was to humiliate him by forced anal penetration."[125] In this case, since the woman is the one manipulating the situation, the man is punished for following a woman.

Figure 1.3. "The *Currutaco* de Sevilla." Text: "El currutaco de Sevilla: . . . in which is recounted the most amusing misfortune that befell a Currutaco at the hands of a Student, for while the Currutaco thought he could use the syringe on the student, the student ended up using the syringe on the Currutaco, along with other details that anyone who is not blind will see." [En el que se declara el más gracioso chasco que le sucedió á un Currutaco con un Estudiante, pues pensando geringarlo el Currutaco, fue geringado por el Estudiante, con lo demás que verá el que no fuere ciego.] (Barcelona: Ignacio Estivill, [18–?]), Generalitat de Catalunya, http://hdl.handle.net/10687/293640.

It is important to note Cristian Berco's explanation that in the Spanish tradition, the accusation of sodomy implied different sexual acts; depending on place and time, sodomy could refer to anything from a broad interpretation of nonprocreative sex to a more specific definition focused solely on anal intercourse. Sodomy and patriarchy, intertwined since early modern Spain, are part of the context of the *currutaco*'s stories; the devaluation of women "meant that their sense of self, the masculinity they so intensely sought in a male-dominated society, was partly defined by what it was not: femaleness."[126] The criticism of foreign knowledge, science, associated with gender reveals the political character of the piece in its rejection of the changes that Spain was undergoing.

SCIENCE, SENSATIONISM, AND GENDER IN THE AMERICAS

Bogotá, Caracas, and Lima were the cities that led the promotion of the new ideas in South America. In Lima the first newspaper in colonial Spanish

America, *Diario de noticias sobresalientes en Lima y las Noticias de Europa*, was published between 1700 and 1711.[127] It was followed in 1715 by *Gaceta de Lima*; but the most relevant publications appeared in the 1790s: *Diario de Lima* and *Diario curioso* (1790), *Erudito, económico y commercial*; *El Mercurio peruano* and *El Semanario crítico*; and *La Gazeta de Lima* (1791). *La Sociedad Académica de Amantes del País* (the Academic Society of Lovers of the Country) was created in 1790 to follow the models of the societies that promoted the Enlightenment in Spain and Europe. The same year it published a proposal for the creation of a newspaper, *Mercurio peruano*, because the "existence of newspapers could indicate when illustration happened in the nations."[128] Many newspapers were already known by the intellectual elites, such as "*Gazeta*, *Mercurio político*, *Diario*, *Semanario erudito*, *Censor*, *Corresponsal del censor*, *Apologista universal*, *Espíritu de los mejores diarios de Europa*," and "*Memorial literario*."[129] This list shows the increasing interest in reading and the news among the minority that was able to read the papers.

A new educational plan designed according to new ideas was created by Francisco Moreno y Escandón (1736–92) in 1774 for the New Granada Viceroyalty, but it failed. It was based on the method of Luis António Verney used in Spain. Modern science and philosophy were introduced in Caracas by the city's native priest Baltasar Marrero (1752–1809) around 1788 as part of the policies of the new Captaincy of Venezuela created by the reforms of Charles III the previous decade. He taught algebra, geometry, Newtonian physics, Kepler, "and the chemical theories of [Charles François] Duvy, [Georg Ernst] Stahl, [Antoine-Laurent] Lavoisier, and [Johann Koch] Cosh," in addition to the theories of Benjamin Franklin and Alessandro Volta.[130] In philosophy he introduced Descartes, Gottfried Wilhelm Leibniz (1646–1716), George Berkeley (1685–1753), Locke, and Condillac—the latter two being the most influential. Andrés Bello (1781–1865), a writer and teacher, was one of Marrero's students, who in 1800 wrote a dissertation on Condillac that used his ideas about language.[131]

As will be discussed, forty years later, while living in Chile as the greatest thinker in the country, Bello wrote a book about grammar that followed Condillac, the author of *Logic*, writing that "a severe logic is an indispensable requirement of all teaching." He also affirmed that "the language of the people [is] an artificial system of signs that in many ways is different from other systems of the same species, from which he assumed that each language has its own theory, its grammar."[132] Also in 1800, Bello met the Prussian naturalist Alexander von Humboldt (1769–1859) who arrived in Caracas at the beginning of his expedition with Aimé Bonpland (1773–1858).

In the neighboring New Granada Viceroyalty, scientific exploration had already begun with the arrival of the Spanish naturalist and priest José Celestino Mutis (1732–1808), a disciple of Linnaeus and a member of the Universalist School already mentioned.[133] As the director of the Royal Botanical Expedition of the New Kingdom of Granada that began in 1783, Mutis met with von Humboldt and Bonpland in 1801, and by the end of the same year had an encounter with another Creole naturalist, a self-taught scientific prodigy from Popayán, named Francisco José de Caldas (1771–1816), who had gained Mutis's attention for his new method to calculate altitude. While in Ibarra (in today's Ecuador) for personal reasons, de Caldas intended to impress the Prussian naturalist so the latter would include him in his expedition. He had Mutis's recommendation and economic support for the position. The plan did not work because in Quito Humboldt had established a close intimate friendship with Carlos Montúfar y Larrea (1780–1816), who, unlike de Caldas, belonged to the colonial aristocracy. Once this relationship became close, Humboldt decided to take the young aristocrat with him for the rest of the expedition, crushing the hopes of de Caldas, the naturalist who believed, with reason, that he was more qualified for the task.[134]

In a well-known correspondence with Mutis, de Caldas denounced his rival Montúfar as a "young currutaco" who was ignorant and dissipated, and criticized Humboldt's choice as a bad lesson in "behavior and morals" by the prestigious European scientist.[135] In Quito, "a Babylon," Humboldt surrounded himself with a group of young libertines who frequented houses where "impure love" was practiced, and he followed them possessed by "this shameful passion of his heart" that blinded him to a point "that could not be believed." Irritation made de Caldas imagine Newton returning from ashes, "the Newton that did not turn into a woman," to tell the young Prussian "in an irate and terrible manner" that he needed to change and abandon the pleasures of a "dark and effeminate life," which contradicted his scientific pursuits.[136] The behavior of scientific men associated with emotion and fashion was criticized both in Europe and the Americas because, as we have seen, different notions of masculinity coexisted inside the universal model of the Enlightenment. Ironically, both de Caldas and Montúfar would work for the same cause in the 1810s when the independence movement united them as patriots fighting for the freedom of their nation. Montúfar became a colonel in the army, an aide of Bolívar, and was executed by the Spaniards in 1816, as was de Caldas.

This denunciation of the homosexual passion that was behind the scientist's preferences was debated in the context of analyzing Humboldt's life, but the interesting detail here is how science appeared linked to a new sociability that led to a manhood that destroyed men morally. De Caldas was of a lower class, had to work, and was passionate about science. His rival was wealthy, well educated, and part of the young elite that took science as part of the new homosocial culture that some believed feminized men of Montúfar's social standing. Regardless of his motivations, de Caldas's complaint about the "Adonis" indicates how science intersected with gender and sexuality.[137] As in France, sentimentalism—the idea that emotions stemmed from physical sensations and served as the foundation of social life—was not limited to the feminine sphere of sociability; these concerns also played a significant role in the primarily male-dominated French sciences.[138]

In 1799 *El Currutaco por alambique*, published in Mexico, enjoyed popularity and was reprinted more than once. Written by a priest who was also a poet and rector of the College of Mines, Manuel Gómez Marín (1761–1850), this satire also denounced the evil example the new sociability introduced among elite youths, which explains the censor's comment that this book "will be an aid to the customs and will maybe repress and solve an indecent excess that makes men effeminate."[139] This satirical poem told the story of a gathering of devils who were bored and decided, with the help of chemistry, to mix all the evil things they could imagine, from which the *currutaco* resulted. "[The *currutaco* was] a hermaphrodite doll / whose dress and figure, / semblance and composure, / presented a person / who created a doubt about whether it was a male or female monkey: / the body seems female, / but the soul is of a macho, and well deserves / the name of hermaphrodite / who being a macho imitates the female."[140] In this version, the devils used science to experiment and once they saw the result they sent the new species to the world, where "the lessons of this monkey and its doctrine" spread all over.[141] Gómez Marín was also a professor of theology who had "introduced the teaching of modern philosophy" in the Real Universidad de Mexico, where his students learned "mathematics, physics, and chemistry."[142] As a naturalist, he was a man who favored the ideas of the Enlightenment, as did some of the *currutaco*'s critics in Spain. All of them questioned the influence of gender on the understanding of the new philosophy and science, because the *currutaco* was the result of an economy of spending and not of ideas and knowledge.

This character was also present in Lima and Buenos Aires. In Buenos Aires, social activities and refined sociability had been elevated by the existence of a viceroyalty court in 1776, which concerned some members of the elite. This process of refinement and good taste changed the city and the cultural attitudes about wealth and consumerism. Over the eighteenth century, what was once a small, modest village evolved into the gateway of a major road network as the capital of a viceroyalty, which transformed the economic and political structure of South America to the advantage of Buenos Aires. This resulted in a swift increase in population and housing development, which transformed the city's landscape. In 1750 Buenos Aires had 13,786 inhabitants. With the progress of commerce, the Spanish Crown promoted migration to the colony, resulting in the arrival of mostly men from Castille, Leon, Asturias, and Galicia. The total number of immigrants was close to 2,000, which increased the total population of Buenos Aires to 24,363 by 1780 and more than 40,000 in 1810.[143] The expansion of commerce began in 1778 with the Regulation for Free Trade (Reglamento para el comercio libre) that left behind the monopoly and, following new economic principles, formed a larger network of commercial exchanges with permission for trade granted to thirteen ports in Spain and 24 in America.

Buenos Aires obtained its first printing press in 1780 when the authorities decided to transfer one from Córdoba so the city could publish its own newspapers and books—the culture of the Enlightenment stimulated the emergence of new publications. In 1801 The *Telégrafo mercantil, rural, politico oconómico, e historiográfico* was published by Francisco Antonio Cabello y Mesa (1764–1814), a Spanish writer and soldier who arrived in Buenos Aires from Lima, where he had been involved in the publication of *Diario curioso* and *Mercurio de Lima*. The new publication was created in imitation of those from other colonies, like Peru, and from Europe. In its application to obtain approval from the censorship authorities, Cabello explained that the goal was to instruct the inhabitants of the River Plate Viceroyalty with "useful news" that enhanced the "immense wealth" of the colony in "the three natural kingdoms" that could "alter the systems praised by naturalists while simultaneously showing the advancement of science, of speculation, and of the fine lectures given by its professors."[144] Cabello y Mesa also stated that concepts would not be abused and that articles would take care to combine "Religion, Politics, Instruction, and principles" using a "soft, thoughtful, censorship" to avoid explosive content (6).

The editor's intention was to turn politics "into a complete system" using "a new method, scientific, clear and advantageous" (11). Finally, he asserted that he would "form a complete Theory of the Science of the proposed Political System," but only the part that was "coherent with our persons, interests, and circumstances, and according to the 'Laws of the Patriotic Literary

Society and Economics Society'" that he was forming (12). The main objective of this publication was to affirm that all philosophers claimed and defended "the desire to be happy" as "the first and only goal of men."[145] It also clarified the position of the editors regarding luxury, which, following sensationism, was not understood as "the number of superfluous things" one could have, but as "the attachment that man has to them, and the influence they had in his happiness."[146] For this reason, the analysis of the viceroyalty's production of wealth is the main point of many articles; in one of them, for example, the area's advantages are described as comparatively better than those that Tyre and Alexandria enjoyed in antiquity.

The colony had "an endless number of other species of comfort and luxury" that made its provinces one of the wealthiest places on earth, and in a position to establish "a strong and powerful commerce."[147] The past lack of understanding of economics valued only silver and gold and considered Buenos Aires and its surroundings as "poor" areas because they did not possess mines. The port's prosperity was linked to the existence of commerce and trade of their "copious and not exhaustible fruits," and since trade meant communication, it had the greatest wealth growth of all.[148] The River Plate provinces and their inhabitants could live "comfortably and in opulence" because of what nature had given them, but they were not doing so. To change, they were urged to start expanding commerce and trade in their natural resources.[149] Slavery was suggested as a possible means of growing wealth if the Crown allowed Buenos Aires to be one of the ports of trade; slavery would also provide the workers needed to increase agriculture.[150] This newspaper also included a note about things to buy or trade, indicating an increased interest in consumerism in the city. One item for sale was "a modern carriage," which indicated the advance of taste in society.[151]

As in Europe and the other Spanish colonies, the increase in wealth and commerce and the expansion of Enlightenment ideas among the youth led to conflicting perspectives on gender and the meaning of sexual difference. The issue of *Telégrafo mercantil* published in November 1801, for example, includes the satirical poem *Definición del currutaco* (Definition of the Currutaco). The description matched the characteristics already discussed: vanity, the desire to look good, and affectation in the way he behaved. The author was alarmed that "our customs have disappeared like light fog," a dangerous event because of the ignorance and superficiality that dominated the new male sociability.[152]

In 1802 another poem was published; it was like those published in Spain and Mexico and written by Cabello under the pen name Narciso Fellobio

Canton, self-described as "an indifferent philosopher." It consisted in a declaration of rejected events and behaviors, including: "I reject [*reniego*] the individual who in order to look like a complete *currutaco* / arranges himself in a way that makes him look / like the Marquis del Berro."[153] In another article, Enio Tullio Grope, whose real name was José Eugenio del Portillo (1760–1843), wrote about the "revolution in the clothing of Peru," in which he noted that there were no laws or outcries from the authorities to prevent it; "a happy revolution" had begun in the way that Hispanic-American women were changing their manner of dress. This event "urged the possible imitation of European women, researching the fashions and the capriciousness of the ancient hemisphere." Peruvian women would soon be adjusting to the "great territory of Argentina or the country of Buenos-Ayres which never knew other clothes than those coming from Spain," and if this situation continued, the differences among the various colonies "would be extinguished" in the future.[154]

Telégrafo mercantil also focused on scientific ideas; for example, in May 1801, it published a letter from Pedro Juan Fernández discussing the use of inoculations during the outbreak of smallpox that was affecting Montevideo.[155] The increasing Atlantic trade resulted in the spread of this disease all over the Americas with terrible consequences. In the United States inoculation was introduced in the 1720s to decrease the number of people who died, which led to a controversy about this practice. Inoculations were the result of knowledge exchanged in conversations between Cotton Mather and his African servant Onesimus, who knew that this prophylactic was used against smallpox in West Africa. Mather convinced Zabdial Baylston to try it, and it became evident that the death rate was lower among those who had been inoculated.[156]

In the River Plate Viceroyalty this treatment was introduced around the 1770s. Fernández's letter reflects on the improvement that inoculations had brought to the viceroyalty, and frustration about the opposition of most mothers to use it on their children during the existing outbreak. Following this letter, an article titled "The Health of the People Must Be the First Law" continued the discussion on mothers' obligation to embrace modern science. Explaining how inoculation had created a war among those who were against it over the past decades, the narrative appealed directly to women whose "sensitive souls" (*almas sensibles*) had been suffocated by ignorance and superstition.[157]

Using statistics that had appeared in Europe during the previous thirty years, the author denounced the character of the "anti-inoculator" for manipulating the facts (85). In parts of the River Plate Viceroyalty, inoculation was common and even the Indigenous population had benefited from it. Appealing to the gendered ideas of the time, the article warned women that because they "idolize their beauty" they would soon say that "smallpox was our executioner: it removed from our faces the sweet attractive features of the Goddess Cytherea [Aphrodite]." The author warned that the future depended on the mothers, who with "a rational complaisance" could cooperate in the conservation of their children "and, as a consequence, the solid happiness of the country" through their patriotism (88). Finally, the disease was racialized, and the introduction of "barbarous Africans" was viewed as an expensive mistake that would cost the lives of many "of our best Americans," which called for the end of the slave trade and the diseases that it transported (88).

Edward Jenner (1749–1823) published his research on the vaccine for smallpox in 1798, thus popularizing its results, and shortly after the article's publication, in 1803, Charles IV organized the *Expedición científica de la vacuna* (Royal Philanthropic Vaccine Expedition). The objective was to promote vaccination in the Spanish colonies, a project that would continue until 1806; Humboldt understood the scientific ambition of this expedition and wrote that the event would "remain forever memorable in the annals of history."[158] After witnessing such an important development and completing his research in New Spain and Cuba, Humboldt continued his travels, arriving in 1804 in the United States, where he met Thomas Jefferson, then president, and another important proponent of vaccination.

REPUBLICANISM, GENDER, AND SCIENCE IN THE UNITED STATES

It is difficult to exaggerate the degree to which the American Revolution changed the views of the followers of the Enlightenment internationally; the writings of Richard Price (1723–91), a supporter of the American Revolution, are good examples of the reaction of revolution advocates in England. In 1784 he published a book explaining the historical meaning of what had happened that captured the views of the extreme liberals in Europe and the Americas. Price was a mathematician, a fellow of the Royal Society of London, a well-known radical thinker associated with Mary Wollstonecraft, and a minister of the Unitarian Church. He explained his "heartfelt satisfaction" originating in the "revolution in favor of universal liberty which has taken place in *America*," a revolution that opened "a new prospect in human affairs," initiating "a new era in the history of mankind." The revolution was the result of the dissemination of "just sentiments of the rights of mankind, and the nature of legitimate government," which "emancipated one *European* country," and was likely "to emancipate others by occasioning the establishment in *America* of forms of government more equitable and more liberal than any that the world has yet known." The American Revolution is described as another link in the chain that included "Newton's physics, the invention of optical glasses, printing, gunpowder, discoveries in navigation, mathematics, and natural philosophy."[159] Political ideas were following the scientific revolution.

Before the existence of the young American republic, the British colonies of North America experienced controversies about luxury, gender, and morality, as the result of a new social culture and the role of science in society. Michael Meranze has explained the extent of a "culture of sensibility" that was evident by the end of the eighteenth century in the transatlantic context of Great Britain and its colonies. This culture "placed great emphasis on educating manners and cultivating empathy and sympathy."[160] Once the American Revolution had triumphed, the figure of the patriot emerged as exclusively related to the creation of republicanism as the result of multiple cultural changes, mostly educational and political, that continued after independence.

According to Leora Auslander, immediately after the republic was created, there were worries "that the New World would be made over in the image of the Old," which could be "bad for the national character and perhaps the nation's very existence, because frugality, labor, and a certain definition of masculinity and femininity were understood to lie at the heart of the Republic." Auslander explains the scandal that erupted around the Sans Souci club in 1784 to illustrate what "was wrong with post-Independence America" for those like Samuel Adams.[161] The protests against the existence of this social club were an expression of the distrust that existed about a masculinity that had moved away from the man/patriot model to embrace the petit-maître as a prototype. As in the cases of the *currutaco* and the *petrimetre* elsewhere, youth with this kind of behavior were understood as contrary to morality because of the luxury and refinement they promoted.

The rejection of this character was also displayed in satires that, like those published in Spain and Spanish America, addressed the gender tensions created by a modernity that oscillated between embracing the culture of sentiment and rejecting it. A satirical play published to ridicule this culture mentioned Richard Price for his outdated vision of republicanism; on the contrary, the youth had a new view that had radically changed old conceptions of honor and virtue. One character in the play, Dr. Gallant, asks Mr. Importance if he has read Price's book, to which the latter replies:

> The Doctor would do to live in Carthage, Sparta, or Rome in the most rigid times; but now those sentiments are wholly imaginary ideas of a Republican's coran [Quran]—they are too antiquated—modern republicanism is of a very different complexion, as contrasted as the severe countenance of Cromwell, and the pretty smooth face of a petit maitre—those of the Doctor's are principles of republicanism of the uncivilized kind—but now we soften them down in the school of politeness, and make them wear a more pleasing garb—the ancient republican spirit is like the old principles of religion—staunch Calvinism—but now we have modernized them, and united them with the court stile of taste and fashion.[162]

Another character, Young Forward, expressed his approval of the fact that not only the ladies but also the men interacted socially "with a particular grace." This happened because of "the happy faculty of blending businesses and amusements" that characterized society. He liked to see that a man threw "aside his professional character" and entered "a polite circle with the Sans Souci air of a *petit maitre*"; he also observed with approval that men were "versed in all tender stile of a billet doux—not calculated only to balance accounts, but to decide with precision the arrangements of a lady's toilette."[163] It is clear through the use of French words that these characters were influenced a foreign culture dangerous for local customs. It was also connected with the materialism of Condillac, which was also becoming a threat to the morality of many in the country.

2

The Patriot

Michael Sonenscher has explained that by the late eighteenth century, Europeans perceived the future as a "race towards ruin" with little expectation of positive outcomes. For example, following the American Revolution, Guillaume-Thomas Raynal (1713–1796) observed that the American colonies were drifting apart from Europe. He believed "that without corrective action, the future would belong entirely to the New World, not at all to the Old."[1] Consequently, analyzing the bonds between government and the governed became essential for a nation's future. The rise of revolutionary politics on both continents brought to the forefront the nature of the relationship between government and its citizens, along with the social obligations of those in power. This marked a shift away from a literary republic rooted in classical culture toward a new conception of power and nation formation, driven by the need to address socioeconomic transformations.

It was in this context that Spain and France experienced a modernization of the concept of patriotism in the second half of the eighteenth century. According to Peter Campbell, new notions of virtue "and the rise of sentiment combined with older currents" produced a popular new idiom in France; the terms "petit-maitre" and "billet-doux," together with "vertu" and "bienfaisance" (virtue and charity), were "central to the changing notions of citizenship, and *vertu* and *patrie* [fatherland] in particular underpin arguments in opposition to both royal policy and the existing social and political order."[2] The linkage of these words also suggests a starting point of reference before the French Revolution, an "imagined community" "in terms of an ideal res publica, a *patrie* with virtuous citizens, seeking the public good, actively engaged in *bienfaisance*, educated by good morals and observing good laws enshrined in a constitution." Both terms are defined in the *Encyclopédie*, and love or devotion "to the *patrie*" was seen "as a virtue," and for several important thinkers it was the first virtue.[3]

Some of these same terms were also used in Spain and its colonies, where they defined the citizen as virtuous and charitable, an individual who also saw himself socially, interested in what was good for all, and helping to achieve it. The measure of honor became attached to the possession of these virtues. As has already been noted, this concept had roots in the Republic of Letters. A speech delivered in 1788 before the Royal Society of Madrid by Gaspar Melchor de Jovellanos (1744–1811), one of the leaders of the Illustration (Enlightenment) in Spain, is a good example of this ideology. He praised Charles III's government, declaring the building of this society, "the home of patriotism," a quality related to virtue.[4] The king was presented as "the father of his vassals" and far from being an absolutist leader (2). The foundation of the kingdom was not God anointing the king but the "men who had associated, recognized a sovereignty, and sacrificed their most precious rights" to assure the possession of property, dragged into this action by nature's selection. The princes had been placed by God "to bring abundance and prosperity to the nations" (5). Charles III was praised as having a heartfelt attachment to philosophy once the throne was his. He was prepared by the teachings he had received in "a long essay on the art of ruling," from which he learned that the "greatest glory of a sovereign is to be supported by the love of his subjects," and that this love was "more glorious when inspired by recognition" (7). He needed to earn the "love and gratitude" of those he governed through good administration and an emotional connection (7).

This king reestablished "people's representation" in order to perfect municipal government, and the "sacred power of the fathers to improve the domestic governance," which was supported in "the patriotic bodies" that assured the king "the title of Father of his subjects!" (8–9). The king's "paternal designs" brought "a new spirit to the nation" by their reliance on "useful sciences, economic principles" and "the general spirit of the Enlightenment" (11). The king's reforms were the result of "the patriotism that inflamed the zeal of some generous Spaniards who meditated much over the public wrongs, and very vigorously asked for reforms: it was then for the first time that the thought of a science that taught how to govern and make men happy emerged" through the study of economics (17).

Accordingly, the domination of Aristotelian philosophy ended because it was based on speculation and, because of poor interpretation by the Arabs who introduced Aristotle to Europe, did not consider experience at all (23). The modern royal subject "abandoned metaphysics" and started "to contemplate Nature, growing the sciences that study it, aggrandizing his being, know[ing] all the vigor of their spirit, subjecting happiness to his freedom" (23–24). Those "citizens" who fought to change society would be favorably judged "by the fatherland" (*patria*) in the future (53). While this was a patriarchy led by a father, the rule of this male authority was founded on love. Authority was emotional, which implied a more personal and subjective bond.

Adding to the importance of the culture of sensibility, Jovellanos affirmed that talent and enlightenment (*luces*) were worth nothing without "goodness and heart's rectitude." This also included women, the "enlightened companions," whose rights were defended by the new government. They were present at the conference of the Royal Society of Madrid and could be members of the society, but their role was "to form the heart of citizens," inspiring in the latter "the tender affections that united goodness with the happiness of humankind." Women needed to inspire in the male citizens "sensibility, this kind virtue" that they had received from nature and men "barely acquired by reflection and study" (54–55). All the actions described in this speech were intended to exalt the patriotism needed to improve the nation.

FREEMASONS AND THE FORMATION OF MASCULINE PATRIOTS

The 1790s was a decade in which the discussion about "patria," patriotism, and citizenship started to develop a more radical rhetoric, which assumed that the monarchy was the real impediment to the well-being of society. In 1788 Charles IV (1748–1819) became king and his wife Maria Luisa of Parma, a student of Condillac, became queen, but their image deteriorated rapidly, in part due to the explosion of the French Revolution, which scared even those who favored the Enlightenment. At this time, a call to rebuild the nation took place, in a narrative that emphasized the values needed to made Spain more modern. The sign of this progress was not the sociability of the salons, but the return to a male virtue that was not assimilable to the presence of the feminine.

In 1759 the French writer Louis Antoine de Caracciolli, Marquis de Caracciollo (1719–1803), complained about the increased interest in materialist positions among philosophers and scientists. He was a representative of the Republic of Letters who worked in the service of the king of Poland and elector of Cologne. Ideologically, he was interested in political reform but rejected the meaning of masculinity among those who had embraced sensationism. A translation of de Caracciolli's book was published in Spain in 1777 by Francisco Mariano Nifo (Nipho) (1719–1803), one of the best journalists and writers of his generation in Spain, to criticize the changes in the nation.

De Caracciolli defined the century as "sterile of virtues, and abundant in extravagancies and horrors," which could only offer to future generations "despicable theater plays, insipid novels, wicked systems, and extravagant opinions." He was upset by how much men "had degenerated compared to [their] elders!"[5] While the fathers of the past had searched for truth, the present generation only appreciated "*illusion*"; the past use of arguments was based on religion while those of the present only used "*systems that contradicted it.*" While in the past men "*consulted their soul,*" the youth of the present "*only [paid] attention to the senses*" and lived to take breaks. The Republic of Letters and its citizen had been replaced by a sensual man. "*The corruption of the heart, and the mistakes of the senses, disguised with the names of Poetry, natural religion, and philosophy*" earned everywhere the greatest applause.[6] In the past, man was "*fearful of sensuality at the expense of losing honor, which contained him within his limits.*" The new masculinity was not attached to honor and morality because of modern science and philosophy.[7]

Feijóo, the Benedictine philosopher already mentioned (see chapter 1), wrote letters expressing his views on materialism. In one of them, written in 1752, he recognized that there were problems emerging from modern philosophy denouncing those "Materialist Philosophers" who pretended "to debase the rational soul, degrading its spirituality" and promoting atheism.[8] On similar grounds, an article published in *Gazeta de Madrid* accused this "*Universal Materialism*" of erasing the special condition of humans in divine creation, "*lowering man to brute condition, deprived from a spiritual soul, and reduced to a shameful need to always seek whatever flattered his self-esteem: it confused all states, all classes, presenting the subordination to the law as barbaric, obedience as weakness, and monarchy as tyranny.*"[9] It was a diabolical sect that denied the existence of God and Providence.

In another letter, Feijoó mentioned the belief of some that another problem threatening religion and the political order was posed by the Freemasons (*francs-masones*). They were becoming a known presence by the 1750s, but he did not consider them very dangerous. Using the word "Muratores," derived from the Latin *liberi Muratores* used by those in the church, he explained that some of these organizations were more interested in fraud than in heresy.[10] Pope Benedict XIV banned Catholics from becoming members of these secret societies at the risk of excommunication in 1738, but this did not stop the popularity of the lodges; in 1751 the Spanish monarch, Ferdinand VI, signed an order to confront "the invention of those who called themselves Freemasons [*francmasones*]" as there were suspicions about their activities against religion and the state, which justified the ban imposed on the creation of these organizations in Spain.[11] The rumored presence of Masons in the military was noted in this king's order, which warned that any of them serving would be ejected.

In 1752 Fray Joseph Torrubia (1698–1761), a recognized Franciscan missionary and naturalist who had spent time in the Viceroyalty of New Spain, in Cuba, and in parts of Asia, organized the publication of his translation from Latin of the pope's Bull and a speech warning people about Freemasons given by Pedro Maria Justiniani, the bishop of Vintimilla in Tuscany. Torrubia dated the origin of this culture to the time of Oliver Cromwell in England; since then, Freemasons had continued their existence "persistently closing their eyes to the light of the Roman Church."[12] It was true that the impulse to create these organizations came from Great Britain; in 1723 the reorganization of London's lodges created what would become the most powerful association because of its extensive network in Europe and the Americas.[13]

In 1782 the English captain George Smith, a Mason and Grand Master for the county of Kent, dedicated a book to HRH Henry Frederick, 1st Duke of Cumberland and Strathearn (1742–1814) and son of the English king, because he was "a protector of freemasons," and Grand Master of the Grand Lodge of England from 1782 to 1790. Smith explained in the dedication that this book was "particularly agreeable" for men "sincerely interested in the welfare of society and of their country," to reflect on the "rapid progress, and general diffusion of the royal art" through almost every part of the habitable world.[14] At this time Freemansonry was "more respectable" in "the vast empire of Germany, and in the kingdoms of Great Britain and Ireland" (2). Those who were "Men of letters" had directed their studies "to their improvement in the masonic art," and books were the fundamental framework of masonic morality and the foundation of this noble art (xiv). Smith noted that violence had erupted "over the whole American continent" and expected that the colonies would become manufacturers once their commerce underwent a revolution (240). This hope was intertwined with a narrative about Spain's decadence that, as mentioned, became more important by the 1760s. While the example of ancient empires embodied the idea of a glorious period followed by decadence and defeat, the decline of Spain by this time was associated with the policies of Charles III (1759–88), who decided to enter the Seven Years' War (1756–63) in 1762, ending the neutrality maintained by Fernando VI (1746–59).

Initially, this was a colonial war between England and France contending for the territories of North America, but it ended with a rearrangement of each nation's colonial possessions: Spain lost Florida and gained the territory west of the Mississippi, including New Orleans; the French increased their control over other islands in the Caribbean, and the English acquired what is today Canada. It exposed Spain's naval inferiority, which instilled a sense of decline, as was mentioned in chapter 1. This conflict also called attention to the presence of Masonic lodges in the American colonies. Since 1733 lodges had been created in most of North America and the Caribbean islands, frequently under the jurisdiction of the Grand Lodge of England, though some followed those of Scotland, France, and Holland. There were 1,247 lodges in Europe, 187 in America, 76 in Asia, and 13 in Africa; Smith estimated a total of 45,690 Masons (243). While England established only one lodge in Spain, the Portuguese had several lodges by constitutions from France as early as 1727, but in both kingdoms the Inquisition made it difficult to form and continue these organizations (204).

Men seeing each other as brothers implied that anyone "regardless of his origins, social standing, or religious background, could become part of this universal community or brotherhood."[15] Following elements related to the occult, the lodges insisted that present knowledge had been introduced since creation and transmitted from fathers to sons since then. Biblical characters were Masons, "the Israelites, at their leaving Egypt, were a whole Kingdom of Masons, well instructed, under the Conduct of their GRAND MASTER MOSES," who marshaled them into "a regular and general Lodge," which established this organization as a foundational social institution.[16]

In 1730 freemasonry's articles were published in Philadelphia where the First Lodge of Boston (St. John's) had members, including Benjamin Franklin (1706–90), who promoted Freemason principles in the press. In 1749 Reverend Charles Brockwell, a chaplain in Boston, justified the forming of these societies because they undoubtedly endorsed "uniting men in stricter bands of love; for men, considered as social creatures, must derive their happiness from each other"; his speech emphasized that each man was created "by providence to promote the good of others," which expanded the homosocial culture cultivated in lodges.[17] Steven C. Bullock estimates that during "the generation after the [American] Revolution, Masonic rituals solemnized dedications of churches, universities, and the United States Capitol. Its symbols adorned quilts, drinking glasses, and tavern signs—as well as the nation's Great Seal."[18] Most of the leaders of the 1776 revolution were Freemasons or had links with them, including George Washington and Andrew Jackson.

The conception of men as brothers introduced a new idea of masculinity that was developed through universalism and patriotism. A lodge member was bound to communication with his brothers who might be living in different parts of the world and in very different circumstances. This openness was translated to gender when the men's interest in their brothers started to be represented as a symptom of abnormal maleness. It also implied the potential erasure of racial differences, which stirred interest in this type of socialization among free Black men. On March 6, 1775, Prince Hall (1748–1807), together with fourteen other free Black men, was made a brother of Lodge #441 of the Irish Registry that was attached to the Thirty-Eighth British Foot Infantry stationed in Boston Harbor. According to Harry Paul Jeffers, for many years "the black churches of America and Prince Hall lodges" were the two strongest organizations in Black communities.[19]

In Latin America the formation of lodges was severely limited by the censorship and control of the Church and colonial authorities. Those from the Spanish American colonies who lived in Europe, though, had more freedom to join. For example, Francisco de Miranda (1750–1816) arrived in London in 1784 following previous stays in Mexico, Central America, Spain, North Africa, the West Indies, Jamaica, and the United States after leaving his place of birth, Caracas, then part of the Viceroyalty of New Granada. Because of Miranda's travels and connections, Alexander Hamilton, George Washington, and Thomas Payne communicated with him. In 1789 he moved to France to fight during the revolutionary conflict and ended up imprisoned in the Bastille. Returning to London he used his personal network to spread knowledge and help those from the Spanish American colonies who were radicalized and thinking about revolutionary violence—which reduced the relevance of the citizen of the Republic of Letters as a masculine model. The ideals of this republic ended with freemasonry, in lodges where women were not accepted. The figure of the patriot and the Mason started to be blended by the second half of the eighteenth century to picture a masculine hero who mainly socialized with men and was driven by politics. The influence of the lodges among politicians and people with power continued throughout the nineteenth century.

REVOLUTIONARY PATRIOTISM IN SOUTH AMERICA

This updated version of patriotism was linked to a modern idea of "good government" that differed from the past. In 1775 a book about the science of government, written by the Frenchman Gaspard de Réal de Curban (1682–1752), was translated and published in Barcelona. Its author was an enemy of the encyclopedists and Descartes, a supporter of natural law and limited government by the people, who had extensive experience in government for his advisory role to the French Crown. The translator was a Catalonian writer, Mariano Joseph Sala, who wrote in the introduction that the definition of "science of government" consisted in introducing "in the vassals' spirits notions needed for them to keep firmly the obedience they must have for their sovereigns." In turn, the king must "soften the yoke in a way in which his power would become "imperceptible," which was the greatest happiness vassals could aspire to obtain. Politics was the "art of making government useful to the people" and was based on the knowledge of laws and politics; the latter involved "knowing the interests of the state" to direct its actions to "the common happiness of its subjects."[20] This was a different understanding of power because it recognized the mutuality of interests that made government possible; at the same time, it maintained support for the monarchy through patriotism.

Complaints about local governments and social unrest had existed since the 1730s in the Andes and had continued in the Viceroyalty of Peru and the Upper Peru region controlled by Buenos Aires since 1776. The Reforms of Charles III made the situation worse for the peasantry and led to the most dangerous insurrection faced by the Spanish Crown in the Americas. Tupac Amaru, a descendant of the Inca rulers, and Tupac Katari, an Aymaran Indian, revolted against local authorities because of bad local government. The former began the war against the Viceroyalty of Peru in 1780 centered in Cuzco; the latter was based in La Paz, in Upper Peru, which belonged to the River Plate Viceroyalty, whose capital was Buenos Aires, in 1781. The viceroy of the River Plate, Juan José de Vértiz y Salcedo (1718–99), born in the Viceroyalty of New Spain, organized the war.

The peasantry who fought in these conflicts shared a political ideology that was anti-colonial, however. As Sinclair Thomson reminds us, several political options existed before the conflict began, and there was no unified project behind the insurgency in the Andes. These views "emerged before [1781] and evolved during an important and lasting process of fighting in the eighteenth century." This demonstrates the presence in South America of a "rich political culture" that was shaping the rejection of colonialism and would have a legacy in the region. Besides the peasantry, "a small group of educated people living for the most part in urban areas, criticized the Spanish government and sustained the idea that a legitimate Inca authority in eighteenth-century Perú" was possible, leading to the project spearheaded by Tupac Amaru in Cuzco, the most influential addition to the political landscape during the era of the great insurrection that began in 1780 and continued until 1783.[21]

Simultaneously, political radicalism was also growing in Spain through the work of Juan Bautista Picornell (1759–1825), a philosopher who graduated from the University of Salamanca in 1778. He was a Mason, a member of an organization originally called the "Spanish Lodge" (Logia España), also known as Matritense, founded in 1728 by the Duke of Wharton, a previous Worshipful Master of the Great Lodge of England, to which the Matritense would be associated until 1779. After the 1751 ban, the secret meetings continued and resulted in the organization of the Great Spanish Lodge in 1767 and the creation of the Matritense Society in 1784: both counted Picornell as a member.

In her analysis of Picornell, Carmen Michelena explains that he participated in the Bourbon movement to reform the kingdom through his educational experiments, and his *Discurso práctico sobre la educación de la infancia* (Practical Discourse on Childhood Education), published in 1786, was associated with the ideas of Locke, Condillac, Rousseau, Montaigne, and Alexander Pope.[22] According to Picornell, family unity, together with education, was the foundation of society, and parents had to be "first in stimulating their children's desire to learn."[23] Similarly to Locke, Picornell explained that boys needed to be treated in the way that laborers treated their own, and that the introduction of luxury and convenience could make them weak and soft. In terms of the content of the lessons, universal history was an essential class in which "the nation, its politics, customs, and the causes that originated the great revolutions" should be studied along with "the progress of the arts and sciences [and] the formation of societies," and how the world had arrived in its present situation.[24]

Picornell also included the "others" who were so relevant in the formation of colonial empires—for him the world was "a great republic"; in teaching about the globe the instructor needed to introduce students to "Chinese, Indians, and Mongols, as their fellow citizens [*conciudadanos*] of this great Republic of the World." Students should be trained to "look with horror on religious craziness" and in general avoid tolerance that endangered the "good constitution of the states, and healthy morals; or their preference for their fatherland over all the other peoples," avoiding "the ridiculous mania of those who had a general affection for all nations," which made them "believe themselves excused of the duties they owed to their own country."[25] This patriotism promoted a universalism based on the idea of brotherhood but in the service of national improvement.

In 1788 Charles IV (1748–1819) became king. Two years later, Picornell translated a book by the French mathematician Charles-Joseph Mathon de la Cour (1738–93) written to illustrate the idea of patriotism as an expression of civic virtue. The translation was dedicated to the Royal Societies of Spain to inspire them to apply the book's ideas to "all the branches of our public Administration" because "the obligation" that the educated members of these associations had "to aid in the goals and tasks of our government, putting in place the precise and practical operations of the public orders in order to work for our happiness."[26] Patriotism was exclusively related to political action and participation in public service, or res publica; according to this model of virility, the patriot was the self-sacrificing individual who was indispensable for the perfecting of society. Defining modern patriotism as "the desire to serve our compatriots, to contribute to their well-being and comfort," Picornell's translation described this new form of political associationism.[27]

The patriot, then, "was always willing to sacrifice his most desired interest, and even his own life" for his Fatherland (*patria*). Love of country was a natural inclination present in all men; patriotism was "a virtue" that needed to be cultivated. It inhabited "the great souls" and by its nature and object "comprehended the happiness of a city, or of one entire nation."[28] In addition, patriotism was the "center of social virtue," which directly elevated the importance of morality, because the man "who did not know how to be a husband, father, friend, or neighbor, will not know how to be a citizen."[29] This model of man demonstrates John Tosh's affirmation that more than a social construction, masculinity should also be considered "as a subjective identity, usually the most deeply experienced that men have," bringing into play "the early formation of the gendered personality in the intimate relations of family life."[30] The development of the patriot's subjectivity was also related to plans for political emancipation and the end of colonialism in the Americas.

In 1795 the revolutionary plot known as San Blas was discovered in Spain with Picornell as its leader; he and his coconspirators were sent to La Güaira, in the Captaincy of Venezuela, to serve their prison term. Documents written in preparation for the revolution explained the motivations of the conspirators: "the education of children in its physical, moral and scientific" manifestations, addressing the "lack of social virtue, the lack of influence many laws had," and supreme ignorance of the "obligations that constituted the relationship between vassals and members of the state."[31] Some of the participants in this plot were also linked to government in the American colonies, such as functionaries located in Caracas, Nicaragua, and the Consejo de Indias. Two years later, the discovery of another revolutionary plot in the Venezuelan port of La Güaira demonstrated how extensively political radicalism had spread in the Spanish Empire.

After the French Revolution and the emancipatory movements in France's Caribbean colonies, many discussions took place about the possibilities of independence and the establishment of the republic.[32] The colonies of Martinique, St. Lucia, and Guadeloupe also had revolutionary organizations, and the Freemasons were very important in the Caribbean. In 1791 a revolution began in the wealthiest colony, Saint-Domingue, under the leadership of Black men, and led to the creation of Haiti. After the creation of the United States in 1776, this was an even more radical example for the Spanish colonies because it would demonstrate that complete emancipation was possible, and the Americas were beginning a new historical era.

In the Captaincy of Venezuela there had been several revolts during the first half of the eighteenth century, including the movement of the Comuneros de los Andes (1781), and the movement of José Chirinos and José Caridad González (1795) in Coro. In this context, in 1796, Picornell and his associates arrived in La Güaira to serve their jail term. Once in prison, young radicals contacted them to share their ideas. Manuel Gual (1758–1800) and José María España (1761–99) were well informed about the San Blas revolution and started to organize a local revolt that demanded the abolition of slavery and the defense of the principles of equality, liberty, property, and security. The movement included white, brown (*pardos*), Indigenous, and Black members.

The insurrection started in 1797, but it was short-lived because the colonial authorities found out and preempted the development of their plan. Picornell was given help to escape, though, and ended up in Santo Domingo, surviving the violence unleashed by the Spaniards against the participants. Gual and España were also able to leave and went to stay in different islands in the Antilles; the latter lived with family in St. Thomas before deciding to return to La Güaira; Gual's base was in Trinidad, and from there he wrote to Francisco Miranda asking for advice about resuming revolutionary action in Venezuela. Miranda, the most prominent revolutionary leader in Spanish America at the time, knew Gual and his family well from his time growing up in Venezuela.

In a letter written from London in December 1799, Miranda replied to recommend taking caution with the governor of the island of Trinidad and leaving for the United States in case of danger. He suggested that Gual introduce himself to Alexander Hamilton, with whom he was friends. He could trust this general completely because "he was the most faithful friend of our liberty and independence in the whole world."[33] More important, he recommended patience in waiting to see whether England would finally deliver on its promises to help the revolutionary cause that had been delayed by the anarchy into which the French Revolution had descended. He did not mention it, but there was also the ongoing Haitian revolution that had radicalized the Caribbean islands and caused ambiguous feelings among those who feared a racial war.

While trying to find support to resume the revolution, España decided to return to Venezuela in January 1799 to reignite the fight, but he was betrayed, captured, and executed on April 8. Shortly after, there was a plan to revolt in Maracaibo.[34] As in the previous attempt, Black and mixed-race people together with French corsairs and Curaçaoan sailors joined forces in an unsuccessful effort to establish a republic on Venezuelan soil. The attempt did not catch on, and it is not clear how the plan developed or who its leader was.[35]

Meanwhile, Gual remained in Trinidad but was killed the next year apparently poisoned by Spanish agents operating on the island. He had described in the introduction to his 1797 translation from French of the "Declaration of the Rights of Man and the Citizen" that the objective was to establish "the people's sovereignty," and bring "to all America the imponderable goodness of a paternal government." Violence was the only way to achieve the goals in order "to save the fatherland" (*patria*), but the destruction of despotism required, at the same time, a "moral and material" revolution, more than just the establishment of another political system. It was crucial to "regenerate the customs to give back to all citizens the recognition of their dignity," together with maintaining "the state of rigor and enthusiasm that had created their revolutionary effervescence" because the latter would disappear once the violence ended if a "positive knowledge of their rights [and] love for their duties" did not exist. The resulting state "would be based on the gathering of feelings and efforts toward the achievement of one single objective," which required the exiling of all those who hated the laws and would oppose the measures taken. The "regeneration of a people could only happen through expurgation" to allow the "enjoyment of the rights of man that constituted the well-being of each individual." This new culture of the patriot would be further developed during the revolutionary movements that led to the independence movements in the next century.[36]

SCIENTIFIC PATRIOTISM, COLONIAL ECONOMY, AND POLITICS

The republic that was revived in the United States, in the words of Allan Potofsky, "was rendered by French commercial ideologies from the end of the Ancien Régime to the Terror as a vast economic experiment for the philosophés and revolutionaries to remake the world."[37] In Spain, the transformation to create a commercial empire was considered "a matter of preservation and existence," and economic reforms and free commerce resulted from "the science of power and preservation being cultivated since the end of the sixteenth century."[38] Urban and rural cultures were also praised or criticized according to their influence on national morality.

The explosion created by the French Revolution in 1789, associated with the triumph of radical sensationalist philosophy and masonic lodges, caused a reassessment of both philosophy and science and their service to the nation. Those who had been skeptical about this philosophy confirmed their objections after the events that took place in the 1790s. It also transformed the image of France as an effeminate nation that existed in the 1770s, in some cases because it was viewed as another example of the decadence that destroyed Athens and was passed to Rome. For example, an article published in London's *Public Advertiser* charged that Great Britain had borrowed "Arts and Fashions from France" but was unable "to cull the useful and improving Vices promiscuously intermingled and imported with them."[39] In Spain, as we saw, there were similar concerns.

The new model of masculinity related to the patriot had many political expressions, involving the Freemason, the revolutionary, and the supporter of the Crown who believed in the inclusion of the educated Creoles. All these types were supporters of a new morality related to politics that eliminated the supremacy of the feminine and repositioned the importance of the citizen in a homosocial environment. In 1791 an article in Lima's *Mercurio Peruano* repeated similar ideas, clearly separating the *currutaco* culture from enlightened patriotism. This publication was created by Jacinto Calera y Moreira, a lawyer and founding member of La Sociedad Académica de Amantes del País (Academic Society of Lovers of the Country) in 1790.[40]

This piece denounced how the "sacred and recommended name of Philosopher" had been attributed, due to "a certain delirium," to "libertines and fanatics, declared enemies of truth and reason," when it only applied to those who "employed their reason [*luces*] toward the achievement of common happiness." As a result, the citizen of the Republic of Letters, a scholar devoted to knowledge, became "a cruelty against the human lineage" because wise men ignored the "needs and misfortunes that surrounded them" without trying to find "the ways to alleviate them." The fatherland deserved this work, which was among the "most important duties of the citizen." The improvement of humanity was the main concern, starting with efforts to solve problems that were visible and local. The meaning of the word "philosopher" was by this time "love for one's country," and it was clear, according to the author, that "in many parts of our [American] continent, the patriotic spirit" was growing; for example, the promotion in Bogota of "the interests of humanity and public instruction" had been reignited. Patriots were in favor of uprooting "abstract ideas" from schools to cultivate "the Useful Sciences" through the creation of patriotic societies that would work "to increase good taste, civilization, industry, agriculture, arts, and commerce," all "true sources of public happiness" that were the final destiny of humanity.[41]

Alexander von Humboldt resided in the Viceroyalty of New Spain from 1803 to 1804, observing a society in transformation. He noted that since the Peace of Paris, signed in 1783 by Great Britain to recognize the independence of the United States, and particularly since the French Revolution (1789), Creoles frequently declared with pride, "I am not a *Spaniard*, I am an *American*!" The example of the United States "and the influence of the opinions of the age [had] relaxed the ties which formerly united more closely the Spanish Creoles to the European Spaniards." The change among the educated elites was closely related to scientific ideas.[42] This naturalist noted that the Creoles living in the city of Mexico were more cosmopolitan and flattered themselves "with the idea that intellectual cultivation [had] made more rapid progress in the colonies than in the peninsula." He affirmed that progress was "indeed very remarkable" in colonial cities such as Havana, Lima, Bogota, Quito, Popayan, and Caracas.

Humboldt indicated that the study of mathematics, chemistry, mineralogy, and botany was "more general at Mexico, Santa Fe [Bogota] and Lima." He observed "a great intellectual ability, and among the youth a wonderful facility in seizing the principles of science." But he concluded that no city of the Americas could "display such great and solid scientific establishments as the capital of Mexico." The School of Mines, the Botanical Garden, and the Academy of Painting and Sculpture were evidence of the advancements made in this colony thanks to the "*patriotism* of several Mexican individuals" and government funding. The best mineralogy work in the Spanish language was printed in Mexico, and he referred to the manual of Andrés del Río (1764–1849) as an example as well as to the first Spanish translation of

Lavoisier's *Elements of Chemistry* that was printed in Mexico.[43] Humboldt criticized a common mistake made in the Spanish colonies because this chemistry was called a "new philosophy" (*nueva filosofía*).

The first director and organizer of the School of Mines, founded in 1792, was Fausto Elhuyar (1755–1833), a Spanish chemist who is known, along with his brother Juan José, as the first to discover and isolate tungsten. He had studied in Paris, and his work was well-known in Europe, where he lectured at several universities, including some in Germany, before being appointed director general of New Spain's mines by the Spanish Crown in 1788. He hired another Spaniard in 1793, the already mentioned del Río, to become the chair of mineralogy at the School of Mines. This mineralogist published the first volume of his *Elementos de orictognosia* (Elements of oryctognosy) in Mexico in 1795, and the second volume in 1805. Trained in physics and chemistry in Spain, del Río moved to Paris and Freiberg to study with Abraham Werner, one of Europe's best geologists and a professor of mineralogy.

It was in Freiberg that del Río first met Humboldt, who was enrolled there to study mining. After stays in Austria and England, the Spaniard traveled to France to work as an assistant to Lavoisier in 1791. Two years later, after the arrest of his mentor during the Terror, del Río fearing for his life left Paris for England, and from there to Mexico. While working in the mines of this colony he discovered element 23, which he named erythronium, though it is known today as vanadium after being renamed by Nils Sefström (1787–1845) in 1830. Humboldt took the sample and the essay describing this discovery when he left Mexico, but it was a long time until the discovery was finally acknowledged.[44]

In the River Plate Viceroyalty, the dominant commercial class in Buenos Aires focused on possibilities that the Atlantic trade could bring to them through expeditions to determine the potential existence of natural resources and to establish posts along the coast to control maritime traffic. In 1747 an expedition to Patagonia was organized to determine the resources available using the knowledge of the natives, which revealed a geography known better to them. In 1752 another expedition to the port of San Julián, located in today's province of Santa Cruz, aimed to determine the possibility of establishing commerce in salt, was financed by a local merchant, Domingo Basabilvaso. Viceroy Vértiz ordered another expedition to the south in 1778 under the command of Manuel Pinazo and Juan de Serdens. By 1783, and due to the expenses of fighting the uprising in Oruro, Vértiz ordered a reduction in establishments in the south because they were considered too expensive with very little resource potential.

The need to fix the borders agreed between Portugal and Spain also stimulated scientific exchanges and research. In 1796 Viceroy Pedro Melo de Portugal tasked Felix Azara (1742–1821) with recognizing the southern border of Buenos Aires at the request of landowners who suffered "the repeated hostilities of the barbaric Indians" who could not be contained through "good treatment, celebrations, or troops placed in convenient places." This problem had been constant since the establishment of Buenos Aires, but by now the expansion of the city because "of the uncalculated wealth for the State and the Royal Treasury" had made contact with the Indigenous populations more frequent, increasing the fear of invasions from the south.[45] Azara was a Spanish naturalist and trained military officer who studied at the Royal Military Academy of Mathematics and Fortifications in Barcelona and had a successful career in Spain before leaving for South America. In 1777 Portugal and Spain signed the San Idelfonso Treaty and Azara helped with the necessary cartography. In addition, he organized an expedition to Paraguay that took place in 1781 to settle another dispute about territory between Spain and Portugal and became the person in charge of these matters and natural sciences in the viceroyalty.

In his report to Melo de Portugal, Azara specified that the frequent violence between the Spanish and Indigenous peoples was due to the intense exploitation of the leather trade, which had left the territory without enough cows, thus depriving the tribes of a source of free food they had since the arrival of the Europeans who brought the first cattle in the sixteenth century. The availability of these animals had been crucial for survival, but the intensification of economic activity by the eighteenth century had depleted accessibility to them for food and trade, forcing the tribes to attack established populations near Buenos Aires. Land had been of little value before the eighteenth century, commerce being a more dominant economic activity in Buenos Aires, but by the 1760s funds had to be assigned regularly to contain the attacks because landowners were becoming more powerful and extending the territory under their control.

Azara recommended continuing to move the border forward because "more leather" would be available for merchants, "meat and bread for the capital, and mules to trade in Peru, securing our current estancias [rural establishments] because the Indians would not be able to enter, and would be killed if they do."[46] While this was described as an area of wealth, it was far from being part of the "modern" culture imagined by the elite in Buenos Aires. The violence and conflict that existed formed a man who needed to adjust to a very simple and practical life, to be knowledgeable in socializing with more diverse populations and cultures, and far from the one imagined as the result of wealth by those who followed the Enlightenment. Instead, over time, this rural man would become the antithesis of the enlightened culture promoted by the elite of Buenos Aires.

By the 1780s communication with Buenos Aires had greatly improved. From this city "six expresses were to proceed annually with letters for Peru and Chili [*sic*], and other provinces of those extensive dominions," all conducted with "so much activity and spirit, that communications have been opened over the famous Cordillera of Chili [*sic*], between the kingdom and Peru, and a regular post for letters kept up, in the most remote jurisdictions; where, before that period, even the very idea or name of a post-office was unknown."[47] European zoological collections had been formed with species coming from the River Plate Viceroyalty, including the great ant bear from Buenos Aires, the Myrmecophaga jubata of Linnaeus, called by the Spaniards Osa Palmera, which was alive in Madrid in 1776. The status of Buenos Aires as a source for scientific specimens was elevated after 1789 when the Megatherium was sent to Spain by the viceroy, Marquis of Loreto. The publication of Joseph Garriga about this specimen reached most scientists in Europe and raised interest around Buenos Aires when the bones were found.[48]

MANUEL BELGRANO, FROM *CURRUTACO* TO PATRIOT

Manuel Belgrano (1770–1820), the general in command of the northern revolutionary army and creator of the Argentine flag during the independence wars (1810–16), illustrates the path that many followed from good Creole patriot working for the economic success of the Spanish Empire to patriotic revolutionary soldier. Born in Buenos Aires into a good lineage, his studies led him to become a follower of the Enlightenment, a movement he got to know well during his education in Salamanca, Spain. It was there that he received the news of the 1789 French Revolution; he wrote that this event "changed ideas and particularly the men of letters" with whom he socialized. The concepts of "liberty, egalitarianism, security, and property" obsessed him, and he focused only on those tyrants who opposed man's enjoyment "of the rights that God and nature" had given him.[49]

Belgrano requested from the pope, and received in 1790, permission to read banned books.[50] At the conclusion of his studies "around 1793, the ideas of political economy multiplied ferociously in Spain" which introduced him to a field that was transforming Spanish thought. His knowledge of this topic led to his nomination to preside at the recently organized Consulate of Buenos Aires, created at the request of local leaders to promote and regulate commerce.[51] He was appointed as its first secretary, took over in 1794, and served until 1809. Belgrano's scientific and philosophical ideas consisted of a mixture of Spanish, French, and English sensationism combined with the economic science that had started to emerge in the physiocratic movement.

As Isaac D'Israeli (1766–1848), the father of Benjamin, noted in 1794, the "ECONOMISTS" were philosophers of prerevolutionary France, who directed their studies to "morals and practical politics" that sought "to render the people happier, by more closely uniting the bonds of society," and teaching man "to apply himself to the study of nature, the mother of lasting enjoyments." The encyclopedists joined the economists to create a new "sect" led by François Quesnay (1694–1774), the author of *Philosophie rurale* (The Rural Philosopher), and the Marquis de Mirabeau (1715–89), the author of *L'ami des hommes* (The Friend of Men). Anne Robert Jacques Turgot, "a practical philosopher," also became part of the group, and these "modest scholars pretended to govern men from their closets, by their influence on popular opinion, the regents of the world," in D'Israeli's view. Intellectually, they were "a monstrous mixture of the French frivolity, and the heavy inconsistency of the English," abusing English philosophy in their simplification of the connection between agriculture and happiness, together with the use of the notion of self-interest, a criticism directed to the use that Quesnay made of Locke's and Condillac's philosophies.[52]

Physiocracy's main aim was "to illuminate the operation of the basic causes that determined the general level of economic activity."[53] Quesnay was critical of mercantilism, and in the interpretation of Max Beer "desired to establish an *ordre naturel* in the place of the existing order, which he regarded as artificial and unsound, morally and materially wrong."[54] He believed that only agriculture was a generator of wealth, an idea that tended to connect morality and natural order with working the land.[55] In Spain, Ramón de Salas (1753–1837), one of the disseminators of the work of the physiocrats, had Belgrano as one of his students from 1786 to 1788.[56] It was during this period that the Creole from Buenos Aires came to know the new science and its main defenders in Europe. Mirabeau saw this new economic science as a solution to the problem of his time, the clash between wealth and virtue.[57]

The young student from Buenos Aires embraced Mirabeau's and Quesnay's ideas, particularly the latter's "Maximes," which he translated to Spanish and published in Spain in 1794. In John Shovlin's assessment, *L'ami des hommes* designated agriculture as "the foundation of the national welfare, and luxury its antithesis," which made it important reading.[58] Belgrano took from these French authors what he needed to speculate about the policies that might allow a harmonic relationship between commerce, agriculture, and morality in the River Plate Viceroyalty. But, while clearly agreeing with everything related to natural order and free markets, he rejected ideas that preserved despotism in any way, which made Quesnay's support of a legal despot difficult to accept. In his introduction to the translation of the French economist's "Maximes," Belgrano explained that if the thought contained in the book was truthful it was because its foundation was "the natural order"; those who had superior intellect knew all the reasoned evidence, whereas ordinary men had "the evidence of their feeling," which reflects the connection between the culture of sentiment and the people who were for him the foundation of politics.[59]

Quesnay believed that society was governed by a natural order, similar to the biological order observed in the human body, which maintained equilibrium among its components and reflected the natural laws regulating the operation of society and the economy.[60] Belgrano published two more books of translations of physiocratic texts that were important to increasing interest in this approach among Spanish readers.[61]

Once he returned to Buenos Aires and took control of the Consulate, in 1794, Belgrano began his patriotic mission of introducing the new ideas into society together with creating policies that favored them. In a report written in 1797, following the example of Campomanes's reports, he expressed his concern for the situation of women in the viceroyalty as women were a "disgraced sex," exposed "to misery and nakedness, the horrors of hunger and destruction" that originated and led them into prostitution, which was "the cause of so many bad things for society." Only by work could the situation of poor women be resolved, and in a footnote he begged readers who did not share his view "to examine at least what means a woman had to subsist" in the colony, and "to what areas of industry she could apply herself." He was certain that those who paid attention would join him in punishing themselves for the "miserable situation of the privileged sex" and would confess that tending to a woman's problems should be the priority given "the needs in which she [was] immersed," and because from the "well-being of the improvement of her situation would surely be born the reform of the customs that will be spread to the rest of society."[62] Similarly, the historian Theresa Ann Smith asserted that Campomanes "had calculated the number of idle women in the peninsula and surrounding islands whose work" could produce growth in the Spanish economy.[63]

In a speech read at the Consulate in 1798, Belgrano asked for the formation of academies and societies that "in all the cultured nations" promoted new ideas to spread knowledge among landowners, peasants, merchants, and artists.[64] The association among those who owned land, merchants, and the representatives of Ilustración (Enlightenment) who wanted to teach the inhabitants of the colony, would lead to progress. "Sweet philosophy, the friend of humanity" was the key to transformation (114). Like all physiocrats, Belgrano believed that the promotion of agriculture was central to society's economic and moral well-being, but due to the dominant role of commerce in Buenos Aires by the 1790s, he tried to avoid the competition between the two, emphasizing the importance of cooperation.

He explained that "agriculture only flourishes with great consumption," for which commerce was needed; the prosperity of one also favored the other (112). The participation of merchants and landowners in the Consulate, ordered by the king, was in the interest of the public good (*bien público*) because "mutual help, the consultation and conciliation of their interests" was the way to change the socioeconomic structure of the colony and the "formation of only one great family" in which all will have "the same object, goal, and means," and will be the source of happiness (104).

D'Israeli also defined "patriots": they were men "who ascending to the source of the Laws, and Constitution of Governments" had shown "the reciprocal obligations of subjects and sovereigns," diving into "the depths of history and its monuments and eternally fixed the great principles of administration."[65] This model applies to Belgrano. He was also a journalist who contributed to various publications. He wrote first for the *Correo mercantil de España e Indias*, founded in 1792, and later for local newspapers such as the *Telégrafo mercantil.*

Young men like him were in charge of government and policies and had to work for the common good; in fighting toward this end men became patriots. This caused friction because not all those who loved their place of birth favored the new science and philosophy or agreed on their meaning. For example, Belgrano's distancing from Cabello y Mesa, the editor of *Telégrafo mercantil* already mentioned in chapter 1, originated partly in their disagreements about the politics of the Enlightenment. It all started with the publication of an alleged translation of *Historia del Dr. Buñuelos* (History of Dr. Buñuelos), written by Boudein. In fact, this satirical piece was written by Cabello y Mesa himself, who made up the surname Boudein, perhaps ironically, since "boudin" means "sausage" or "pudding," it might have been created as a mix of both.

According to Estanislao Zeballos (1858–1923), the author of the poem was Cabello y Mesa and the fictional Dr. Buñuelos was Belgrano.[66] Written in first person, the protagonist described his incessant travels around the world, starting with his birth. He described himself as "extraordinary" and a "prodigy," born in the "previous century," at the time of "the Enlightenment," into an illustrious and aristocratic family.[67] He was "blonde like the air," very well educated in philosophy and law, but "with nothing in the brain."[68] His superficiality and affectation were proof that he was truly not a man of ideas.

In another section of the poem the protagonist explains that he had been in Boston to attend a congress dealing with "this new phenomenon called the *Currutacos*." Among the experts "there were terrible debates about the true sex of these rare specimens, made of male and female animals," but since in them "the feminine sex predominated" the authorities ordered that their clothes be removed to be exhibited to the public." Dr. Buñuelos continued his travels around the world until he died and was buried in a mausoleum. His epitaph was written by a cook who "made porridge when he wanted to make stew," interesting wording considering that "mazamorra/porridge" also meant confusion of ideas and "puchero/stew" was a typical Spanish meal.[69] The portrait of Belgrano made by François-Casimir Carbonnier while he was in London depicted him wearing European clothes and an affectation in pose indicates the origin of Cabello's satire.

Belgrano, the young secretary of the Consulate, had his chance to take revenge, since Cabello y Mesa's newspaper was canceled after the publication of articles that got him in trouble with the authorities in charge of censorship. One of them, "Politica," was published anonymously and contained an analysis of the conditions in Buenos Aires and the Malvinas Islands, and how to improve them. It was originally written as a report to the Crown by the Spanish explorer Juan de la Piedra on his appointment as commissar superintendent (*comisario superintendente*) of part of Patagonia in 1778. Cabello took the liberty of editing and shortening the original to fit his own beliefs and exposed the differences of Belgrano's generation's morals and the culture of consumption.

"Politics" claimed that the capital of the viceroyalty was a perfect place for lazy men because the abundance of food, which, together with the "sweet way in which women socialized [*trato dulce de las porteñas*]," and "many other

Figure 2.1. Portrait of Belgrano. *Retrato del general Manuel Belgrano*. Fortunato Fontana, 1941. Oil painting. Reproduction of François Casimir Carbonnier, "Retrato de Belgrano [ca. 1815]." Courtesy of the Museo Histórico Nacional, Buenos Aires, Argentina.

circumstances" produced in the city "the vile behavior of most European men."[70] These men encountered free horses and free hospitality because the "owners of the hacienda were very happy" to have them for any length of time on their properties. More importantly, "the fair sex had a singular affection for any European man," and there were so many available that "each man will have a dozen women," most of them "filled with a thousand charms and graces that were difficult to resist." This abundance of women meant that "one out of thirty got to marry" while the others "lived in perpetual celibacy or were corrupted" (23). The young men who immigrated from Europe did not marry; they lived with "a freedom that had no limit" because they only wanted to make a fortune and return to their country to find a proper wife in their hometowns.

Unlike the women from Buenos Aires, "petrimetras" who spent hours in their boudoir or sitting in conversation, European men wanted women who used "the spindle and the spinning wheel" and dressed in a humbler way. Other men were less successful, "abandoned themselves," and lost the "desire to return to Spain." Plenty of men left their towns to travel to the River Plate Viceroyalty but were corrupted in the colony and "unable to execute their thoughts" (24). The author continued emphasizing the surplus of single women, stating that they generally "were very poor and without much luxury" that "none wanted to present herself looking less decent than the other females, using ways to accomplish this goal that deserved compassion" (25). This problem could be solved by forcing the men to either marry their women or return to Spain.

Racially, according to this publication, Buenos Aires was no better. "A multitude of free blacks and mulattos" hid and protected the enslaved people who escaped. Those should be settled in the Malvinas Islands because "they were susceptible to culture and reproduced very well in spite of their bad reputation" (27). The allegations that Buenos Aires's women were prostitutes and that racial divisions were not enforced met with indignation among those who lived in the city. Still, as we will see, the ideas about the colonization of the islands continued until the British took them and renamed them the Falkland Islands.

Telégrafo was replaced by *Semanario de agricultura, industria, y comercio*, edited by Hipólito Vieytes (1762–1815), another devotee of the new economic science including the work of Adam Smith. The prospectus of *Semanario*, written by Vieytes to get the censors' approval, explained how he and his collaborators saw the present as a time in which "the spirit of domination and conquest was not anymore the principal passion that fed men's hearts."[71] They had by then "abandoned the sword" to occupy their armies in activities that improve the economy, such as agriculture, and there was no longer "the horror of those fields" from which "destructive men destined to destroy their own species" emerged. The "blood of the warrior" was succeeded by the "sweat of the peasant and the horrible sound of the military trumpet by the pastoral flute" (viii). This coincided with Campomanes's desire to change the culture that had characterized Spain; something that those like Cabello, an ex-soldier, might not have favored.

Vieytes wanted to create "Societies of Economics [*Sociedades económicas*]" and "chairs in Agriculture and Chemistry" to "spread to the general public the knowledge of essential principles for the peasant and the artist" (iii). Moreover, "it was about time that the voice of those with wisdom . . . be heard distinctly in our modern populations" (iv). Knowledge and education were universal and "nothing could contribute more efficiently to this objective than the publication of a newspaper," which was precisely what was being done by the creation of *Semanario* (iv). Regarding the newspaper's content, Vieytes promised the publication of articles dealing with agriculture, commerce, moral education, domestic economy, professions and arts, the sponsorship of peasants and artists, the elements of chemistry that were more useful, the economics of rural areas, and the best productions of trades and arts (vii).

The first issue published an article written by the philosophy professor Manuel José de Labardén under a pseudonym, Juan Anselmo de Velarde, to deny what was claimed in the problematic article "Politica," starting with the idea that "the majority of the Europeans that arrived in this province [Buenos Aires] were debased for living in it."[72] This led to a discussion about the abundance of women, which the article denied, as it did with respect to the desire for luxury as the result of the expansion of consumption related to the commercial character of Buenos Aires. In this city "there had not been, there is not, and would not easily exist" luxuries, because "its happy constitution" was opposed "to the accumulation of money in a few hands" (46). The impression travelers had of Buenos Aires's women was the result of a "cleanliness that suited their gender, reflecting their dedication, showcasing their economic knowledge, enhancing their appeal to lovers, and captivating their fiancés." Their appearance was not the result of a love of luxury, but an example of "their particular taste in choosing designs" that once completed resulted "in a dress that seems brought from Paris" (47). If in the city there were many "courtesans," as they were called in the *Telégrafo*, there were also plenty of "angelical virgins"; thus, debased women were not what characterized Buenos Aires (47).

The plans for economic development used a model of complementary roles. A September issue, for example, noted that economic growth depended on agriculture because "it was able by itself to increase the opulence of peoples [*pueblos*] to a degree almost impossible to calculate."[73] Actions intended to "defend the moral, political, and physical preeminence of agriculture over all the other professions derived from the luxury and depravity of society" were excused because it was beyond dispute that they were "the creator[s] of science and states," the engine and foundation of civilized societies.[74] Buenos Aires was placed in the "center of the commercial world" and protected by a wise and enlightened government, aware that changes could lead to luxury and moral degradation, but that this could be diminished by the example of those men working in commercial agriculture, which required more physical sacrifices and intense manual labor.[75]

Another important economic activity was industry, needed "to increase the population and consumption" that erased poverty and promoted "science, arts," and the renewal of nations. All these economic sectors should assist each other and complement the different needs of society. Those devoted to agriculture should not see the merchant as an enemy, and the latter should not compete with industry for economic control. Vieytes, like many thinkers at this time, was trying to understand and define the meaning of modern work, and its value in stimulating labor in a society that, like Spain, limited jobs according to social hierarchy until 1783. On March 18 of that year the king declared the work of blacksmiths, taylors, shoemakers, carpenters, and all those of similar kind to be "honest and respectable" (*honesto y honrado*); working in these professions did not "degrade" a family or the people who practiced these trades, and for this reason they could now participate in governmental positions.[76]

Campomanes had suggested this to the king to improve economic productivity and administration in Spain, where the nobility and honorable people, a significant part of the population, spent more than what they produced because they did not work in professions considered below their social standing, and those who were productive economically were not able to hold office or enjoy privileges related to social prestige. Since the desire to work as needed in a modern society was not natural but artificial, the abundance and diversity of natural products made life so easy that men did not need to work in the colonies, which could only be stimulated on the American continent by a culture of wants generated not by need, but by desire.

Benjamin Franklin wrote about this problem in the United States, and it is not a coincidence that Campomanes sent him a copy of the royal decree of 1783 in their correspondence. Franklin believed that men were inclined to have a life of less effort that was related to maintaining the precious freedom praised by new political ideas. This idea was informed, as was common in the Americas, by coexistence with Indigenous populations who were not attracted to the European way of living, thus creating conflicting perceptions about the kind of freedom that modern societies allowed. The problem concerned how to promote artificial needs that violated natural demands in a balanced way that did not affect the right to be free or the morals of society.[77]

For Vieytes, similarly, the science of economics explained how to develop the need and will to work, both in men and women, through growing "the numbers of things people desired" beyond just food, with the aid of industry and commerce.[78] Besides silver, only "leather with hair" was exported to Europe; the abundance of silver decreased the need for work and made everything more expensive.[79] The solution would arrive when "the citizen-patriot" devoted himself to transforming the society and its values.[80] In another article, Labardén called for drastic economic change, which should be the mission of the illustrated (i.e., enlightened) and patriotic men in the River Plate Viceroyalty, but always considering that economic expansion also presented dangerous outcomes; trade "brought all those things that are useful not only to satisfy our needs" but "also to supply different tastes, and amazingly increase luxury," which could degenerate society because it led to effeminacy.[81]

Semanario also had a naturalist and scientist as one of its writers. Pedro Antonio Cerviño (1757–1816), born in Spain and living in Buenos Aires since 1774, was interested in economics and the improvement of government with enlightened ideas. He left to study mathematics and fortifications at the Royal Military Academy of Barcelona, returning to the capital of the viceroyalty in 1781. He was friends with Félix de Azara and assisted him in his expeditions, drawing maps for border disputes and expeditions like the one to the south. Once the Navigation School was created (*Escuela Náutica*) he became its first director; he also headed the Academy of Mathematics created in 1812; two years later he completed the topographic map of the city of Buenos Aires that was printed in London in 1817.

Cerviño represented patriotism through scientific work in favor of society, a perfect image of the modern patriot already discussed. In 1803 he wrote articles explaining how the emptiness of the Pampas was the result of the lack of economic activity; cows, lambs, and ships that reflected the importance of meat, wool, and leather commerce were absent, making the observer suspicious that "the love of the fatherland and the glories of the nation were extinguished" from its inhabitants. It seemed that the entrepreneurial character that distinguished the Spaniards from the rest of the nations had disappeared.[82] He also described a plan to maintain borders with the Indigenous population that promoted a "soft" way to treat them, including promoting the development of trade and agriculture in the territories under their control. The idea was to turn them into "useful" vassals of the king through an agreement to move the border to Rio Negro, in Patagonia, far from the city of Buenos Aires.

In a second article related to this topic, Cerviño explained that rural work made "men stronger" and accustomed them "to work outside and fatigue." Permanent exercise aimed at carrying arms during times of war and peace was a "modern invention; previously, armies were only gathered when there were enemies to fight, and the men returned to the fields and workshops after the conflict was over."[83] The model of masculinity Cerviño promoted required men who had military discipline and the will to populate the southern territories with an owner's mentality that made him enjoy and seek to possess property. This included attendance in church on religious holidays and presenting himself with "decency and cleanness," the markers of "good behavior and presence" necessary to become prominent in society.[84]

In 1804 the economic articles continued, including one in which Cerviño explained that the desire to possess was innate in men; the existence of those who did not own property was related to the lack of means to obtain it; and if those like him "satisfy this desire" through good governance and assistance, in the future there would be an excess of population in the areas that were empty. They needed to "transform [those who did not have property] into useful vassals to the king and the state." The creation of a list with all the names of those living in poverty was needed to "remove them" from a condition created by "their education, lack of enlightenment, and remaining inactive." Eliminating leisure time (*ocio*) was the solution to ending "the corruption of morals, vices and crimes" that concerned families and altered the social order. Property "would make them produce whatever they can produce," but their "lack of capital" and the "accumulation of capital" by the wealthy were serious problems that led to the corruption of society. But he wrote, "the love of the fatherland when well directed found resources for everything." This sentiment would provide "abundant funds" that if well distributed would make miracles.

Semanario de agricultura also published articles about the terrible depiction of those born in the Iberian colonies in geographical texts published in Europe. For example, an article demanded, "Stop suffering the despicable epithets of lazy and indolent" that were given to those living in the colonies of Spain and Portugal in America by a "modern writer that made us heirs of the vices, but not of the virtues of our ancestors."[85] This was a reference to William Guthrie's *A New Geographical, Historical and Commercial Grammar*, first published in 1770—the most popular book on geography that had forty-five editions by 1827.

The first French version of this book, *Nouvelle géographie universelle*, appeared in 1798, translated from the English edition published the same year, making it accessible to those who could not read English, which explains its influence in Spanish America. Buenos Aires is described as "a beautiful town surrounded by superb buildings," and the most commercial place in Spanish America.[86] In 1804 the French edition was translated into Spanish, making the book accessible to those who did not read French or English.[87] But, while sections changed, or were added, the association between the feminine and degeneration in the Spanish and Portuguese colonies remained in all the versions printed since 1770, indicating why Vieytes and the followers of the Enlightenment were so frustrated. "The Creoles have all the bad qualities of the Spaniards, from whom they are descended, without that courage, firmness, and patience, which makes praiseworthy part of the Spanish character. Naturally weak and effeminate, they dedicate the greatest part of their lives to loitering, and inactive pleasures. Luxurious without variety or elegance, and expensive with great parade and little conveniency, their general character is no more than a grave and specious insignificance. From idleness and constitution their whole business is amour and intrigue; and their ladies of consequence are not at all distinguished for their chastity or domestic virtues."[88] The portrait of the manners and customs of the Portuguese in America, by the most judicious travelers, is very far from favorable. They are described as a people, who, while sunk in the most effeminate luxury, practice the most desperate crimes.[89] Scientists also reproduced these same concerns; Cornelius de Pauw claimed that stagnant waters, the product of a deluge, were the cause of the lack of progress in the American population, its degeneration, and the decay suffered by those Europeans living there.[90] More important, de Pauw's characterization of syphilis as an "America's disease" (*mal d'Amerique*) made a most feared sickness the continent's fault.[91] Although the Spanish American priest Antonio Sánchez Valderde (1729–90?), born in Santo Domingo, wrote a strong rejection of this characterization, the image of the tropics as the epicenter of primitivism and sickness would not change.[92]

COLONIAL PATRIOTISM AND VIRILE WARRIORS

The beginning of the nineteenth century presented a different set of challenges for those ruling the River Plate Viceroyalty. Napoleon's control of French and European politics, first through the Consulate (1799–1804) and later with the creation of the empire (1804–15), radically transformed the image of French culture as effeminate and corrupt. During his rule the patriot-soldier became the most important social figure, enhancing values related to honor, patriotism, and a martial and virile masculinity.

Patriotism "in the armies of the Consulate and the Empire associated the honor of the soldier with the honor of the nation."[93] This meant that France, instead of being portrayed "as a beacon of freedom, enlightenment, and civilization," came to be characterized as "a warrior nation that needed to maintain its prestige and position in Europe through military conquests."[94] The importance of "Revolutionary virtue" was replaced by "Imperial virtue," which emphasized the "displays of martial and sexual prowess."[95] This was quite different from the model that promoted peaceful and cooperative association.

If Vieytes rejected "the spirit of domination and conquest" to embrace a new era of peace organized around agriculture, trade, and commerce, this plan would soon be compromised, both culturally and politically. The previous conception of free trade as the origin of a new world of cooperation and interdependence was shattered by the European wars that emphasized competition and destruction instead. The virilization of French culture, the end of French republicanism, and the creation of a conquering empire shattered some of the Enlightenment expectations in Europe and the Americas. The Napoleonic Wars gave commercial warfare much greater significance than it had before because they were intimately related to the expansion of industry and commerce and the ability to control national and international markets.

The plan to create an integrated economy between Spain and its colonies was mortally hurt by the defeat in the Battle of Trafalgar against England in 1805, which crippled the control of the seas needed to develop trade. In 1806 the confrontation between the British and French took a new turn when, in May, Britain "declared a blockade of the Channel from Brest to Elbe," which allowed Napoleon "to pose as the defender of national rights against British economic and commercial tyranny." The Berlin decree of 1806 denounced the blockade as a violation of international law, and as retaliation France blockaded British ships and did not allow them to stop in Europe's ports. The Milan Decree followed in December "declaring that all neutral shipping using British ports or paying British tariffs was to be regarded as British and would be seized."[96] This measure directly affected the United States, forcing it to retaliate by closing its own ports to European trade in 1807. None of these events supported the possibility of cooperation among nations.

The war among the European empires, though, gave some hope to those in the colonies who wanted emancipation. Francisco Miranda had already tried to start a movement in his native Venezuela in 1804 with the support of the United States, but upon failing he accused the United States of treacherous behavior.[97] He was also unsuccessful two years later when the English refused to give him more support. According to Miranda, if England "had [earlier] given support to the South American colonies, they would already have been emancipated from Spain," in part because the European wars meant that they were the bargaining chip of any possible negotiation, which enhanced local patriotism.[98] In June 1806 British troops led by Sir William Beresford invaded Buenos Aires, and a second deployment followed a year later, directly involving the local population in the European hostilities.

Buenos Aires established its militias to successfully expel the invaders, accomplishing something that Spain had failed to achieve. The triumph also inflamed the patriotism of the viceroyalty outside Buenos Aires. In the north, for example, the city of Tucuman quickly organized a demonstration to support the capital of the viceroyalty, which was recognized by the Cabildo of Buenos Aires in a letter stating that its inhabitants were the first "to give proof of loyalty and Patriotism, promising to send to the coast a certain number of people to defend this soil." The reason for this action was "the cause of God, the King, and the Fatherland [*Patria*]."[99] This patriotism was not expected by the British because the importance of commerce in Buenos Aires seemed to make the city friendly to them since they led the cause of free trade in the world.

The invasion also created concern among other Viceroyalties in South America; on September 12, 1807, a British Foreign Official Paper was published in the *Weekly Political Register*; taken from the *Peruvian Minerva*, it contained the manifesto sent by the viceroy of Peru to the inhabitants of the viceroyalty after the English capture of Buenos Aires. In it, England is accused of "using every means it can invent" for "cementing and propagating its detestable tyranny," using vile practices that set aside "the most sacred principles of the rights of man, and trampling upon all usages and customs, for many ages universally received and observed amongst civilized nations," which was related to the British Atlantic blockade, and how it had become the main obstacle to free trade.

One of the acts the English committed immediately after taking over Buenos Aires was "to lull to sleep the understanding of the inhabitants of Buenos Ayres, with hopes of imaginary happiness"; lies that covered "the iron chains which their oppressors have prepared for them; to stupefy the native energy of those Spaniards; to alienate from their hearts, if it is possible, the love, the fidelity, and the gratitude they owe to the most benevolent and just" of all kings. The intention of these actions was to deter Buenos Aires's inhabitants from "retrieving their lost honor" offering "immediately to free their commerce from the heavy duties and imposts to which it has been subjected." The offer also included a promise to respect the Catholic religion "and its holy Ministers," and the preservation of "local laws and national customs" under British rule.[100]

The manifesto ended with two important reminders to the population of the Viceroyalty of Peru. First, the support that the English had given to the revolutionaries of Haiti, in what is described "as the most atrocious rebellion recorded in the annals of nations." Second, a call to the Indigenous people: "Indians! You are such interesting objects of the tender care of our most amiable monarch!"; "Indians! Wheresoever the English nation has gained a footing, your's [*sic*] have been enslaved, reduced, and destroyed without mercy."[101] Just a few years after the horrible repression of the Tupac Amaru movement, and with slavery still accepted in society, the Spanish government presented itself as the defender of the universal rights of men and civilization.

The same reasoning was present in Belgrano's proclamation in March 1806 as the secretary of the Consulate, when it was clear that Buenos Aires could be attacked by the English. He wrote a declaration to request support for the Spanish authorities, presenting an enlightened version of the nature of the king's power and the way it treated his subjects who were "citizens," which implies a different kind of colonial rule. These citizens were in debt to the monarchy because of its efforts to "eradicate from the American soil" barbarism in order to replace it with "religion, laws, sciences and arts," based on the "love for the loyal vassals" who "lived with the most tenderness in the heart of Charles IV," the king who governed through "charitable intentions" and never used "his illustration [*luces*] and his power" for a purpose other than the "happiness" of those he governed. He had a "liberal hand" and was the "tireless agent" who would allow those in the viceroyalty to "live comfortably and securely" under the "extremely soft yoke of its domination."[102] Citizen-patriots needed to gather to keep the English away from Spain and its colonies.

Alexander Gillespie, one of the men who had been part of the invasion of Buenos Aires, and a major in the Royal Marines, recognized in 1818 the possibility that the local elite might have "espoused revolutionary objects" but England's hostility had "revived in them all their national animosities, and drowned every other feeling but the thoughts of our extirpation."[103] The emergence of a new subject after the first invasion is captured in a letter directed to Buenos Aires's City Council (Ayuntamiento) and signed in August 1806 by Martín de Alzaga, a leader of the cavalry and a supporter of the Spanish king. Together with other notable men from the city, they wrote that it was "astonishing" to see how much "the courage inflamed by the enthusiasm of honor, loyalty, and patriotism" could achieve. It had "transformed the spirits," creating "a new generation of warrior men," and in what was "a prodigious metamorphosis," it even changed the "weak sex" that had substituted "the sweet power of the insinuating Venus" for "the fury of terrible Mars," fighting together with "tender children mixed with the strong male," to expel the invaders.[104]

The English defeat did not end the influence of the European wars in the American colonies, though; in 1808 a dispute erupted about the succession to the Spanish throne. Support for the Spanish monarchy was split between the unpopular Charles IV and his more popular son Ferdinand, which ended with the Aranjuez mutiny that forced the abdication of the father in favor of the son, who became Ferdinand VII. Napoleon used the crisis as an excuse to invade the Iberian Peninsula and kidnap the new king, and on May 25, 1808, the French emperor announced that he was planning a constitutional monarchy to replace the Bourbon dynasty. The first step was a call for representatives from provinces and cities because he wanted to know the "desires and needs" of those who lived in them. Once these were known he would appoint another king who would rule according to a constitution, which would reconcile the "authority of the sovereign with the liberties and privileges of the people."[105] In June 1808 he appointed his brother Joseph as the new king, whose rule began in July after the passing of the Bayonne Constitution.

This document was prepared with representation of those born in the colonies who were living in Spain at the time and introduced changes that had important consequences; first, the territories outside Spain were no longer called colonies, and now became Spanish provinces. The Creoles had the same rights as those on the peninsula, including the freedom to cultivate, create industries, and engage in commerce. Freedom of trade was also granted, and concessions and privileges for import and export were prohibited. The provinces also had the right to have representation in the courts. Despite these reforms, the French lacked solid support in Spanish America for many reasons, among which were the failure of their own republicanism, patriotism, and loyalty to the figure of the king, even among the Indigenous population, coupled with the rejection of the French liberalism they associated with excesses following the 1789 revolution.

The Iberian Peninsula remained occupied by French troops while part of the Spanish army continued supporting Ferdinand. The Junta Central claimed to hold both the monarch's authority in his absence and legislative power through regional representatives that constituted the nation. Formed in 1808 in Aranjuez, it relocated to Seville that same year, and by 1810, it was conducting sessions in Cádiz and León Island. At this final meeting, representatives from the American and Asian colonies participated, though none were present from the River Plate Viceroyalty. The colonial representatives advocated for the return of the Jesuits to Spain, stating that it was "of the utmost importance to cultivate a taste for the sciences." Furthermore, they emphasized the necessity of the Jesuits for the "progress of the missions," which were instrumental in spreading and nurturing the Christian faith among Indigenous populations.[106]

The presence of two competing authorities complicated matters in the River Plate. Santiago Liniers (1753–1810) was appointed interim viceroy in 1807, in recognition of his heroism during the British occupation. As viceroy, he received conflicting messages following Ferdinand VII's abdication in 1808: one from Joseph I, declaring himself the new king of Spain, and another from the Junta of Seville, claiming it held the monarch's authority; both demanded recognition of their authority. Liniers was French, and rumors about his disloyalty circulated together with accusations of malfeasance. The Banda Oriental, today's Uruguay, rebelled against Liniers's authority and established a Junta in Montevideo in 1809 under the order of the Spanish governor Francisco Javier de Elío y Olóndriz (1767–1822). Liniers was replaced in 1809 by Baltasar Hidalgo de Cisneros (1756–1829), an officer in the Royal Army born in Cartagena, part of the New Granada Viceroyalty, after his designation for the position by the Spanish Junta.

Rumors continued about how the European colonies would align after this crisis; it was suggested that some were plotting with the Portuguese and their ally, England, to give the sovereignty of the Spanish American colonies to Carlota Joaquina, the sister of Ferdinand VII, queen of Portugal, by then living in Rio de Janeiro because the whole court had moved to Brazil to avoid becoming prisoners of Napoleon. Those who supported the Enlightenment were divided; for some, patriotism meant the defense of Spain, and for the most radical, it meant the creation of a new republic. In addition, there were political conservatives who supported the traditional colonial organization, and many who were just defending their place of birth amid the disintegration of Spain.

The chaos enhanced the importance of the local militias, whose power grew, another factor being the militarization of Buenos Aires as the capital of a viceroyalty to maintain territorial integrity. The troubles of the period were recorded in *Correo de comercio*, a newspaper edited by Belgrano and, by early 1810, published weekly to replace Vieytes's *Semanario* after it ceased publication in 1804. It was the initiative of "a group of Patricians" because they were "ashamed that the great capital of South America, now deserving the attention of the civilized world," no longer had this kind of newspaper.[107]

Publication began on March 3, 1810; an article in the third issue affirmed the importance of education as "the true origin of public happiness," though in the River Plate Viceroyalty the schools were in terrible shape, thus condemning the population of certain areas to a life without "Law, King, or Religion," a situation that promoted the laziness and abandonment that dominated rural territories.

The kings are depicted as rulers who "constantly had pushed for the illustration [*ilustración*] of their peoples," and as a result "no colony in the whole universe" had a poor class that "received as many benefits" as the Spanish ones. Followed by a terrible period in which the state had stopped promoting education, there was now a new time in which "wisdom had come to defeat the rays of barbarism, promoting all that was useful, advantageous, and in particular the promotion of knowledge so virtue could occupy the place it deserved," and the "Nation" could acquire the enlightenment that would benefit those who were part of it.[108] The objective of education was "the expansion of knowledge" and the "formation of the moral man" who was needed to govern under the new principles.[109]

The weakened power of Cisneros as viceroy was obvious when he changed positions related to free trade, which led to his removal from power. On May 25, 1810, a junta led by its conservative president, Cornelio Saavedra (1759–1829), assumed the provisional government of the River Plate Viceroyalty, but with members divided about the future plan; the more radical members were from Buenos Aires: Juan José Castelli, Miguel de Azcuénaga, Manuel Alberti, Belgrano, and the two secretaries, Juan José Paso and Mariano Moreno. The debates about how to organize political power and to whom men should be loyal resulted in many disputes and political struggles, particularly about the role of Buenos Aires as the capital of a vast territory that had divided interests. In these circumstances, another newspaper, *Gazeta de Buenos-Ayres*, started circulating to promote a radical turn for the political movement that had started in May. Mariano Moreno (1778–1811) was behind the publication and its most prominent writers were Juan José Castelli, Manuel Alberti, Bernardo de Monteagudo, and Belgrano.

The first issue, published on June 7, 1810, included an epigraph taken from Tacitus, capturing the times that inspired this publication, which was not unique; David Hume's *A Treatise of Human Nature*, opened with the same reference: "It is the rare happiness of the times when you can feel what you want and say what you feel."[110] One article illustrates how the culture of sensibility influenced the description of the ceremony organized for the swearing in of the members of the junta. During this event the "man-philosopher" appreciated more than anything "the spectacle of people electing without incident" who would oversee government. In this public event, the "sensitive souls fainted" in front of "a very sweet impression to which they were not accustomed": the image of numerous troops that "instead of the ferocity that had distinguished them in combat" were now showing "tenderness."[111] These public displays of the new representatives of the people produced in addition "trust and more secure hopes, and elevated the youth's souls, and made the oldest cry out" about "such a glorious day."[112] This emotional political spectacle announced a new era.

In the third issue, an article requesting freedom of speech and press introduced a much more aggressive political and emancipatory stance. It was an affirmation that without freedom of thought "the arts and useful knowledge would not advance," and society would continue "respecting the absurd ideas consecrated by our parents and authorized by time and custom."[113] A veiled attack on the more moderate political forces was a clear rejection of "our outdated opinions" together with a call for less attachment to what it considered "ours" in order to "introduce the Enlightenment." This implicit bashing of colonial culture was a typical expression of a new generation's belief that they represented a new historical time.

The Seville Junta had failed, and a Regency Council replaced it in 1810, requesting that representatives of the "Spanish and American Spains" gather in Spain. This resulted in the elevation of "Spanish Americans" (Españoles Americanos) "to the dignity of free men" after the movement to reject the French invader had begun.[114] They "were not the same" since the revolution had started because they were no longer under "an intense yoke while away from power, regarded with indifference, abused by greed, and destroyed by ignorance."[115] This patriotism was displayed to unify the goals of the citizens of the empire and make it clear that after the king's return the monarchy needed to govern under liberal principles.

In a similar fashion, in the *Comercio*'s issue of June 30, 1810, Belgrano introduced a discussion about the best political system that was made by "patricians," those members of the best families born in Spain and Spanish colonies, because they were all part of the same "*Nation, and the same Monarchy, without any distinction in our rights and obligations.*"[116] This nation needed education and good government supported by the right philosophical system, especially through the teaching of logic, "the art of thinking with exactitude, or of using intellectual faculties conveniently, analyzing and reasoning" (207). It was for this reason that the first step needed was the evaluation of how useful the new ideas coming from this discipline were, what advantages men could "obtain from knowing if sensations are thoughts or knowledge," if the essence of things or "their existence can be the object of the idea; and other things of this kind" that were taught in what was called modern logic (207).

The solution was the adoption of a method that put "reasoning constantly on trial through experience, conserving only the facts that are truths given by nature," the latter being the "only teacher that cannot deceive us" (210). It was needed to "reduce reasoning to simple operations" (210), and short judgments, so it was never "out of sight of the evidence that must serve as guidance" (140). The logic that Belgrano recommends is Condillac's, which was "recommended by the patricians from the European and American Spains," and already used to introduced government reforms (141, 211).

Regarding the idea of nature, in the issue published on June 2, 1810, Belgrano understood it as the creator of diversity because the "talents of man" were diverse and nature had created this distinction to reward talent.[117] The same idea is repeated in September: the "infinite providence, whose work is nature, has wanted the variety with which it has adorned it, making men dependent on each other." The "Supreme Being" has formed links in nature in order "to lead peoples to conserve one another," and with the intention of being praised for its creations. This Being "manifested its love and greatness through the knowledge of the marvels with which it had populated the universe." For this reason, the passions unleashed by "the reciprocal dependence that men had for the variety of merchandises" also constituted a natural force—one that economic policymakers needed to recognize. In the same way, the productions of industry were infinite, and manufactures were the result of perfecting what was given by nature. Industry and work transformed nature to make it better, to diversify it economically.[118]

After the creation of the junta in 1810 came a period of intense interest in the promotion of scientific knowledge, with the organization of new institutions, and a new curricular design for classes at the elementary level and above. This same year Jean-Jacques Rousseau's *Social Contract*, translated by Mariano Moreno (1778–1811), was published for the instruction of American youth, though the parts on religion were suppressed. In the prologue, Moreno detailed how disappointing it was for his generation to see the Spaniards' failure to form "a government that deserved" the people's trust, and to write a constitution to end the anarchy, even when the heroism of those fighting was admirable. This debacle had taught those in the River Plate Viceroyalty that the victory of liberalism would not come if the nation's population was not educated. The responsibility of the patriot "citizen" was to communicate his "illustration [enlightenment] and knowledge" that were virtues of patriotism. All "classes, all ages, all conditions" would share the benefits of Rousseau's efforts. [119] Spain's incapacity for creating an enlightened government had left the colony alone in the process of creating a good government that for Moreno was the republic.

REPUBLICAN BROTHERHOOD AND NATION-BUILDING

The year 1812 was important because the pressure to declare independence and organize a new form of government increased together with the conservatives' fears of making this commitment. By this time, the United Provinces of New Granada, led by Simón Bolívar, had collapsed, demonstrating the problems of establishing a republican system and the weakness of revolutionary politics led by a minority. In this same year, the Spanish priest and theologian Rafael de Vélez denounced those who called themselves "*Philosophers* because of their professed love for the sciences and their desire to find the truth."[120] They also called themselves "*Strong-Spirits*" and "Liberals because they easily renounced their previous opinions and followed the new ones that were more enlightened." They believed themselves superior to every other human and "their fatherland [was] all the world: their compatriots all the men," even the worst, calling themselves "the true Cosmopolitans." All over Europe they were known "by the names, Enlightened, Materialists, Incredulous, Libertines, Masons, Atheists."[121] The formation of a liberal elite was viewed as treasonous by conservatives because of their openness to foreigners and their lack of interest in continuity and tradition.

The ex–River Plate Viceroyalty was building a liberal revolutionary side that pressed for a more radical path to independence, but a conservative Triumvirate had removed the more radical individuals from power, including Mariano Moreno and Bernardo de Monteagudo, which then led to the

Revolution of October 8, 1811, supported by General José de San Martín and Carlos María de Alvear, both military officers recently returned from Spain. They forced the resignation of the previous members and supported the formation of a new Triumvirate that represented a liberalism favoring independence. Important measures were taken in 1812, among them the creation of the National Public Library and a prohibition against introducing more enslaved people. The new authorities called for an assembly to take place in January 1813 with the purpose of declaring independence and passing a constitution, but these objectives were not achieved due to the distrust of some provinces toward the supremacy of Buenos Aires, a capital identified as favorable to commerce.

The Free Womb law was passed, making all the children of enslaved people born after this legislation free, though the institution of slavery continued until its abolition in the constitution of 1853. The Inquisition was banned, and the Triumvirate was replaced with the creation of a Supreme Directorship serving a term of two years. The primary inspiration for the emancipation of enslaved people stemmed from the historical writings of Abbé Guillaume Raynal, which gained widespread popularity in Europe and the Americas. As noted earlier in this chapter, Raynal was pessimistic about Europe's future and focused on the actions of European conquerors and the Catholic Church in colonizing America and Asia. In 1770 he published, with other contributors, an anonymous book titled *Histoire philosophique et politique*. Later editions included the name of the main author. The 1780 edition featured sharp criticism of colonial slavery, predicting the rise of someone who would lead enslaved people in a revolt against tyrants, including the Spanish, Portuguese, English, French, and Dutch. The book was banned by the Inquisition, which is not surprising; nevertheless, copies and translations circulated.[122] Mariano A. Pelliza wrote that Raynal's book was even more influential than Rousseau's *Social Contract* and it had prepared those who would be identified as the most radical revolutionaries in Buenos Aires: Mariano Moreno, Pedro José Agrelo, Juan José Castelli, and Bernardo de Monteagudo among them.[123]

Those who favored revolutionary ideas began associating themselves with a lodge that prioritized the emancipation of enslaved people after creating independent nations out of the colonial system in America. The modern masonic lodges had their origins in the seventeenth century, with a textual base that combined "Magic, Cabala and Alchemy as the influence making for the new enlightenment."[124] The main concern of this movement was "striving for illumination, in the sense of vision, as well as for enlightenment in the sense of advancement of intellectual or scientific knowledge."[125] The colonies were also suspected of having secret societies that were instrumental in organizing the plans for emancipation. There is evidence that in 1804 Masonic lodges were already operating in Buenos Aires; a year later the Portuguese Juan de Silva Cordeiro organized a lodge named "San Juan de Jerusalén de la Felicidad de esta parte de América" connected with a lodge in Baltimore. Already a member of Madrid's "Matritense" lodge and a supporter of liberalism, he had traveled to the United States and Brazil before coming to Buenos Aires to escape from being captured under an order from Lisbon for his activities in the secret society. Miguel de Azcuénaga provided a house for the meetings, and many prominent men joined, according to Gervasio Antonio de Posadas.[126]

The Logia Lautaro (Lautaro Lodge) gathered the elite of Spanish American emancipators and was the most prestigious and effective. According to Jules Mancini, it was Miranda who, through Jesuit priests born in Spanish America, organized the foundation of a secret association, the "Junta de las Ciudades y Provincias de la América Meridional."[127] Antonio Zúñiga agreed that Miranda was the main figure behind the creation of these societies once he arrived in London and contacted local lodges. After Logia Lautaro he created two others, "Rational Gentlemen" (*Caballeros racionales*) and "Great American Reunion" (Great American Meeting). Among the members mentioned by Zuñiga were "Saturnino Rodríguez Peña, Servando Teresa Mier, Santiago Matiño, Benito Lizarraga, Olavide, Pozo, Sucre, Lord Melville, Lord Marduff, Count Fife, Sir Home Popham, Sir David Bair" and some English gentlemen.

Lautaro followed the Masonic rules with some modifications; the first grade was Apprentice, and "besides the introduction to the symbolic aspects of the liturgy" the new member needed "to commit his life and possessions to work for American independence." The second grade was the "Companion" (*compañero*) who had to "profess his democratic faith"; the third grade was Master (*maestro*), and the affiliate needed "to work in civil propaganda in favor of the new ideals." The fourth level, Rosencrantz (Red Cross), required the brother to exercise influence to support revolutionary action; those in the fifth grade, Kadosh, worked on the military action of the revolution.[128]

The Supreme Council was based in London, where Miranda personally introduced the political leaders of the American colonies to the organization, including an initiation ceremony and an oath never to accept any government as legitimate unless it had been elected by "the spontaneous free will of the people," and also to affirm that "the republican system [was] the most adaptable to govern for the Americas," which demanded that the members use all the means available "to make the people choose it."[129] Prospective members were carefully investigated and at the beginning the goal of moving the revolutionary movements toward republicanism was ignored.

The Logia Lautaro's ideology was synthetized in the words: "Union, Force, Virtue," in the belief "love your neighbor" (*amor al prójimo*), and in the "freedom of the human race" (*libertad del género humano*).[130] By 1808 this society had chapters in Buenos Aires, Chile, Venezuela, México, Nueva Granada, Ecuador, Peru, Caracas, and Chuquisaca.

San Martín and Alvear opened the new chapter in the River Plate with the objective of "working for the system and plan of American independence and its happiness, working with honor and proceeding with justice." Its members had to be "Americans who were distinguished by their liberal ideas and their patriotic fervor."[131] One of the lodge's political leaders was Bernando de Monteagudo (1789–1825), a radical liberal close to San Martin, the military leader soon to be liberator of Argentina, Chile, and Peru. Monteagudo was from the northern province of Tucuman, and not a member of prestigious colonial lineages based in Buenos Aires. In fact, he was accused of impure blood, due to suspicious gossip about his origins, including the possibility of a free mulatta (*parda*) mother. Another leader was Vicente Pazos Silva (1779–1853), or Pazos Kanki, a priest from the north region of the viceroyalty, who was a descendant of a prestigious Aymara lineage of which he was very proud. He had completed his doctoral studies at the university in Cuzco and in Chuquisaca, where he studied law and finished his religious studies to become a priest. He was fluent in Quechua, which he taught in Cuzco. Later, while living in London, he translated the New Testament to Quechua.[132]

Pazos Kanki lived in Buenos Aires in 1810, where he became friends with some of the leaders of the republican movement. He explained how Napoleon's invasion of Spain resulted in the flooding of ideas about independence and political freedom among educated youth such as himself, whose imagination was exalted by the reading of Rousseau, Voltaire, Mirabeau, and other philosophers, which prompted them to political action. He was one of the writers for *La Gaceta de Buenos Aires* in 1811, and the next year he founded *El Censor*, a very influential publication. Political frustrations forced him to leave for London, where he decided to leave the priesthood and marry a local woman. Returning to Buenos Aires in 1816, he brought the equipment to establish a print shop where he continued publishing in favor of republicanism.

Monteagudo and Pazos Kanki had in common being foreigners to the culture of Buenos Aires. The former, a typical radical of his generation, was a member of the Patriotic Society founded in 1812 to support the political ascendancy of emancipatory liberalism, indicating the emergence of a new form of patriotism delinked from the needs of the empire. He defended the new philosophy and its culture of sensibility in his articles published in *Mártir o libre* (*Martyr or free*) the same year. In one piece, he explained that when a man turned into citizen, he should give the "fatherland [*patria*] generous love and all the sacrifices he was able to provide." He should be "obedient and respectful of the laws and judges; and give the other citizens help, protection and fraternal sensibility together with honor, dignity, and virtue." These duties came from "the natural instinct that inspires them and the reciprocal convenience" that develops them.[133]

In the same spirit, the reaction of the people to the election of Juan Martín de Pueyrredón (1777–1850) as a member of the Assembly was described by Monteagudo as "so emotionally sensible" as to justify its unanimous approval.[134] This speech made clear the political objectives of those like him and the role that the culture of sensibility had in it. His universalist historical narrative started with the primitive man and moved through different eras of human development until arriving at the American revolutions led by "the sensible hearts' good desires" that had awakened the lethargic colonial society.[135]

Society should be dedicated to "promote the Enlightenment" that "guaranteed the happiness of a state."[136] This change was the result of "the education of the people" and the "formation of customs" to be followed. The latter was the creation of the appropriate environment through "good legislation linked to the nature of climate, the characteristics of its inhabitants, and the passing of time that was sometimes able to form a more or less moral people through the impression *of virtue*."[137] Nature gave rights, among them the right to be happy and to perfect humanity through knowledge.

Monteagudo understood the nineteenth century as the one in which "the monopoly on scientific knowledge was abolished" allowing its spread from the "Artic circle to the Pyrenees." This facilitated experiments "in the physical and moral sciences" that had "deduced practical consequences, which had "influenced the happiness of the human race in ways that were still not felt."[138] The objective of this science was to let men know their "true relationships with what exists, and the advantages the masses of organized beings could get from knowing them and the means to obtain them."[139] Scientific knowledge was the foundation of politics.

While the speech conveyed ideas like those of the colonial patriots, it introduced a key difference: the focus shifted from economics to other sciences.

This change reflected the need for a new philosophy to organize the emerging social and political order, enabling the break from Spain. This form of patriotism tied the republic's future to the integration of all aspects of life with scientific principles for national development. At that time, however, there was no comprehensive philosophical system that could seamlessly merge politics and science to shape a universal republican citizen. It was only with the spread of French ideology that a new model of the republican male began to take form.

3

Man of Ideology

Elizabeth Colwill has explained the French Revolution "as a watershed in the multiple discourses of womanhood of the eighteenth century," and the same can be said about manhood after the May Revolution of 1810, when shifting ideas of men's roles were debated in the River Plate Viceroyalty.[1] By the end of this decade the adoption of the first system of knowledge applied to all aspects of the republic began and helped advance a model of man based on property, republicanism, and science. Connecting the virile citizen of the revolutionary wars with the liberal man of property and science, French ideology created a comprehensive template for republicanism based on Etienne Bonnot de Condillac's sensationism.[2] Ideologues devised an economy organized by the division of labor and the development of a civil society according to the division of gender and social class. The influence of this philosophy was felt across the American continent, and was defined as the science of ideas, or science of man, according to the writings of Antoine Destutt de Tracy (1754–1836) and Pierre Jean Georges Cabanis (1757–1808), both professors at the National Institute of Paris.

Ideology was also connected to the development of physiology, which was transformed by the interactions of the distinct disciplines that would later be known as part of biology.[3] According to Michael Gross, Cabanis explained sentiment as related to "anatomically conflated nerve and muscle fibers by describing the latter as a combination of nervous pulp and cellular tissues," and every part "which acted—which meant every living fiber in the organism—did so because innervation made it sensible and contractile."[4] Unlike Cabanis, other scientists, such as George Cuvier (1769–1832) and Jean-Baptiste Lamarck (1744–1829), defended the idea of a vital force, and the physiologist François Magendie (1783–1855) defined phenomena of living organisms as related to nutrition and movement.[5]

In her analysis of the ideologues, Colwill explains how, for the most part, they ignored the vast majority of nonelite women who were compelled by necessity to participate in the public sphere to support themselves.[6] While Manuel Belgrano promoted the need to create jobs for poor women, the focus of the new philosophy was on those who were part of the elites. Under the influence of this philosophy, a woman became a man's creation, being both the product of his imagination and the mother of his children. She was expected to express her love for the man in a way that affirmed his individuality and status. Yet the obligations and responsibilities brought into men's lives by marriage contradicted the ideals of autonomy and individualism caused by commercial expansion, the emergence of professionalism, and the politics of the revolution.[7] By the 1820s, French political evolution indicated that men would zealously defend property and politics, which symbolized their authority, from anyone who might challenge them.[8]

PHILOSOPHY, PHYSIOLOGY, AND GENDER

The work of Isaac Newton and other European mathematicians inspired a desire for rationalization in all fields, which initiated a search to establish a new foundation for those sciences that had already a "mathematical form." In other fields, for which the conceptual structure was less secure or coherent, the establishment of foundational principles progressed more slowly. For example, electricity and magnetism moved from narrative theories and experimentation to mathematization. Chemistry, medicine, and natural history adopted systematic models of information organization.[9] The social sciences were developed by the influence of the natural sciences' fundamental concepts and methods. The result was a theorization of humans and their societies in both Europe and the Americas. Method was understood as referring "to a double movement of analysis and synthesis by which the phenomena of a field are reduced to their elements and then restructured into a true whole that can be known" by reason.[10]

In 1801 Destutt de Tracy wrote that one could only have an "incomplete knowledge of an animal if its intellectual faculties were not known," particularly among humans who had the capacity to think.[11] His book was "part zoology" and part a history of man based on the philosophy of Condillac. Robert J. Richards explains that this philosopher also "believed one could resolve all ideas into the sensations that produced them and thereby test their soundness."[12] Destutt de Tracy himself explained that his *Projet d'éléments* was basically a synthesis of the philosophy of Condillac, an attempt to fix a problem the latter created when he failed to write a text for students explaining his complete doctrine, linking his discoveries in "the art" of speaking, reasoning, and teaching. The purpose of this text was to provide "an exact and detailed description of our intellectual faculties, their main phenomena and most relevant circumstances" in a useful manner to contribute to the study of science.[13] In this interpretation truth was not exposed but discovered through a precise procedure.

Destutt de Tracy used many examples from political developments in the Americas to support his political philosophy. For instance, he affirmed that the most successful nation in the world was the United States, whose citizens "every twenty-five years, double their culture, their industry, their commerce, their riches, and their population," unlike the "miserable" Spaniards "notwithstanding all their advantages."[14] At the same time, regarding science, he disagreed with Georges-Louis Leclerc Comte de Buffon's idea of America as the cradle of degeneracy. He explained that "when a part of mankind has been accustomed to abuse its power, we may well say that they have degenerated," which led some to believe that "they are no longer born the same—that nature has changed—that the race is depreciated—that they have no longer force or courage," which was all "very false."[15] Those who were able to access enlightened education were able to transform themselves through ideas that were based on the observation of experience.

The physiological side of this philosophy was developed by Cabanis; in 1794–95 he wrote *Coup d'oeil sur les révolutions*, which he finally published in 1804.[16] This work stated that "in proportion as the boundaries of the Sciences are extended, it becomes more requisite to improve the methods of arrangement and instruction" (1). He used an analogy to explain the problem brought by the rapid accumulation of knowledge. It was as if an "inquisitive traveler who on his way to collecting all that interests him" saw his baggage growing "at every moment" and was frequently forced to examine it. As a result, he needed to "either get rid of unnecessary or duplicative items or to arrange them in a better order" that used less space to ease their transportation and use (3).

At a time when all the branches of sciences were being renovated, those physicians who were interested in philosophy needed to regard it "as a duty to unite their efforts, in order to accomplish this great reform of their science and their art" (5). The time had come for "placing medicine on the level with the other sciences, and for determining with precision, their mutual relations" (6). Cabanis also indicated that when principles are carefully derived by comparing and analyzing all relevant facts, the resulting system or overarching doctrine is no longer a mere speculation. Instead, it forms a clear and accurate representation of the science, as far as current knowledge allows: "and the new discoveries, that may be made, will be easily referred to the general principles, to which they relate; whether they tend to confirm them, or contradict them, and to render some changes or modifications of them necessary" (1–2). In this attempt to transform the medical field it was clear that "medicine and moral philosophy form but two branches of the same science, which, when united, constitute the *Science of Man*" or what was known as anthropology by the Germans (304). This would become the most important science for the process of nation-making over the remaining nineteenth century.

Physiology was the solution to all the problems and confirmed "all their speculative and practical truths." Humans' sentiments, passions, virtues, and vices depended on the "physical sensibility of the system, or the organization that determines and modifies it," and for this reason "metaphysics and moral philosophy" were equally necessary for medicine (305). A physiologist needed to be educated in the diseases of the body and the mind, the latter requiring great observation and sensibility. Cabanis declared that the "errors of the imagination, of those of the passions and desires" were evidently "the causes of a great proportion of the miseries of man." These were diseases that "generally depend on his own errors, or on those of society," and were always "liable to aggravation from the depraved state of his moral constitution" (384). The morality of the nation could be measured by the diseases it transmitted.

In terms of gender, the roots of ideology were in Condillac, who, according to Catherine Bobbs Peaden, saw the "feminine at one pole, the point of origin, and the masculine at the other." The difference was more in degree than in kind, and the feminine determined the masculine "as a seed [determines] the future plant." The masculine might appear more powerful, but "it must always return to the feminine to replenish that power" through "an aesthetic experience, which returns us to our physical and emotional roots."[17] As Jacques Derrida notes, Condillac's work does not contain a simple articulation of binary difference, but the ideologues changed this approach.[18]

Destutt de Tracy's ideas were different because he argued that while "women had the same rights and probably the same capacities as men, they were most usefully occupied in domestic functions, in which they excelled." Unlike women, men were the "natural" representatives of families in public affairs. The sexes were not "unequal," simply endowed by "nature" with different functional roles, which resulted in a division of labor.[19] Building a new social order required careful leadership that should be in the hands of those who were more educated. So, while he defended the rights of all men to participate in elections regardless of considerations of lineage and social rank, he also devised intermediate steps that would place the final decisions in the hands of a few men educated in the ideas that would build the new order.

Women were important because they would not only give birth to future men but also educate them accordingly to form a citizenry respectful of the new morality and laws. In this sense, Destutt de Tracy agreed with other philosophers of the time who posited a "direct relation between physical sensation" and thought production. One way to offer "evidence of this relation was to show how sensory systems differ between men and women."[20] These naturalized differences formed the division of labor based on gender.

Unlike Destutt de Tracy, Cabanis appears more confusing regarding gender. In her work on art, Ewa Lajer-Burcharth notes a contradiction in his ideas; on one hand it seems that nature "brings forth man with fixed organs and faculties; but art can increase these faculties, change and direct their use effectively creating new organs." But, on the other hand, "it is the woman's anatomy that determines her social destiny." Women were mothers, and any other aspiration was against the obligations of her sex. Lajer-Burcharth characterizes Cabanis as confused between physiological and moral dimensions, but ideology could not separate the two—as Cabanis had mentioned—and concentrated on the confinement of the feminine body within the conceptual and social reality of the family.[21]

In terms of the French politics under the Directory (1795–99) this position was related to the "ideological shift towards the family" with marriage becoming the regulator of the affects and drives related to men's primitive and natural organization.[22] Sexual desire was the result of physiological transformation that both men and women underwent during puberty, though with very different results; socially, women became coquettish, experts in the arts of seduction, but sexually they were dominated by a modesty (*pudeur* in French, and *pudor* in Spanish) that made them uncomfortable with the practice of sex.

Women were different from the educated men because the latter needed to be aroused and excited, distracted from their political duties, to have sex, which placed woman's stimulation of male desire among the moral obligations of motherhood, but controlling women's sexual behavior was politically necessary to protect the proper raising of future citizens.[23] Her existence could not be conceived outside the needs of men, as Cabanis concluded in his writings. Since the value of females would always rely on "the impression" they left on men, those men who "truly love them" could not find "pleasure in seeing them carry a musket or marching in formation," and even less on a platform where "the interest of the nation" was debated.[24]

Women's moral obligations were delineated according to the impression they made on men and how this would shape society, which was very different from the ideas defended by the Marquis de Sade (1740–1814), who around the same time understood the republican subject as a subject of desire. He supported the idea that the new republican order should follow the uncompromising logic of desire, disregarding all the rules and barriers established by societies in the past. The bedroom (boudoir) was viewed as a space where the desiring subjects—both male and female—enjoyed equal rights. On the contrary, Cabanis viewed the bedroom as the reproductive place of the nation, regulated by the social and political needs of the state.[25]

Women created impressions through a visual order that organized social values and played an important role in the self-transformation of men. This view entailed that new medicine would consist in bringing together the physical and mental to create the science of man, according to which the human nature would be controlled and disciplined.[26] At the end of the eighteenth century Cabanis introduced the idea that the intervention of the body through physical exercise and habits could helped change perceptions, which altered men's character (62). Ideology made a distinction between physiological ideology, related to sensitivity in the body, and rational ideology, concerned with mental and moral phenomena; both would be relevant in Argentina because they were seen as the new foundation of the republic by the leaders of liberalism in the 1820s. Cabanis's work relates to gender and masculinity because it emerged in a context in which emotions and sentiment were also studied physiologically, which medicalized the culture of sensibility.

As George Mosse explains, by 1752 physicians such as Albrecht von Haller had developed an interest in the sensibility of the nervous system, and in this context, the phrase "to be nervous" emerged (61). Nervousness remained "a principal foe of manliness well into the twentieth century," and control of its related diseases became a main goal of medicine (61). The research of Philippe Pinel (1745–1826) became the source for studies of these disorders and their gendered meaning in France. Pinel's *Nosographie philosophique*, published in 1797, was based on his medical work at L'Hospice de la Salpêtrière as a professor at the Paris School of Medicine. He studied nervous diseases as issues intimately connected to the history of human understanding and moral philosophy, areas that were also crucial to creating a happy society.

Pinel's method followed Condillac's in his understanding of systemic medicine. Ideologues "conceptualized man as the product of his environment: by changing it, they sought to improve the human condition; they were eager to apply their 'science of man' to concrete social situations and to assume the role of public servants."[27] They also rejected "the traditional belief in a soul" and adopted concepts "of sensitivity and irritability as the only bases for sensation and movement." Feelings "emanated from matter" and should thus be researched.[28]

Men and women had clear physiological differences in Pinel's model; he wrote about the "masculine energy" that characterized talented men. He also differentiated diseases suffered according to their organic constitution using historical references from ancient history: for example, Spartan culture helped men to understand the "advantages of a masculine institution that allowed young men to strengthen their bodies," and the "enslavement of young boys and girls to regular exercises, coarse food, and laws of austerity" observed during marriage, "which were all powerful ways" to promote "love of the fatherland [*patrie*] and heroic devotion" making clear how education and costumes could transform the nation.[29] He defined "neuroses" (*névroses*) as "lesions of feelings and movement" that originated in the brain, cerebellum, and spine, where the nerves were located.[30] Neurosis could be the result of dietary and aphrodisiac (*aphroditique*) behaviors, and ophthalmological, auditory, and affective conditions, indicating that feelings, like matter, could suffer injuries that needed to be medically diagnosed and treated. Some of the diagnoses he analyzed were hypochondria, melancholia, mania, hysteria, spasms, hydrophobia, convulsion, and nervous disorders.

Men who studied too much, were deprived of sleep, and lived an unbalanced life could experience mental problems, as was the case with those who studied without a method, exercising "more the imagination and memory than the reason," which could "degenerate into a manic state," based on Pinel's observation of patients. But it was not the role of philosophical medicine to answer in all its breadth the question about what were "*the most suitable ways to develop [male] scholars' talents and natural aptitude for the sciences without harming their health or contracting diseases.*"[31] Medicine, like all sciences, was gendered. For example, mathematics was deemed a masculine mental pursuit, as Newton and French revolutionary leaders in 1799 believed. The reading of Condillac also gendered the imagination, as discussed in the first chapter, because it enhanced "sensibility, nature, and cultural production", as Mary Sheriff has explained.[32] In her view, this philosopher introduced a "feminine" combinative cross-dressed imagination, "like a man bedecked in woman's jewels and finery."[33] On the contrary, the ideologues introduced a virilization of Condillac's philosophy that matched Napoleon's virile culture in France.

Pinel thought that women were victims of unbalanced passions, which led to hysteria, a disorder more common among the young who had an "ardent constitution," women of all ages "who refrained willingly or by force from sex," young widows who abandoned themselves to "obscene readings," and married women exposed to "a long absence" of their husbands.[34] Because they possessed "much sensibility, energy of affections, and maybe also the unstoppable vivacity of their imagination," they were the most exposed to the same nervous sicknesses that affected men, but "complicated many times over by hysteria in a greater or lesser degree" (40). The number of "alienated women" (*mujeres enajenadas*) was double that of men in public hospices, indicating that women's higher predisposition toward sensation made them more prone to mental disorders (8–9).

The evolution of psychology was closely tied to the culture of sensibility. Jessica Riskin has concluded that this culture shaped the predominantly male French scientific community, which played a pivotal role in the development of French psychology, marked by a rejection of Cartesian principles in favor of what Riskin describes as "sentimental empiricism." As a result, scientists and philosophers began to focus not only on sensory experiences but also on individuals' emotional responses to them. In Argentina, this emerging culture of sensibility merged with the efforts of Romantic writers who delved into themes of subjectivity and emotion, as will be explored in the next chapters.[35]

IDEOLOGY IN THE AMERICAS

On November 3, 1803, the French economist and ideologue Jean-Baptiste Say (1767–1832) wrote to Thomas Jefferson (1743–1826), then president of the United States (1801–9). In his letter, Say praised American citizens, noting that although "the offspring of Europe," they surpassed their forebears in merit. He described Europeans like himself as "elderly parents, raised with foolish prejudices, constrained by outdated obstacles, and swayed by childish notions." Say believed Jefferson's generation would show Europeans "the true path to liberation, for you did more than win your freedom: you asserted it." He praised the American Revolution because it had defended a personal liberty that was "compatible with maintaining community." People were no longer able to "sully noble causes with excesses." Civil liberty was the "true goal of society" and political liberty "a means to that end."[36] Similar sentiments appeared in letters to Jefferson from Destutt de Tracy, who wrote hoping Jefferson would translate and publish his writings, as Napoleon's opposition to ideology had led to the ban of such sentiments in France.

In 1811, Jefferson was instrumental in the publication of Destutt de Tracy's work, a critique of Montesquieu's *Spirit of the Laws*; he translated and edited the book, including an appendix with writings on the same topic by Condorcet and Helvétius.[37] At a time when debates about government administration and the economy were central to politics, France offered a unique perspective: its failures. In a short period, the French people had transitioned from an absolute monarchy to a parliamentary system, then to regicide and various forms of republicanism, culminating in Robespierre's terror. These lessons were vital in the United States, where institutions lacked a solid republican foundation and citizens needed education on a coherent, scientifically grounded republic, which explain why this book on Montesquieu was used to teach principles of civil government.

Jefferson also translated Destutt de Tracy's *A Treatise on Political Economy*, published in 1817, for use as a textbook on the subject. He hoped to place the book "*in the hands of every reader in our country*."[38] It offered "*sound principles of political economy*" that would protect "*public industry from the parasite institutions now consuming it*," which partly referred to economic and moral values.[39] This case demonstrates that while ideology had evolved because of France's troubled republicanism, at this time it was developed through a broad intellectual community across Europe and the Americas.

In another letter to Jefferson, Destutt de Tracy included a postscript in which he explained that he "must boast" about an excellent translation of his work in Italian, this "fourth volume, along with my *Principes logiques*, was similarly translated into Spanish and published in Madrid with all the requisite permissions." Indicating how his ideas were spreading in Europe he explained that the Spanish translator "was appointed to a chair of political economy created for him in Malaga." Such chairs also already existed in Madrid and Barcelona. The king of Württemberg had just established one in Stuttgart, and the French philosopher believed that soon "they will be all over the world before we see a single one in our unhappy France." This philosophy supported both a cosmopolitan model for men and a nation based on universal ideas, rights, and principles.[40]

Pierre Jean Georges Cabanis also had connections with prominent figures in the United States. As a medical student in France, he served as secretary to Madame Helvétius, whose salon hosted notable intellectuals. There, he met Benjamin Franklin, who asked Cabanis to revise the French translation of his essay "The Whistle" in 1779. Cabanis completed the revision, which was later published in collections of Franklin's works. Thomas Jefferson, also part of this intellectual circle, met Cabanis at the same salon. In 1802, the physiologist sent Jefferson a copy of his newly published *Rapports du physique et du moral de l'homme* (Relations Between the Physical and Moral Aspects of Man), a compilation from lectures delivered at the Institut de France that explored the relationship between physiology and psychology.[41] Cabanis was a member of the American Philosophical Society in Philadelphia, the city where his work *An Essay on the Certainty of Medicine* was published in 1823. The recognized physician René La Roche (1795–1872) translated it.[42]

Also in 1817, the Spanish translation of Antoine Destutt de Tracy's *A Treatise on Political Economy* was published in Spain. Cabanis's *Coup d'oeil* Spanish translation was released in 1820 after much struggle with censorship by the Church, which initially denied permission to print the book, deeming it obscene and antireligious, among other objections.[43] The Spanish edition of Destutt de Tracy's treatise was translated by Manuel María Gutiérrez (1775–1850), an economist and the chair of political economy in Málaga mentioned in Destutt de Tracy's letter to Jefferson. He was also a translator of Say and a passionate defender of free trade who promoted ideology in Spain with arguments like those circulated in Spanish-speaking areas in America. He believed that the science of political economy had not progressed because it lacked "in its study the same method of all the other areas of human knowledge."[44] Ideology provided a clear idea of "*property*" and willpower, "because all our power consists in the use of our physical and intellectual forces" (xxviii). The employment of human forces, or *work*, was our "only primitive wealth" and true society started when the goal and will were "to increase the power" of each individual (xxx).

According to its translator, this book was a study of man's organization, "his natural and innate propensity for empathy, the need for reproduction and the dependence on the being reproduced," which noted the importance of reproductive politics in this philosophy (xxxi). Humans were "naturally driven" to live in society, it was "our necessary state," which completely refuted the mistake made by Hobbes who understood men's natural state as one of continuous isolation and war (xxxi). The prosperity of a society was the result of the will of individuals to accumulate wealth, which Gutiérrez described using Destutt de Tracy's example of "the picture of a nation perfected" in which we can see "the immense accumulation of goods possessed; the result of prodigious changes in industry" caused by the "reciprocity of services" and the great number of alterations that had occurred. The three greatest benefits of such an interpretation of the meaning of society were: "*the gathering of forces, accumulation and conservation of knowledge [luces], and the division of labor*" (xxxiii). All these were possible because of the empathy that made humans prone to collaborating and helping each other, including the desire to reproduce and continue.

This interpretation naturalized labor, accumulation, marriage, and the existence of a desire to increase one's wealth. It was for this reason that work was productive or unproductive, the true foundation of political economy. Two principles could be deduced from this. First, "the true producers" were those who studied "the laws of nature and learn[ed] their use and what must be done with them" to make things useful or increase their utility. Second, there was no division between agriculture and industry; those who worked the land also transformed what they received from nature, which resulted in an agrarian industry. Productive human labor was understood as constantly transformative and always introducing change, perfecting what was received from nature.

As a result, there were two types of industry based on factories and commerce. Value was determined by "the intensity of our desires, or the importance and extension of the sacrifices that we are willing to make in order to acquire the thing that we want," which makes the generation of desires an essential part of economic growth—desires and wants resulted from teaching citizens what was valuable. The wealthiest nation was the one that had "the most educated and devoted workers, or the one that produced more value in a specified amount of time" (xxxix). Productive individuals worked to accumulate capital and things, and sensibility stimulated the will to acquire them.

Because not everybody had access to accumulated capital, this process of change also created a division of labor, which led to the birth of the differences that had been established among them, "some ruining themselves and others getting rich" (xli). The salaries of workers were determined by the owners who needed to recuperate their initial investment in addition to making a profit, which explained social relations—the common opposition noted "among businessmen; among salaried workers and businessmen; and finally among consumers, businessmen, and salaried workers" (xliii). More importantly, because "not all individuals have the same degree of intelligence and strength," there was no equality, so " whatever is done to establish it [would not be] useful, because art cannot destroy the works of nature." What could be done was "to protect the weak from the powerful" (lviii). The means of existence were those that prevented individuals from "uncomfortable and bothersome experiences in life," which reintroduced the problems of luxury and consumption that already existed. As a model of man, the ideologues defended the bourgeoisie by naturalizing inequality: a model of productivity related to the creation of a perpetual perfecting of nature connected to sensibility in service to the desire to consume and produce.

Reproduction, and by extension sexual roles, were exclusively the work of nature, which occupied itself with the conservation of species, taking less care "of the individual," which led to the conclusion that "population should always be proportional to the means of existence." As a result, the "population of ignorant and barbarous savages must stay the same or be reduced, if they lacked the means of existence." Furthermore, according to Gutiérrez's understanding of the author, the latter agreed with Malthus in thinking that the population in civilized nations tended to grow too much in relationship to the means of existence. It was "barbaric to increase" the number of civilized people too much, and it was in the "interest of man" to "diminish the effects of his fecundity" (lxii). This corresponded to the model of a man who kept his desires in check and was not controlled by his passions and sexual urges; desire was concerned with accumulating capital for buying things, which is a part of this philosophy that greatly influenced Karl Marx years later.[45]

The belief in the perfectibility of humans through the study of science and the arts supported by ideology was also important for Spanish American writers. In 1823, for example, Andrés Bello (1781–1865) and Juan García del Río (1794–1856) were living in London where they published *La Biblioteca americana*, which included articles on art, music, science, and literature. An article about pedagogy for teaching music addressed a method that consisted in applying "to the science of music" the analytical and philosophical path traced by "Bacon, Locke, J. J. Rousseau, Condillac, Destutt [de] Tracy, Cabanis, etc.; the same that, fortunately, [was] being adopted in all the sciences and arts to facilitate their study and accelerate their progress." Ideology was the first system to synthetize the previous ones.[46]

The Marquis de Lafayette, who was Destutt de Tracy's friend, traveled to the United States from 1824 to 1825, and his secretary, Auguste Levasseur

(1795–1878), wrote a book describing his experiences in the country. In one passage, he wrote, "Mr. Salazar, Chargé d'affaires from the Republic of Colombia for the United States," proved himself a man "instructed in republican institutions, and entirely devoted to them." One night he "gained" everybody's hearts by "giving a toast to Mr. Destutt de Tracy, and by proclaiming the happy influence produced by his writings in both hemispheres," which indicated the extensive network created around this thinker.[47] In the case of Buenos Aires, Destutt de Tracy's philosophy acquired political power during the years in which Bernardino Rivadavia (1780–1845) was governing after pursuing diplomatic work in Europe for more than five years (1815–21). He was close to Jeremy Bentham (1748–1832) with whom he corresponded, and knew Destutt de Tracy. In 1817 in a letter to Jefferson the Marquis de Lafayette wrote that he had lately "seen a Commissioner from Buenos Ayres [Rivadavia] who Assures me His Commonwealth is more advanced in political knowledge and practice than we are aware of."[48] The agents operating in favor of the independence of the River Plate Viceroyalty promoted the country's potential.

Klaus Gallo's analysis indicates that other thinkers were also influential in the thought of Rivadavia. In his visits to London from Paris, he established contact with Jeremy Bentham and James Mill (1773–1836), the former being the one who convinced him to support a republic and not a monarchy.[49] While this is true, it is also demonstrable that the mixture of ideology and utilitarianism was considered the philosophy of republicanism at the time in Spanish America. For example, in Colombia, the ideologues Bentham and Mill together with Claude-Frédéric Bastiat "formed the trilogy of high education in Colombia" during the 1820s. Bentham and Destutt de Tracy were in fact the official philosophers whose work by 1826 defined the "study plan" (*plan de estudios*) in Colombia; two years later, a group of young men radicalized by the English philosopher attempted to assassinate Bolívar after he became a dictator. The Liberator banned the teaching of Bentham as result of this attack. Still, by the 1830s, the mixture of utilitarianism and ideology returned to education; according to the philosopher Rafael María Chasquirrilla (1857–1930), it was useful for promoting science and finding "a social and rational order" based on the study of the ideologues, grammar, and logic.[50]

BUENOS AIRES AND THE POLITICS OF IDEOLOGY

In Argentina, ideology renewed Locke's and Condillac's culture of sensibility after years of political conflict. Between 1810 and 1820 "there were six revolutionary governments: the First Junta (May to December 1810), the Big

Figure 3.1. The nation as a woman of arts and sciences. "Triumph of American Independence." Original printed by Rudolph Ackermann. London, 1825. Courtesy of the Museo Histórico Nacional, Buenos Aires, Argentina. "Explanation: The genius of the American Independence begins its triumphant run crowned by the hands of Prudence and Hope and carrying in its own hands the symbol of Liberty. Six horses pull its cart in representation of the republics of Mexico, Guatemala, Colombia, Buenos Aires, Peru, and Chile. Temperance and Justice direct them. The genius of the Arts and Sciences decorates this great and interesting spectacle while Abundance and Commerce offer, with the emblem of Eternity and Union, the happy presage of America's future luck."

Junta (January to September 1811), the Conservative Junta (September to November 1811), the First Triumvirate (November 1811 to October 1812), the Second Triumvirate (October 1812 to January 1814), and the Directorio (January 1814 to February 1820)."[51] Finally, in 1816 there was an agreement to declare independence from Spain, and the United Provinces of the River Plate (Provincias Unidas del Río de la Plata) emerged as another new nation on the American continent: its representation was gendered—the nation was a virtuous woman.

A manifesto of independence written by the representatives of the United Provinces and circulated in Europe was published in London in 1818. It justified this radical action by criticizing Spanish bans on foreign entry, the "extermination of natives," "depopulated cities," monopoly, lack of self-governance, and economic neglect of mines, which showed "none of the progress evident in enlightened nations." It also condemned the prohibition of scientific education, limiting instruction to Latin grammar, ancient philosophy, theology, and canonical jurisprudence.[52] It is clear in the text that banning the access to science was a justification for emancipation.

In March 1819, Destutt de Tracy wrote to Jefferson about the United Provinces of the South; while he and others respected the "very prudent" silence of the United States regarding the former River Plate Viceroyalty, "the friends of liberty, numerous in Europe, eagerly await your open recognition" of the new nation. The English government supported this cause, "earning their gratitude," and the French government would soon follow. Support from the United States would bolster the nation's future.[53] Jefferson replied, reaffirming his support for South America but noting that "immediate acknowledgment raises other considerations" regarding European involvement and reactions from Spain and England. Jefferson stated that "every generation must pay its own debts as it goes," and the United States had fulfilled its obligations. By then, his country received and protected "the flag of South America in its commercial dealings with us." However, he cautioned that the new nation's future could be "problematical," as its capacity for self-government was unclear, potentially leading to "wise governments or military despotisms."[54] The events of 1820 proved his assessment correct.

The primary challenge facing the former River Plate Viceroyalty after independence was how to determine an appropriate form of government. By this time, European powers viewed republicanism as related to political radicalism. To secure European recognition, Manuel Belgrano proposed an Inca monarchy that blended precolonial lineage with a system acceptable to

European sensibilities while moving away from the values that characterized the independence process in Spanish America. This proposal failed and republicanism ultimately prevailed.

The second challenge was to draft a constitution that balanced the interests of diverse provinces while addressing tensions over Buenos Aires's ambitions to dominate the former viceroyalty's territory. This struggle manifested as a conflict between Unitarians—liberals advocating for centralized executive power based in Buenos Aires—and federalists, conservatives who championed provincial autonomy. In 1819 the Unitarian-backed constitution established a strong central executive, which many provinces perceived as a threat to their independence, and Buenos Aires landowners saw as detrimental to the city's commercial interests. The constitution faced widespread rejection, and its fate was sealed in 1820 when a coalition of provinces associated with the landowning class of Buenos Aires defeated the proponents of this document at the Battle of Cepeda.

These events led to the dismembering of the nation, as each province had complete autonomy and only weak connections among themselves.[55] As a result, General Martín Rodríguez (1771–1855) was elected governor of Buenos Aires this same year for a four-year term; he appointed Bernardino Rivadavia as the minister of foreign affairs and government (*ministerios de relaciones exteriores y gobierno*) and Manuel García (1784–1848) to head the Ministry of Economy. They were powerful ministers, particularly after the governor left to fight against Indigenous incursions, which had resumed on the southern frontier.[56]

In 1820 Rodríguez signed the Miraflores Treaty with some of the Indigenous chiefs to curtail raids (*malones*) and to establish the border south of the Salado River. This agreement did not last, and the conflict continued until the end of Rodríguez's term after three military campaigns that mobilized thousands of men. His successor as governor of Buenos Aires, Juan Gregorio de las Heras, finally signed a peace treaty in 1825. While this was going on at the southern border, Rivadavia initiated his plan to implement the new philosophical and scientific ideas. He was not a military man as leaders usually were, he did not command hundreds of men ready to fight as landowners did, and he belonged to a culture that was understood as much less masculine than theirs.

Rivadavia and his allies saw themselves as cosmopolitan men who had lived for years "in foreign nations performing public service for the country, which made them independent from the factions that divided the city." They had closely observed what had happened in Europe, especially France's troubled republican attempt, which helped them understand their own problems. They established a new government under this principle: "*All theories must be related to the country's organization, using them in a practical way to demonstrate their efficacy*" (24). This echoes ideology's understanding of government as ruled by empirical laws that were followed in the passing of several decrees, which signaled a new beginning: an emphasis on the rights of property, even for foreigners and during war, which attracted foreign capital; approval of a decree mandating that the government publicize its actions; and granting amnesty to all those who had left because of the civil conflict.

In May 1824, the British representative in Buenos Aires, Woodbine Parish (1796–1882), wrote to Rivadavia to request a report that explained all the useful information needed in London to assess what had happened since Rodriguez became the governor. Ignacio Nuñez (1792–1846) was assigned to answer Parish's demand. Nuñez was a writer, public servant, soldier, member of the Lautaro Lodge, and supporter of Rivadavia, who had taken him to Europe while he was a representative of the government. He was also one of the writers of the 1819 constitution. The resulting document detailed the state of the country around this time with the intention of convincing the British to recognize the independence of the United Provinces. It explained that the "first ten years of the revolution [had] been spent in the perpetual fight" between the different factions that wanted to shape the nation (24).

In Nuñez's narrative, Buenos Aires had transformed itself in the past three years, and everything was "order and prosperity; and the freedom compatible with each of them," and was enjoyed as much as the "shedding of the blood of two hundred thousand souls" had been in the past (53). The new institutions had established deep roots, and under the leadership of Buenos Aires the "National Body" had been reconstituted except for Alto Peru, now Bolivia; the province of Montevideo (Uruguay) occupied by Brazil; and Paraguay, which had seceded (81). The success of Buenos Aires was very important in the report because Nuñez believed that restoring and strengthening public order in Buenos Aires would, once achieved, extend stability across the fragmented territory. Invitations to a National Constituent Assembly were sent in 1824, with meetings planned for early the following year to discuss a new constitution. The aim was to organize the country practically, drawing on "our own experience" while avoiding past theoretical speculations (30).

The new administration significantly improved public education. The University of Buenos Aires was created in 1821, and "a school of moral sciences, another of natural sciences, and another of religious studies, where the young men of Buenos Aires and other provinces were educated." Scientific societies were organized and funds given to improve the Public Library. Together with the general improvement of education, the administration of justice was changing because of independent judges. There were plans to establish a correctional code, which would be the first in the country, and the frontier of Buenos Aires was organized by judges (31). This was important because the ongoing conflict was affecting the property of many ranchers (*estancieros*). Nuñez clearly asserted that the Indigenous population of the country would not be integrated according to the civilized ideas of the government of Buenos Aires.

A meeting was held in 1822 with the main chiefs (*caciques*) of the "*Aucaes* and *Pampas*, *Huiliches*, *Tehuelches*, and *Ranqueles*" to convince them to sell the land extending from the south of Buenos Aires to Patagonia, but this did not work due to the "resistance of the Ranqueles who belong to Chile" and constantly resisted peace, "by their courage, influencing all the other Indians," who had rejected the deal. Due to this failed attempt the government had no other choice but to "follow the example of the United States, albeit in better circumstances because Buenos Aires will defeat these 'barbarians' more easily; they did not exceed eight thousand men in total, and their weapons were bolas [*boleadoras*] and spears, their only advantage being mobility due to their mastery of horse riding" (170). The conflict would continue until another expedition exterminated the Indigenous population of Patagonia by the 1880s.

Regarding slavery, the introduction of supporting documents included letters exchanged between Great Britain and the government of the United Provinces after the passing of the emancipation law of 1813. First, the hypocrisy of the British government is revealed; Britain supported Brazil in its complaints about the abolition law of 1813, which, following the example of Haiti, had granted freedom to all the slaves of foreign countries that had set foot on the territory of the ex–River Plate Viceroyalty, raising concerns that those in neighboring Brazil would escape to the emancipated territory. Lord Strangford, the representative of the Crown in Rio de Janeiro, justified his support for Brazilian slavery based on the fact that in Great Britain, emancipatory principles were the "simple and natural result" of the British constitution, established centuries ago, and not because of a recent law created to specifically regulate slavery.[57] In its reply, the government of Buenos Aires mentioned that this provision had been suspended as requested; those who fled would be returned to their owners, but it also stated that the United Provinces did not want "*to attract a population that [was] not wanted, moreover a population that had been pushed away from this soil through the banning of the introduction of slaves*."[58] The wording seems to imply that the end of the slave trade was related more to a determination to avoid an undesired population than to emancipatory ideas.

Once Rivadavia had political control of the government of Buenos Aires more drastic measures were approved. In 1823, for example, a decree placed funding of the church under state control the under the same conditions that regulated other public institutions; also, all unused church property was placed under government control.[59] Such measures renewed the religious tension that had increased after a legal opinion issued in 1827 by José Eusebio Agüero allowed Protestants to marry Catholics and to maintain their religion without the church's dispensation. José Manuel Avazoa questioned this opinion based on several considerations. He started by mentioning that Argentine women, "as usually happened everywhere with this sex," did not have "a solid instruction in matters of faith" and lost interest in the dogmas of their religion once they married. Even many men born in Argentina, "and many ladies who are supposed to be of the upper class" considered "it an important part of their education and culture to despise the Catholic religion, and pretended to be philosophers, or protestants, looking at persons of faith like us as fanatics." Englishmen were the masters (*amos*) of the nation, and "many magistrates boast about despising our Catholic dogma" in public.[60]

According to Avazoa, even in cafés, at social gatherings (*tertulias*), and while strolling (*paseos*), "our dogmas, mysteries, sacraments, precepts, rites, and ceremonies were ridiculed." The Catholic temples were victims of "profanations" by "those who called themselves philosophers, protestants, or carefree [people]" (*despreocupados*), who behaved this way because they knew they had impunity.[61] Catholics felt this lack of security because the schools "have been teaching, and maybe still teach, impiety," while promoting a philosophy that formed libertines. At this time Catholic liberals had not yet developed their own ideas with respect to how these politics of religion could work in their favor. The work of Charles Forbes René de Montalembert (1810–70) would make this possible later.

IDEOLOGY AND EDUCATION IN BUENOS AIRES

In 1812 Bernardino Rivadavia, as a member of the Triunvirato, requested that Manuel Pinto, who was going to London, hire "two professors of mathematics, one of experimental physics, one of chemistry, one of mineralogy, one architect who could also draw, a printmaker, and a professor of political economy."[62] In 1815, the Spanish engineer and soldier Felipe Senillosa

(1790–1858), who left his country after the political conflict as a result of the return of Ferdinand VII to the throne, arrived in Buenos Aires. While he was in London, a close friend who was from Buenos Aires convinced him to move to his country, Argentina.[63] Immediately after his arrival in Buenos Aires he created a newspaper, *Los amigos de la patria y de la juventud*, which he published until May 1816, shortly before the independence declaration of July 9. In this same year he became the director of the Academy of Mathematics in charge of changing the academic programs. In 1817, to provide study materials, he published *Gramática española* (Spanish Grammar), which he had written in 1813, and it had received the approval of Destutt de Tracy. Senillosa had intended to publish it in Paris, but the war in France changed his plan and the manuscript remained unpublished until he moved to Buenos Aires.[64] In 1818 he published another text for students, *Tratado de aritmética dispuesto en XXIV lecciones* (Treatise on Arithmetic in Twenty-Four Lessons), which updated the teaching of young men of the elite following the principles of ideology.

The first issue of Senillosa's newspaper included a letter from Francisco de Paula Castañeda (1776–1832), a Franciscan priest. He was part of the liberal clergy who had supported the independence movement and was the editor of several publications. According to this priest, the objective was to turn "*each house into a school, each neighbor into a teacher, and each teacher into a man of wisdom.*"[65] This would lead very soon "*to convert coffee shops into schools and [playing] cards into books, because even the professional gamblers that infest our loved fatherland, will be ashamed of being ignorant, lazy, and lost.*" He ended his letter by asking Senillosa to state what was to be done because he, Castañeda, was "born a slave of those who promoted the illustration [enlightenment] of the peoples." The young generation spent too much time having fun and socializing in the wrong places; they needed to be disciplined and educated to be citizens ready to serve the nation.[66]

In addition, Senillosa introduced a new way to understand the roles of men and women. In an article published in *Los amigos de la patria*, "Sobre mujeres" (About Women), men were clearly depicted differently from those of the past, and women needed to adjust to this transformation. The ideal woman was no longer just beautiful, but "one that even when she was not very beautiful had a candid and good soul, treated people with kindness, had a sincere heart, was hardworking, knowledgeable, prudent, [and] discreet. " But unlike in the past, women were not the ones who controlled sociability and men; now women needed to "become the delight of all men who had good sense" and enjoy this happiness to unite their lives to them. It could be dangerous if in addition to all these qualities women were also beautiful, because with "so many reasons for admiration that enchanted the men's souls," even those who were the most inflexible would "lower themselves at the cost of their good sense, making a hard sacrifice of their own freedom."[67] Men's independence and autonomy from women's charms were needed for the men to become true citizens focused on the building of the nation; the authority of the best men should be the foundation of society.

In the same issue an article on the army and the need for soldiers' discipline was published. It contained an explanation stating that since violent actions were part of the soldier's life, and this violence was "outside the natural order," soldiers had to be compelled to commit such acts through "violent and extraordinary" means and should be led by "an aristocratic order" guaranteeing that the army never fell into "anarchy." These leaders should provide an example of "obedience, love for work, and constant firmness." The latter, above all, should be accompanied by "rectitude," a moral example that exhibited virtue.[68] The soldiers needed to offer blind trust to superiors whose actions they followed without question. Addressed in the same issue were scientific matters, including medicine, geography, and mathematics, which were very important because they "rectified reasoning; teaching how to reason," and were essential for the youth.[69]

The first issue also included a supplement with two readers' letters and the editor's replies to them, all addressing the problems created by women who abandoned their children. The second letter mentioned concerns about the "physical education and morality" of the children whose mothers placed them under the "fatherland's pity."[70] They existed as the result of "perfidy, weakness, and even need," and the generous protection of this part of the population indicated the "moral civilization of the peoples" of the world. On the contrary, the death of babies abandoned in the streets of Buenos Aires was shameful and required reform and the creation of a philanthropic society led by "Ladies [*señoras*] who were friends of humanity," devoted to fixing the problems faced by their own kind. The solution was first to establish a seminar that would shape the children as students; second, women from the higher social class would become the teachers and have the opportunity to "exercise the tender crafts that naturally lead the sensibility of their hearts."[71] It is unclear whether the letters are real or Senillosa was proposing his social plans through invented letters, as was frequently done, but the emphasis on physical and moral education, honor, republican virtue, social class, and gender differentiation clearly coincides with the ideas of the ideologues.

Senillosa favored the creation of an institution led by the best women in society. He suggested the wife of the director of the state as its possible president, to make women "agreeable to the eyes of the Lord" and those of men, who would see in them "true mothers of the people" who deserved admiration for their virtue. Men would "crown women" and everyone would be grateful for their contribution in building the nation.[72] Unlike in the past, women were now confined to domesticity and learning how to demonstrate the necessary republican virtue. The need to create schools, academies, philanthropic societies, and charitable institutions is defended as the foundation of sociability. A political body, or a state, "could be compared to any other living body" and it had many parts that made it work.[73] The first step for men who were leaders was to form "good morals, keep passions in check, using reasoning" and fulfill their obligations through their own actions to contribute to the "general movement of the machine" that was society.[74]

In 1817, at the celebration of a school opening in Mendoza, Toribio de Luzuriaga (1782–1842) told the audience that "the country invites you to share the ideas of the Enlightenment [*las luces*]. Minerva's temple opens to all without any exclusion." He continued, emphasizing that "the soldier, the writer, those who govern, the religious man, the man who practices agriculture, and the liberal arts constituted the happiness of any state." Scientific instruction was not only "an adornment but also a gift needed by the military." Ending his speech, he appealed to parents to "inspire in your children the generous desire to learn science at an advanced level" to help the country.[75] This was the climate in which the adoption of ideology occurred.

The first controversy regarding the study of ideology took place around the work of Juan Crisóstomo Lafinur (1797–1824), who was well-known for his poetry and considered by some to be the precursor of Romanticism.[76] He fought in the Independence War and was a student in the Tucumán Academy created by Belgrano. In 1819 he published a logic book, and another with his class notes for post-elementary education. He taught Condillac's work, translated into Spanish in 1784, and spread the ideas of Locke, Bacon, and Descartes together with the works of Galileo and Newton. He was a follower of the ideologues and an avid reader of Destutt de Tracy. In his book he introduced the study of mathematics as a tool for understanding physics and recommended the purchase of a lab for the education of the youth.

Lafinur was clear on the need to replace the Scholastic method "with the inventions and knowledge of the modern school" but not everybody agreed with him, and he faced strong criticism. Cosme Argerich (1758–1820), a well-known doctor, wrote a letter in his defense to the newspaper *El Americano*.[77] In it, Argerich praised Lafinur for his scientific work and "good taste," and mentioned that the new philosophical ideas being discussed were based on the philosophy of Cabanis and Destutt de Tracy.[78] Since the work of Cabanis was considered materialist, there was mistrust about his approach, and concern about its consequences. Materialist philosophy was associated with modern ideas that were against Scholasticism and Catholic dogma.

Lafinur replied to Argerich's letter with one of his own, in which he tried to clarify his philosophical beliefs and denied that he was reducing everything to matter. According to him, sensibility was not the result of matter, but "of the presence of an immaterial being to an organized body" (51). He also explained the phenomenon of the moral man and "of a sensibility given by those who followed Newton" in order to analyze what happened in the physical world (51). More important, he believed that "the attention of the soul directly modified the well-being of the organs," because of the impression left by sensation. Certain smells "make some people leave the room, while others feel attracted to it." This occurs because the soul integrates diverse feelings into a unified impression, which proves that "the soul has an energetic and active nature, in fact an immaterial one" (52). This interpretation included a defense of spiritualist positions, which were connected with the metaphysical component present in ideology.

According to Lafinur, "Our internal impressions, the ways of being that are commonly called *feelings, or pains of the soul*, such as enjoyment and sadness, trust and fear, weakness and strength, activity and idleness, etc.," were simple products of sensibility. The definition of feelings as "pains" resembles Cabanis's use of the word "injury," and it is probably the origin of the definition. In Lafinur's view, the French physiologist had demonstrated that "there are no ideas in the soul, nor *innate principles* either theoretical or practical."[79] The development of a society, then, was the result of a series of *continuous changes*. A change was "a transaction that was to the advantage of all parties."[80] Lafinur also taught that "commerce and society were one and the same thing," associating the development of science and socioeconomics in a single bundle.[81]

Lafinur also taught about the importance of morality, and the virtues that were needed to develop a perfecting society. He followed the work of the Abbé de Mably (1709–85), a defender of republicanism, whose writings were introduced to students in an 1815 translation likely done by him.[82] The French philosopher's best-known work, probably written in 1758, "offered a representation of political life in which the fabric of law was dissolved into the exercise of the will." For this reason, "no form of government could be defended" if it had not originated in "the active will of the people." Neither "a primitive contract, nor historical prescription, nor tacit consent, could properly be regarded as depriving people of the exercise of its will."[83] Such thinking gave a minority— enlightened educated men—the justification to impose their ideas and destroy the old colonial system to create a republican society.

According to these ideas, Lafinur explained the existence of four fundamental virtues needed to affirm political will: clemency, gratitude, patience, and love of glory and fatherland.[84] This morality was also related to instinct, the internal pulsation that, following Cabanis, he classified as superior in species that were less rational.[85] "The capacity to feel was the spirit's primary ability, and all other actions are connected to it."[86] Man was "a moral being" who reigned over nature since "feeling" was the force in him that determined his thoughts and elevated his designs (99). This would be important in the future when the culture of sensibility became a way in which men understood their own place in society. The man of feeling would be a very different model for the youth than the natural man, who did not possess the ability to elevate his ideas for the benefit of society. In terms of progress and improving, Lafinur believed, as did most of his contemporaries interested in the subject, in the chain of being that connected the most developed creature, obviously man, to a descending chain that included beings of the lowest species (106).

He also understood movement in terms of space and time. One could perceive space seeing the visible objects that occupied it; and one could also perceive duration according to the extent of a person's ideas and sensations (93). In 1820 Lafinur organized a famous public examination for three students who were charged with giving one speech each. The first, by Luis Belgrano, required him to demonstrate "with philosophical reasons, the divinity of the Christian religion." The second, by Ignacio Martínez, dealt with "the history of the physical and moral man." Finally, in the last speech, Lafinur himself criticized Jean-Jacques Rousseau's idea that the sciences had corrupted customs and made man worse.[87] Those responsible for the evaluation could also select any piece by any writer, "old or modern," and ask "for an analysis of its beauty."[88] Due to the achievement demonstrated by these examinations, some described 1820 as the "beginning of modern teaching" in Argentina.[89] Coincidentally, it was in this same year that civil war exploded because of the divisions brought about by the 1819 constitution, which threw the country into anarchy.

The instability created by the new ideas, and the turmoil that dominated Buenos Aires, did not help Lafinur. Conservatives closer to the church accused him of negatively influencing the youth, leading to his forced departure from Buenos Aires. He relocated to Mendoza, where he faced further conflicts and, in 1822, moved to Chile. He died tragically in an accident in 1824. His work, and that of the intellectuals mentioned in this section, allows us to understand a transition related to the "transformation of the "metaphysical" problem of the soul into the "analytic" question of the origin of human knowledge, a development that would continue in the following years.[90] It also helps us understand why the philosophy of Victor Cousin (1792–1867) would be adopted in the 1830s to solve the conflicts and contradictions that were apparent by this time.

The ideas introduced by Rivadavia during his tenure as secretary of state of the Buenos Aires province, certainly pleased Jeremy Bentham who considered him a disciple. In a letter sent by Rivadavia to the English philosopher, on August 26, 1822, he mentioned that one of his main objectives was "to protect commerce, the sciences, and the arts," through the action of the government. He suggested that he was adopting Bentham's ideas in his governmental plan, but his deferential tone warrants cautious analysis; Jonathan Harris warns against accepting the Argentine's enthusiasm at face value, and this seems more plausible.[91] As seen in the correspondence between Jefferson and Destutt de Tracy, these exchanges reflected dual dynamics shaped by each side's distinct interests in politics: while Bentham and Destutt de Tracy were marginalized in their countries, unable to sway their governments, American leaders held power. Their aim was to establish their new nations as significant within enlightened culture and to advance revolutions to an institutional stage recognized by Europe. Such acknowledgment was vital for expanding international trade and adopting European philosophers' frameworks to express their nation in scientific and philosophical terms. Ultimately, they sought to transform the Republic of Letters into a national endeavor. Moreover, philosophers offered cohesive knowledge that unified the nation and educated citizens to ensure its continuity.

In connection with Bentham, Rivadavia supported the method of John Lancaster (1778–1838) that allowed large groups of students to receive basic education through a monitorial system by which teachers trained the best students to teach other students. Bentham promoted this method through his support of Hazelwood School, located near Birmingham, and Rivadavia sent his sons to study there.[92] The Lancasterian method gained popularity because it provided a low-cost way to educate many students, requiring fewer teachers. The creation of the British and Foreign School Society in 1814 further facilitated the expansion of this method in the Americas. Lancaster himself moved to the continent in 1818, first to the United States, where his work was already known. In 1825 he moved to Caracas at the invitation of Simón Bolívar but returned to North America two years later after his relationship with Bolívar soured.[93]

In 1820 James Thomson arrived in Buenos Aires as a member of the British and Foreign School Society. As in Great Britain, his teachings focused on enabling boys to learn to read the Bible as part of a nonsectarian effort. In a book he published in 1827, detailing his experiences in Buenos Aires, Chile, Peru, and Colombia, Thomson described how warmly he was welcomed in the capital of the United Provinces, noting that a priest became his closest friend. He also mentioned the school he established, which enrolled one hundred boys who were taught reading, writing, arithmetic, and basic grammar. Shortly after his arrival, the representative of Chile hired him to introduce the Lancasterian method in his country, leading Thomson to leave Buenos Aires a year later.[94]

In 1822 the Lancaster school manual was published and adopted not only in Buenos Aires but also in Tucumán, Santa Fe, San Juan, Patagones, and Mendoza, where a Lancasterian society was formed with Lafinur and the Scottish botanist John Gillies (1808–89) as members. Gillies, also known as Juan Guillez in Argentina, had arrived in Buenos Aires from Edinburgh in 1820 to complete scientific work. He was responsible for the introduction of the silkworm in Mendoza, chemical analysis of this province's metals, and studies on the botany of the Cuyo region that ended in the discovery of several species, including a medicinal plant that was part of the *Amaryllidaceae* family named *Guilliesia* in 1826 by the English botanist John Lindley (1799–1865), a member of the Royal Academy of Medicine of London. In this same year Gillies returned to Scotland.[95] Alexander Caldcleugh (1795-1858), another English traveler who was in Buenos Aires in the early 1820s, was quite impressed with the number of European books that were circulating in the city, and the interest that they provoked among the local population.

> A large public library was established some years ago by the endeavors in part of some English merchants; it consisted, at first, of about 12,000 volumes, but has been since considerably augmented; it is under very good regulations, and by the printed return, it appears, that between the 21st March and 31st December, 1822, it was visited by 2960 persons, of whom 369 were foreigners. Some curious MSS. are in the hands of private

> persons, who were willing to sell them but the price was extravagant. The best private collection I visited was that of Dr. Segurola, before mentioned; it was chiefly enriched by the spoils of the extinct company, and shewed to perfection the beauty of their system, and how well it was adapted to the order of things. Books are permitted to enter free of duty, and the quantity of French books that have arrived in the country, and been conveyed a considerable distance inland, is surprising.[96]

Schools that adopted Lancaster's method flourished in South America, Mexico, and Central America, becoming the preferred way to educate boys. As Thomson wrote, the work of the British and Foreign School Society united "all the nations of the globe, by brotherly ties of science and religion" because "all true and enlightened patriots" across the continent cried, "*Give us education*." This urgent need to promote widespread education to sustain the new nations explained the popularity of the Lancasterian system in the Americas.[97] Science was the crucial piece of this political system even among those who were religious.

Rivadavia, following the work of the ideologues, also introduced new ideas about gender as part of his policies. In 1823 he wrote a decree to create the *Sociedad de Beneficiencia* (Charity Society) led by women from the city's best families. The text began by stating that "the social existence of women" was "too vague and uncertain" at the time. Everything related to them was "arbitrary" because what applied to some did not apply to others, "beautiful and good qualities sometimes harm them, while the same defects" were formerly useful. This was an imperfection of "the civil order" that had to be addressed because "the physical perfection of a people" emanated equally "from the beauty and sanity of both men and women," because it was also "their moral and intellectual perfection." Nature had given women a "different destiny and means to serve" compared to those of men. Nature had provided women with qualities of heart and spirit not possessed by men. Even if men tried to perfect their qualities "they would move away from civilization if their ideas and feelings were not associated with those of the precious half of their species."[98] These complementary roles also affirmed a clear difference between the two, natural qualities that separated the sexes.

In addition, the decree established that it was "useful and just to give serious attention to the education of women, the improvement of their customs, and the means to provide for their needs, in order to establish laws that fix their rights and duties," which would assure "the share of happiness that belonged to them." But the way in which this prosperous situation could arrive was through "the public spirit of the ladies who, because of the distinguished situation they had obtained and the gifts of their hearts and spirits, led their sex, proving their aptitudes." Accordingly, the government was giving them "their first chance" because it believed that "the only way to create permanent social and political relations" was "to illustrate and perfect men and women," and individuals and people, which is the justification behind the association. This was not work, it was charity, an activity that was viewed as closer to women's moral and spiritual attitude.

This institution would oversee the direction and supervision of girls' schools, the management and inspection of orphanages, houses for "public and discreet childbirth," women's hospitals, and "all public establishments directed to the well-being of individuals of this sex."[99] The government was responsible for providing the funding from a reserved account. Rivadavia, charged with implementing the plan designed by the decree, personally named the first group of women who constituted the society, all members of the city's most prominent families

The society opened on April 12, 1823, in a ceremony attended by Rivadavia, who explained in his speech that the institution had three main objectives: first, "the perfecting of morality"; second, "the cultivation of the spirit in the beautiful sex"; and third, "the dedication of the spirit to what is called industry, the result of the combination and exercise of those qualities" that characterized women.[100] The secretary stated that while nature had given the sexes common qualities, it had also provided some that appear only in men or women; among women it was "sensibility and the gifts of the heart that the beautiful sex possessed in abundance, attributes that contributed decisively to the formation of the moral." These qualities explained the influence of women over men: "She communicated sweetness, affability, and all those qualities related to the heart," which built character and made grow "in men the same feelings that women possessed in a superior manner."[101] Rivadavia reminded those in attendance that the "force of the revolution in the country had greatly corrupted morals," because men were not influenced by women while they devoted themselves to "ambition and their partial interests," which caused them to quarrel with each other.

Finally, he recommended that cultivation of the spirit should be the main duty of the society's members. The revolutionary ideas that "attempted to put women on the same level with men" had created a great injustice because "nature had given them greater aptitudes." Proof of this superiority was that the wisest men who created the most sublime productions "had always descended to the sphere in which the sweet treatment of enlightened women" had placed them.

The society's actions were evaluated in Rivadavia's message to the Representatives of Buenos Aires in 1824. In it he explained that the society had "completely satisfied public expectations: the effort and intelligence of the ladies" who were part of it resulted in "progress in the education of the girls" a crucial marker of progress.[102] While the differences between men and women were considered "natural" in the colonial society, the meaning of nature was now different. Liberal men defended a model that implied the ability to change, perfect, and transform, which in terms of gender required that women occupy a permanent place that did not allow them to erase differences or challenge men's power, as had happened among radicals who were part of the French Revolution.

This stability was even more important at a time in which Buenos Aires was experiencing a great transformation because of the economic growth created by the reforms introduced since 1820. It became a more cosmopolitan city, which included an expansion in consumerism and the introduction of new European fashions. The growth of commerce attracted international businesspeople who also completed scientific work while in the country. For example, in 1824 Jean Théodore Lacordaire (1801–70) arrived in Buenos Aires as a businessman and started to complete scientific expeditions in his free time, leading to the creation of an important entomological collection. He stayed for four months during which he initiated his pursuits as a traveling naturalist. George Cuvier called him to work with him, and he returned to France. He wrote for the *Annals of the Entomological Society* and the *Revue des Deux Mondes*, mandatory articles for most Argentine liberals in the next decades. His writings about his stay in Buenos Aires would become very important by the 1840s, as we will see in chapter 4. In addition, two travelers and naturalists from Switzerland, Johann Rengger (1795–1832) and Marcelin Longchamps, wrote about their impression of the city after seeing it again in 1825, six years after their arrival in 1819 on their way to Paraguay where they had stayed to complete their research.

> But it was Buenos Ayres particularly, which we had some difficulty in recognizing: the streets had been paved; public promenades laid out; and, as it may be said, a second city erected by the addition of several stories to the houses; a marketplace, with several halls, and surrounded by a gallery, had just been completed. The animation of commerce was visible everywhere; a forest of masts covered the roadsted [*sic*]; the streets were filled with carts,

> conveying merchandise, and on every side were seen stores and workshops. Several establishments for the instruction of youth had been formed and were in a prosperous condition. And looking to the literary resources to be found in it, we would have supposed that it was quite a European city. The dress, manners, and customs were also European, and the number of foreigners of all nations was so great that it was difficult to say which was the indigenous language. To give some idea of the increase of population, it may be sufficient to mention that the number of French alone, established at Buenos Ayres, amounted, it was said, to six thousand.[103]

They added that this transformation was due to Rivadavia whose actions "bear testimony to the wisdom of his views, and to that elevation of sentiment and extent of knowledge" that characterized him.[104] This modernizing impulse began to decrease with war and the blockade of the port, which reduced trade, and the cost of the conflict shrank the economy. In 1825 a war was declared to recover the Banda Oriental, which was occupied by the Empire of Brazil, and, at the same time, General Rodríguez ended his term creating a dispute among those who wanted to get more power. A new constitution was ratified in 1826 by an assembly originally convened in 1824, and while the provinces participated there was distrust about the role of Buenos Aires, which increased after Rivadavia was named president of the country in the same year.

The new president remained an unpopular figure and an object of ridicule for his customs and behavior, and this was not only mentioned by locals. The English Esq. J. A. Beaumont offered his impression of a meeting he had with Rivadavia in 1827 to discuss the failure of a colonization plan that the president had promoted in the past. Arriving in the middle of the war, the colonists lacked the promised resources and Beaumont, who had known Rivadavia in England where he visited his father, requested a meeting with the president. In London, Beaumont "frequently shook [Rivadavia] by the hand" and "joked with him" at his father's table, and he had not felt, "as perhaps [he] ought to have done, the awfulness of the presence!" In a detailed description, he implied that the president was a ridiculous figure estranged from the reality of his own country. Once the door "opened with solemn slowness," the president of the Argentine Republic, "gravely advancing, and with an air so dignified that it was almost overpowering" revealed himself. He had been transformed.

> [Don Bernardino appeared to belong] to the ancient race which formerly sojourned at Jerusalem; his coat is green, buttoned *à la Napoleon*; his small

> clothes, if such they can be called, are fastened at the knee with silver buckles; his whole appearance is not very unlike the caricature portraits of Napoleon: indeed it is said, he is very fond of imitating that once great personage in such things as are within his reach, such as the cut or the color of a coat, or the inflation of an address. His Excellency slowly advanced toward me, with his hands clenched behind him; whether this, too, was done in imitation of the great well-known, or to gain something of a counterpoise to the weight and bulk which he bore before him, or to guard his hand from the unhallowed touch of familiarity, it might be equally difficult and immaterial to determine; but his Excellency slowly advanced, and with a formal patronizing air, at once made known to me that Mr. Rivadavia in London, and Don Bernardino Rivadavia, President of the Argentine Republic, were not to be considered as one and the same person.[105]

The depiction of Rivadavia as a ridiculous figure, shared also by his rivals in Buenos Aires, contrasts with the brilliance that prompted Bentham to cite him as a model for nation-building. In an 1824 letter to the Greek provisional assembly, the philosopher included a regulation that had "guided the legislative assembly of the Republic of Buenos Aires in South America" for three years. Rivadavia sent it in a letter dated August 26, 1822, which Bentham had received only recently. Proudly, Bentham noted meeting the Argentine in London, where he was handling business, "still governing through disciples he trained and the distinctive reputation he earned."[106] Bentham viewed Buenos Aires as the only nation emerging from Spanish and Portuguese colonies with a "stable and prosperous foundation," earning the English government's "unequivocal esteem." The contradictory depiction of Rivadavia as both a vain fool and a prominent intellectual was a common practice among liberals of his era. While they promoted universalism, dissenting compatriots favored the autonomy allowed by federalism, and a distinct culture reflective of their identity.

John C. O'Neal has asked, "Does the mind in the writings of these theorists [ideologues] refer to man, and the body and sensation to woman?"[107] The answer to this question in the Argentine context is quite complex; gender instability brought about by the cult of sensibility challenged the limit of sexual definitions. In Europe, as Anne Vila has shown, the situation was similar since intellectuals even when they exalted sensibility "also expressed a deep anxiety about it: moralists and physicians alike viewed sensibility as a potentially dangerous quality that could lead to emotional excess, moral degeneracy, and physical debilitation." The transformation of medicine after Cabanis situated sensibility "somewhere between enlightenment and pathology."[108] The idea of liberal masculinity as a pathology would be argued during the years the conservatives took power from the 1830s to the 1850s.

SCIENCE EDUCATION AND PUBLICATIONS

In 1821, the year the University of Buenos Aires was founded, political economy began to be taught. James Mill's *Elements of Political Economy*, published the same year, was translated and was the basic book for students. In the preface, Mill announced that this was "a schoolbook," so the choice is understandable, particularly when he also explained its objective "to detach the essential principles of the science from all extraneous topics, to state the prepositions clearly and in their logical order, and to subjoin its demonstration to each." He recommended that people of either sex should read the book attentively.[109] The creation of a chair in this field was considered so important that Buenos Aires's Legislature announced in 1823 that political economy was taught at the university "and the expansion of its lights will give our fatherland [*patria*] intelligent administrators" who would lead the country in the future.[110]

The teaching of medicine received a renewed impulse that incorporated the areas of physiology that were making great progress as demonstrated by Cabanis's work. The medical school was founded in 1799 and by 1821 it had six professors. There was an attempt to hire the French explorer and botanist Aimé Bonpland who was living in Buenos Aires to teach, but because of complaints about hiring without competition the appointment was canceled. In 1822 the government created the Academy of Medicine to gather all the experts, including foreigners. In his speech at the reception, Rivadavia stated that the destiny of the nation was assured, "the Enlightenment [*luces*] and civility [were] developing and making gigantic, majestic strides" in the country.[111] The United States had already recognized Argentina's independence and the country's scientific progress would help move Europe in the same direction. He expressed his satisfaction about the reception of "professors of science on whom Buenos Aires places its hope for the support and spread of the Enlightenment, without which our nation cannot progress."[112] Publication of the *Annals* of the Academy began in 1823; by this time, twenty-one students were enrolled in classes, and studies on vaccination had also begun.

SCIENTIFIC TEACHING AND PUBLICATIONS

Interestingly and reflecting changes introduced by physiology and the influence of debates about ideology, chemistry was taught for medical students by

Manuel Moreno (1782–1857). An article published in 1822 observed that as students learned the new medicine, "the ideas of Magendie, Bichat, Richerat, Alibert, Pinel, Tenard, Orfila, etc." were introduced in classes.[113] Moreno, a graduate of the University of Maryland and member of the Massachusetts Historical Society, taught the principles of chemistry, "the general forces that produce the chemical phenomena, to intervene actively in them." These forces were of two kinds, "according to what we know: the forces of attraction and forces of repulsion."[114] In a speech Moreno delivered to introduce his chemistry class, he mentioned Philippe Pinel, who wrote that this field "had given a method to medical matters that previously contained so much inexact and delusional knowledge." Modern chemistry had helped "physiology and anatomic pathology," and its object was "to know what made composite bodies formed by the combination" of elements; it was to study the nature of the power that originated these combinations.[115]

Moreno was aware of the new chemistry derived from the work of Antoine Lavoisier (1743–94) and of the opposition that Jean-Baptiste Lamarck (1744–1829) had expressed to this remarkable transformation of the discipline, which earlier did not have the large reach that it acquired later due to how it changed philosophy. The doctrine of the gases had been discovered between 1772 and 1785 when experimental philosophers led the renovation of scientific thinking, of which Lavoisier's discovery of oxygen was an example. In Moreno's view, "La Mark" had ruined his reputation with his attempts to destroy the new science. The main problem was that he was a synthetic philosopher who based his chemistry on speculation and not on experimental work, as was done by the younger generation. Carl Barus, a professor at Brown University, wrote that the French naturalist Lamarck's work of physical and chemical speculation was still related to the old alchemistic philosophy that had been abandoned because by then philosophy and mathematics were structuring scientific studies and creating a different type of synthesis.

A decree signed in 1826 established the courses that were needed to complete the introductory preparation for entry to the university. The following order had to be followed: Greek and Latin, philosophy, arithmetic, geometry, algebra, and, finally, experimental physics. Students also needed to be prepared in drawing, French, and English.[116] The *Revue encyclopédique*, an important publication for liberals, published two articles on Argentina in 1827 recognizing the importance of the educational reforms. The second article described the composition of the faculty teaching at the University of Buenos Aires. In 1827, an article in the newspaper *Crónica política y literaria de Buenos*

Aires affirmed that "of all our recent institutions, the one on which foreigners would focus their attention someday" would be the laboratory of physics and natural sciences that was being established in the Convent of Santo Domingo with equipment coming from Europe.[117]

It was under the supervision of Pedro Carta Molina from Torino, where he had studied medicine. His involvement in the politics of Italian unification sent him into exile, so he knew the scientific institutions that existed in Spain, France, Switzerland, Germany, and England. In London he met Rivadavia, who invited him to move to Argentina to be appointed a professor of experimental physics at the University of Buenos Aires in 1826. Three years later Carta Molina devoted himself only to the teaching of medicine. The laboratory allowed the teaching of physics, mechanics, hydrostatics, hydraulics, pneumatics, electricity, acoustics, and light. The equipment was also very up to date and included the machines needed to study all these fields. An abundance of publications dealing with science, philosophy, arts, and culture started appearing during the period, marking the highest intellectual point of the first half of the nineteenth century. Influenced by Rivadavia, Ottavio Fabrizio Mossotti (1791–1863) also came to Buenos Aires to work in astronomy, replacing Carta Molina in teaching experimental physics after he had dedicated himself only to medicine. In 1830 Arséne Isabelle (1806–88) arrived in Buenos Aires to complete scientific research, meeting with Bonpland, Mossotti, and Carlos Ferraris who oversaw the conservation of the Museum of Natural Sciences, and Joseph, the son of the famous painter Jean-Auguste-Dominique Ingres (1780–1897), who was a well-known merchant.[118]

The union of science and sensibility was reflected in the publication of *La Abeja Argentina* in 1822. Mariano di Pasquale defines the journal's role as the "spreading of a new medical and physiological modernity that came from the French medical school," which included support for the creation of a school and a National Academy of Medicine. In addition, public hygiene, vaccination plans for the province of Buenos Aires, identification of diseases, and climate studies were some of the topics discussed in this journal's articles.[119] Artistic forms that stimulated sensations were also included to promote good taste and the emergence of a proper sensibility that facilitated students' learning of modern ideas. This monthly was published by the *Sociedad literaria* (Literary Society), founded in 1821, and it was the fourth attempt to create an institution that promoted Enlightenment ideas. In 1811 a "Club" was formed, then in 1812 the *Sociedad patriótica* (Patriotic Society), followed by the *Sociedad del buen gusto* (Society of Good Taste) in 1818. The Literary Society also published a weekly, *Argos de Buenos Aires*.

As Jorge Myers has explained, the *Argos*'s writers, not more than a dozen, were promoting the ideas of the Enlightenment, but the audience was not large. The sophistication of the articles, and the engagement with the most recent scientific and cultural novelties made them difficult to understand for those who did not know the appropriate context.[120] There were twelve members in the Literary Society, and they belonged to the most important families of Buenos Aires. Cosme Argerich, Juan Antonio Fernández, and Vicente López were the ones mostly responsible for writing about medicine and science.[121] In spite of its small circulation locally, *La Abeja* gained renown, even though it was not published for long. Alexander von Humboldt, for example, cited it in his *Personal Narrative of Travels*, and in 1835 it was mentioned in the *American Journal of Science and Arts* published in the United States.[122]

The first issue of *Abeja Argentina* explained that the union of science and art was very important to improving the nation, and for this reason the mission is described as "generalizing all types of knowledge" coming from civilized culture. The sciences are represented as "the image of movement," the needed impulse which transformed societies.[123] In the case of medicine, the transformation came through "the rules of public and private hygiene," which explains why the journal contained a section on sickness and prevention, and a popularization of scientific ideas.[124] The creation of the University of Buenos Aires in 1821 and the school of medicine shortly after were presented as examples of the new era of development in the country, since science was not able to exist and expand "without the influence of the government."[125] Due to the influence of physiology, developing the study of this discipline was a priority.

Other articles clearly noted that by the end of the eighteenth century an important change had occurred. This transformation helped establish that "observation united to experience was the foundation of all the arts" (*artes*). This new approach was translated to politics through "a new science" that belonged to the "moral class."[126] For this reason, the highest authority was "the professional, or scientist . . . the authority of those men who make a profession out of an art or a science."[127] In this way "the sciences apply to the arts, and the arts to the public work."[128] Music, for example, was part of the union between science and art.

The daily *Argos of Buenos Aires* featured music criticism together with articles on politics, arts, and science. An article praised Juan Pedro Esnaola, who had perfected his skills as a composer and pianist in Paris and Madrid, for his performance at a social gathering. The critic noted that on this occasion "everything that we saw and felt" was so appealing, leaving no doubt that the "school of music must increase the civilization and culture of the American family."[129] This emphasis on music was the result of the efforts of Juan B. Alberdi (1810–84), to whom we will return in chapter 4. Upon his return to Argentina in 1822 he founded a music academy that was supported by the government of Buenos Aires. On November 11, 1822, the first monthly concert demonstrated the progress made by the music academy. "Eleven of the most advanced girls beautifully played several pieces of the Italian and Spanish song cycle. Three male students helped with the choir together with the young teacher and director." Influenced by ideology, the review mentioned that artists were able to make nature beautiful through the study of philosophy, since the latter allowed the understanding of the "immutable laws of universal harmony," a unity in which "the moral and physical world" rested.[130]

On the first page of an issue published in 1823 an ad appears seeking subscriptions for the reissue of a translation of a work of "Destutt de Tracy" published in 1798 because "reading it must be popularized and be placed in the hands of all for the good of the country."[131] The publication would be cheap, three reales, to make it accessible to everyone. Among the sciences, "the study of the natural sciences" was the "only origin of all human knowledge," and on the latter man should always base "the foundation of his concepts and works."[132] The function of the recently created science commission was thus "to classify the physical and mathematical sciences in a natural order"; to determine the limits "that their relations have among them, and with the arts and fine arts," in order to "perfect the moral man, and to enjoy true freedom, and, as a consequence, happiness."[133] Morality in republicanism was taught through science and the arts, which developed the aesthetic sense that was a distinctive feature of the new society.

Unlike the ideas defended by conservatives, whose positions were closer to a Hobbesian perspective, natural law was understood by liberals as the relationship between man's extended knowledge and nature. Truth was "the essential principle of the natural law"; justice and liberty resided "in the perfect agreement of the positive and natural laws, and in the meeting of truth, "justice with liberty and morals," which was "the origin of the political and social sciences." On the "perfect knowledge" of these sciences depended "the perfecting of the human species," and on its perfecting depended "its happiness." But this perfection was only the result of "experience," experience acquired over time.[134]

These ideas were reflected in a program for the study of post-elementary education that started with classes in "drawing, which teach us how to imitate," followed by classes in natural history and languages. The second stage covered the study of math, and the third the study of the "general sciences, being those literary or political."[135] These three sections followed the educational plan put in place in France after the 1789 Revolution, but with a larger emphasis on the sciences. The first section taught natural history, one foreign language, and Latin or Greek; the second section introduced aesthetics and geometry; and the third prepared students in physical sciences and mathematics (geometry, calculus, physics, astronomy, chemistry, and natural history), literary sciences (literature, ideology, and grammar), and moral and political sciences (natural law, philosophical history of people, and political economy). The educational plan demonstrates that by the 1820s the union of philosophy and science reflected the interests of the nation and its educated population in a way that formed a republican sensibility.[136]

An article on statistics indicated the importance of these kinds of studies for the ability to govern a modern country. A crucial issue in the accumulation of data was the measurement of population growth, and it was here that the promotion of a "sober life without tumultuous passions" was recommended. A prosperous nation needed to increase its number of inhabitants at the same pace as existing resources, and "moral and political causes" influenced the increase of births. A poor economy decreased the number of marriages, "and only through marriages could the state expect an increase in births." Marriage was the foundation of a good society, and it was "demonstrated" that those who lived "in matrimony produced more children than those devoted to corruption." It was a fact that "women's communities, polygamy, divorce" were some of the causes of a stagnant or decreasing population. The figure of the husband and father became the foundation of modern society.

CHAOS AND CIVIL WAR

Bernardino Rivadavia took power while the war against Brazil was still at the center of the country's politics, and despite Argentina's defeat of the empire's forces, a treaty was signed by Manuel García ceding the Banda Oriental (today Uruguay) to the Brazilians. There was an immediate scandal in Buenos Aires; both Rivadavia and the Congress rejected what was signed, the president resigned; the National Congress was dissolved, and the provinces returned to governing themselves. In such a critical situation the city and province of Buenos Aires were under the command of the federalist Manuel Dorrego. He was left with the burden of assuming the representation of foreign affairs and dealing with a civil conflict that was believed to be leading the nation into chaos. Finally, in August 1828 a treaty was signed ending the war, creating a new republic, called Cisplatine at first, which is Uruguay today. Brazil evacuated its troops from the new nation, and Dorrego was able to turn his attention to the internal problems of his country.

The war between Unitarians and federalists continued, and the role of secular ideas was very much part of it. In 1829 the Unitarian general Juan Lavalle (1797–1841) fought the government of Buenos Aires, and after one confrontation took Dorrego by surprise and made him his prisoner. Immediately, he ordered the execution of the governor without a trial, on grounds of his illegal behavior. Despite pleas from the United States and British representatives, the order was carried out. The failure of two constitutions and the senseless killing of the leader of a federalist faction, seemed to announce another conflict, which led to the stalling of the educational reforms in the 1830s. In 1834 the exiled liberal priest from Jujuy, Juan Ignacio de Gorriti (1776–1842), wrote that the civil war disgracefully "had destroyed everything, or almost all the scientific institutions of the country"; the schools of ecclesiastic and natural sciences had received a "hit" that painted a very dark picture of the future. Teaching at the University of Buenos Aires was canceled, and the "topographic institute was also closed down." Defending the need of a modern state for the teaching of sciences, particularly natural sciences, Gorriti lamented that the new authorities "were accepting of the fact that savage styles and the dirty manners of their allies the barbarians" were providing "the norm of sociability everywhere in the country."[137] Gorriti also complained about the materialism of Destutt de Tracy and Holbach (Paul-Henri Thiry, Baron d'Holbach), and blamed their ideas as being partly responsible for the failure of the liberals.

Materialist philosophy was "a true disgrace" for the new American nations. Gorriti noted the incredible work that "these alleged philosophers put into spreading their ideas; all of them as pernicious as the men who defend them." The main problem was that in all the new Spanish American republics the new materialist ideas "had inoculated the youth." According to Gorriti, the purpose of these intellectuals was "corrupting" the youth through their "depraved" ideas, teaching young people to enjoy the pleasures of life while alive, "because with death it is all over for us." This lack of restraint made the young generation attack government and laws, "satisfying their brutal appetites like beasts, with which they pretend to equate our being" (180). Gorriti's criticism located a crucial problem that continued in the future, the risks of materialism's supremacy in philosophy and science.

His writings also demonstrate the different ideas that were part of what we call a "liberal" movement. On one side, they defended "civilization" and science over barbarism, but the emergence of new philosophical systems that questioned metaphysical and "spiritual" explanations to create the foundation of the new nation divided those who defended republicanism. The reception of a wave of ideas coming from different understandings of how modern republics should be organized created a maze of possibilities that contradicted the supremacy of reason associated with the Enlightenment. Gorriti, for example, despised Voltaire, Diderot, and d'Alembert, as well as d'Holbach and de Tracy, but he was a supporter of the moral teachings of Condillac, Mably, and Manuel Tesauro (185). It is very curious, though, that Gorriti's view of the world is dominated not by optimism in progress, but by an almost utilitarian view that promoted religion, or spirituality, as a way to survive the horrible conditions present in humanity. "They [materialist philosophers] made the pains of poor humanity even worse, and without consolation. All the sophisms of materialism will never protect men against life's adversities. We live constantly struggling against the elements, against men, and against ourselves. Natural elements bring great calamities, and misery is felt everywhere in a horrendous way. Thousands of individuals suffer from scarcity of resources, all events that take away from us our friends [and] those who favor us, and [their lack shows] us a very sad future. And where is the consolation? There is not one at all" (181).

This quote is helpful for understanding the dilemmas faced by liberals in Argentina at this time. On one side, the emphasis on the environment provided a helpful way to imagine a civilized future led by education, but the ideologues presented a perspective that was quite materialistic in its understanding of scarcity and the relation between man and nature. It also reduced everything to biological developments, including in the brain, which enraged Christians. The need for metaphysics and a spiritual conception of progress was then related to the concerns about an overtly materialistic philosophy such as the one that had failed. In 1810 Belgrano had already called attention to the importance of metaphysics; he mentioned his struggle to determine whether speculative physics should be the foundation of education or metaphysical beliefs. He ended up choosing the latter because it proposed "to know the divinity on which we depend, and the soul, the most noble portion of ourselves." Following Locke, Belgrano distinguished two types of metaphysics: first, the most ambitious one that attempted to know "the essence of beings and the most hidden causes"; and second, a more moderate perspective that recognized "the weakness of the human spirit" and was not concerned about "what cannot be reached," respecting limits that were evident. Schools needed to "move away" from any proposal that claimed knowledge of the absolute and speculative and instead should embrace the teaching of religion because it was the "main and indispensable foundation of the state" and "the firmest support for the citizen's obligations." He proposed laughing at the "moral virtues that were not based on our saintly religion. Reason and experience teach us this lesson constantly." Gorriti continued this kind of thinking.

From the point of view of masculinity, ideology challenged the past worship of women and the sharing of a social space by both sexes that led to fears about effeminate men. Men had a clear role as fathers of and providers for the family, owners of property, and devotees of politics, while women's role was to support men through their own sphere of spiritual action and motherhood, made possible because of a division of labor according to sex.

4

Eclectic and Romantic Man

Bernadino Rivadavia left for Europe after he resigned as president, leaving the country in chaos, a situation far from unique among the Spanish American nations during 1827. In Mexico a pro-Spanish conspiracy was organized by Joaquín Arenas, a priest; a coup in Chile led by Colonel Enrique Campino was defeated; Bolívar lost control of Peru after a mutiny and became a dictator by 1828; in the United Provinces of the River Plate the war against Brazil was not over, and economically and politically the country was devastated. Only in 1828 was the peace agreement signed with Brazil, and it established a new republic in the Banda Oriental, which was the origin of today's Uruguay. By this time, it was clear that the Enlightenment ideology was chaotic, full of contradictions, and far from being the organized rational system defended by liberals. Not only had it led to wars and violence, but, because of the emergence of new fields, the sciences also seemed to be less certain. Economically, the promises of capitalism and free trade were also more doubtful after the economy crashed in 1825 when the Bank of England was nearly destroyed because of speculation with the debt of Latin American nations, and a year later the stock market crashed in the United States due to fraudulent financial practices.

These events were followed by calls denouncing materialism, selfishness, political radicalization, and the lack of morality in the Americas and Europe. As a result, the 1830s was a decade of generational renewal, in which those born around 1810 started to reevaluate their faith in sensationism and utilitarianism to build the nation. The intellectual conflict was defined by the split between spiritualists and materialists; the former, as revealed by Larry Sommer McGrath, "revolutionized the metaphysical and theological study of the self on the basis of the very sciences that appeared to cast doubt on the authority of philosophy and religion."[1] Materialism was associated with the belief in the eternity of matter and the denial that the spiritual realm existed.

The failure of the liberal republican project led by Buenos Aires was covered in the international press. For example, one article that appeared in the *Revue des Deux Mondes* in 1835 was translated and published in Mexico in the same year.[2] The author of the French original piece was the Belgian entomologist Théodore Lacordaire (1801–70), who published several stories about his experiences in South America from 1824 to 1830 for the *Revue des Deux Mondes*. The translation and the introduction were written by José R. Pacheco (1805–65), a writer and Santa Ana's future foreign minister. He stated that his article was notable because it helped reflect on the fact that reform was more important than change; France and the Spanish American nations had all passed many constitutions following revolutions, but this had not solved the problems; internal conflict and civil wars had only increased the number of mercenaries who fought, and paying for the service of militias had taken most of the state's resources that could have been invested in more productive ways.

The collapse of the United Provinces of the River Plate exemplified how a nation could be dissolved, and the dangers faced in the process of nation-building even after applying the best ideas; the case of France was even worse, "seven constitutions have been ratified in ten years," only to end in strong despotism.[3] The dissolution of the United Provinces in the 1820s was so significant that Lacordaire's article was published "in many newspapers of the United States and Europe, revealing to the world the meaning of revolutions and pronouncements" (5). Mexicans "could be ashamed after finishing the reading" because it was a warning about their own past and present actions. In "political, economic, and generally scientific matters," there could be no pronouncements guided by "the general will because this involved infinite numbers of people who were ignorant because of their lack of education on these [modern] subjects" (1). Popular desires had become an obstacle to improving the nation, a view that became more dominant among the educated elites of Europe and the Americas.

Pacheco paid close attention to the fact that Lacordaire was writing "at the height of the admiration that was aroused in Europe for the yet unfinished war of the Spanish colonies against their mother country" (*madre patria*); this scientist was obsessed by what was happening, and "had a hot head" like many others, so he decided to go to the "Eldorado of freedom," Buenos Aires. In the translation, anarchy was clearly a perpetual danger: "Everyone aspired to be president, and each one, in turn, held the office"; the republic was accustomed to these activities (7). But tired of the repetition of four revolutions one after the other since he had arrived, and with the threat of another looming, the traveler decided to leave the city and relocate to the capital of the province of Entre Rios, a small rural town. Its inhabitants formerly lived reasonably well, but since the turmoil and the various political plans put into effect, they were mostly poor, and they were surviving by stealing. The governor, a man described as an imbecile, complained about the foreigners who were arriving "all the time in the country." He said that the country "was good" but not "ours" anymore, it was "owned by foreigners." They took the leather and the cattle, increasing prices and making the locals poor (11).

The story ended with another popular uprising, this time in the area that the traveler had imagined as peaceful. The main character concluded that all the criminals and turmoil of Spanish American countries were presented as "popular pronouncements, of popular will" when in fact they were the expression of ambitious fools without merit, "calling people a rabble [*chusmas*] who did not know what happened" (25). The idea of representation of the "people" that had been so ingrained in republicanism lost relevance after the failures of new nations in the 1820s, and the majority of the uneducated population was maligned as responsible for the political mistakes made by the educated class.

The inability to establish societies grounded in modern ideals, despite efforts like those of Bernardino Rivadavia's administration, highlighted the necessity for a new science of humankind. This discipline would investigate the principles and dynamics shaping social structures to comprehend how societies develop. It would emerge from a universal philosophical framework that applied consistent principles and laws to the analysis of specific societies.

THE EMERGENCE OF A NEW GENERATIONAL WORLDVIEW

The main thinkers analyzed in this chapter shared a generational worldview held by men educated under enlightened principles, regardless of where they were born. The proliferation of knowledge was a problem that continued in the 1830s and required a political adjustment to new ideas that were emerging in philosophy and science. The French socialist Henri de Saint-Simon (1760–1825) had feared that "in dividing and subdividing, science would create communication problems leading at best to the duplication of effort and at worst to intellectual anarchy."[4] This situation stimulated an interest in the creation of a unified philosophical and scientific worldview that was developed by Romantic ideas and French eclecticism. It is difficult to define exactly what Romanticism was because ideologically it contained many contradictions that are hard to explain in a comparative manner.[5] For the purpose of this book, Michael Löwy and Robert Sayre's understanding of this movement as a "worldview, that is, a collective mental structure" that can be expressed in "quite diverse cultural realms," such as literature, philosophy, sociology, history, and so forth, is very helpful.[6]

Scientifically, Charles Darwin (1809–82) belonged to this generation, as did most men born around 1800 to the 1810s. As a student at the University of Edinburgh, Darwin's education was influenced by Scottish philosophy and the two tendencies that emerged there – "realistic intuitionism" and "objective idealism" led by his grandfather Erasmus Darwin (1731–1802) and Thomas Brown (1778–1820), respectively, both claiming their concepts to be the foundation of "scientific work in physiological psychology."[7] At this time, Edinburgh was the center of transformist and evolutionist ideas in Great Britain.[8] As was the case in the eighteenth century, Scottish philosophers and naturalists had many contacts on the continent at this time, particularly in France through the work of Jean-Baptiste Lamarck (1744–1829) and Étienne Geoffroy Saint-Hilaire (1772–1844).[9]

Robert J. Richards understands Darwin as having a Romantic conception of life, and his sources were some of the same ones that were influential among his peers in the River Plate area, starting with his admiration of the work of Alexander von Humboldt (1769–1859), whose partner in the South American expedition, Aimé Bonpland, lived in the Littoral area of the Argentine confederation by the 1830s.[10] The importance of Romantic science was that it allowed for a unification of ideas that stemmed from the continuous transformations experienced in the economy, science, history, and philosophy since the 1780s.[11] As we saw in chapter 3, the influences of utilitarianism and ideology were understood as materialist interpretations of life, and, according to their critics, viewed as deprived of moral concerns. The Romantic view was against such an assumption; Charles Darwin's nature was not only intelligent but also moral, very similar to Friedrich Wilhelm von Schelling's (1775–1854) and Humboldt's.[12] The followers of Romanticism in Argentina shared this view in a literary sense but they rejected the nature and people populating their country because their sociability was not framed by the modern ideas that they defended.

The same experience characterized France in 1830, and the July Revolution of that year was related to a philosophical revival led by Victor Cousin (1792–1867) and the Romantic movement that coexisted with it. Larry McGrath describes him as a philosopher aiming "to create a just middle ground that would restore France following the 1789 revolution." Cousin promoted his ideas between 1830 and 1851 in his roles at the Ministry of Instruction and the University of Paris, where he taught. His significant influence established psychology as a central element of the philosophical curriculum in the lycées, where the state educated the elite. This prominence cemented this philosopher's status as one of France's most influential intellectuals. Cousin converted psychology into a vehicle of the Christian faith based on God's design of the psyche as a means to search for the Absolute.[13] This proposition enabled intellectuals to address one of the most urgent questions of the time regarding how reason and culture related to the biological or natural in the formation of selfhood.[14] After Cousin, masculinity acquired a different role based on the spiritualization of the nation and the need for contemplation and reflection.

An article about Victor Cousin published in 1890 by the *Edinburgh Review*, originally written in 1869 by the historian François Mignet (1796–1884), explained that by the 1810s and 1820s the "philosophy of freedom" had moved away from France and the United Kingdom to be developed in Germany. This region was "the European center of attraction for all intellectual and aspiring spirits. Its culture both in literature and philosophy had attained a climax of achievement" by the 1830s.[15] Both the Scottish philosopher Sir William Hamilton (1788–1856) and Cousin coincidentally traveled to Germany in 1817 to learn about what was considered the main philosophy of the time, and they would remain friends and correspondents for the rest of their lives. Cousin and Hamilton wanted to return to Descartes to correct the problems that empiricist positions had created in philosophy.[16]

Hamilton himself explained in 1829 that following the decline of Descartes's and Malebranche's philosophies, and after Condillac expanded Locke's limited principles by reducing all knowledge to sensation, "sensualism" (sensationism) emerged as the dominant philosophical theory in France and was nearly unchallenged until Cousin's philosophy became known. Cousin's work was novel for its attempt to combine "the philosophy of experience, and the philosophy of pure reason, into one."[17] Influenced by Scottish philosophy, his approach maintained the interest in experience, but it analyzed consciousness into a broader and more significant set of elements than those recognized by Condillac's school. It demonstrated that certain mental phenomena could not be reduced to sensory modifications. It established that intelligence relies on principles that, as preconditions for its activity, cannot be products of that activity. Furthermore, it identified innate, necessary, and universal concepts within the mind that could not be explained as generalizations from the contingent and particular experiences of the external world (195).

It was also influenced by the ideas of Johann Gottlieb Fichte (1762–1814) and Friedrich Wilhelm Schelling (1775–1854), who, in Hamilton's view, were part of the new German school, which argued that experience, being limited to the phenomenal, transient, and dependent, cannot serve as a true foundation for scientific knowledge or certainty, as it lacks inherent reality (197). Hamilton explained that there were two elements of intelligence in Cousin's philosophy: the unconditioned (expressed as unity, identity, substance, absolute cause, infinite, and pure thought) and the conditioned (expressed as plurality, difference, phenomenon, relative cause, finite, and determined thought). These are correlative and interdependent. The unconditioned, though absolute, is not conceived as existing in isolation but as an absolute cause that must manifest in the conditioned (its effect). The two are linked as cause and effect, each realized through the other, forming a third element of intelligence: their connection (197).

Concerning the concept of selfhood, Victor Cousin posits that individuals are mere fragments of a broader humanity, fully realized only through the balanced development of all its core principles. History of philosophy is essential for understanding transformation, with each era distinguished by the partial emphasis on one aspect of human intelligence within a significant portion of society. Given that there are three fundamental elements, human history is divided into three epochs: reason (rationalism), sensation (empiricism), and sentiment (idealism and intuition). This is the new science of man (201). Hamilton critiqued Cousin's claim that the unconditioned is positively known in consciousness, arguing that the unconditioned and conditioned are conflicting notions, and that the absolute and infinite are inconceivable within the frameworks of time, space, cause, or substance.[18]

Politically, the philosophies of Schelling and G. W. Hegel (1770–1831) revitalized the view of freedom, including an updated analysis of what colonialism and the revolutions that freed the American continent had brought to European identity. Interest in India's religion and mysticism found these two rivals on opposite sides; Schelling used sacred texts from India to defend pantheism; Hegel defended Christianity as part of a distinct and different European tradition. The declaration of Haitian independence in 1804 marked a pivotal moment, enabling philosophers focused on freedom to envision its universal realization through the abolition of one of human societies' most enduring institutions: slavery. The fact that enslaved people themselves had brought this end about strengthened the belief that humanity was entering a new era characterized by emancipatory politics. But by the 1820s Hegel had changed his mind and accepted the continuity of slavery, and his views became more conservative until the time of his death.[19]

Cousin befriended Hegel during his visit to Germany, but their philosophies diverged, as the French thinker focused on reconnecting seventeenth-century French philosophy with contemporary political needs. His work was a regenerative effort following the catastrophic collapse of Napoleon's empire and the Bourbon Restoration (1814–30). In contrast, Spanish America, despite frequent political crises, largely maintained republicanism. For France, the abandonment of the revolutionary ideals born in 1789 was a profoundly traumatic experience. Paul Janet (1823–99), a spiritualist, wrote a biography of the author of *Cours* after his death and in it quoted the memories of the writer Charles de Rémusat (1797–1875) of Cousin's 1815 philosophy class and his calls to "take courage and lift up your souls," since nothing sacred could be lost after Napoleon's defeat, which made an impression on the "humiliated" youth affected by catastrophic failure and British ascendancy in European politics.[20] Cousin demanded that students return "to the doctrines that find the true origin of reason in contemplation of the necessary truths, returning to its prerogatives at the same time as its laws." This would resume the interrupted work of the French Revolution "by purifying and strengthening its principles." It was time to inaugurate a philosophy that could serve free men.

As a result of Cousin's teaching, the young hearts "vowed to consecrate themselves to the cult of the just and the true," which "was to be [the mission] of their whole lives."[21] Due to the political changes in French academia, Cousin was marginalized in 1820. Still, eclecticism increased in popularity and in 1827 Théodore Jouffroy "proclaimed in the *Globe* the death of Condillac's empiricism, while Jean-Philibert Damiron hailed eclectic spiritualism."[22] A year later, Cousin returned to his chair in philosophy at the University of Paris and a magnificent event convened "two thousand attendees" who, captivated, listened to doctrines "incomprehensible to most." The philosophical discourse sparked "unprecedented interest in Paris and France, unmatched since Abelard's era," highlighting the profound significance of this novel philosophical approach for those engaged in politics.[23]

Joan Scott understands Cousin's objection to Condillac's philosophy as based on the rejection of the "notion that a self was merely a collection of sensations"; instead, he offered a "substantial self that, he said, preexisted its experiences and could be found by conscious introspection, a science of internal observation." He argued that metaphysics and science worked together to uncover the truth, which was exactly what the ideologues tried to destroy. Accordingly, for Cousin and his followers, the self was real, endowed with free will and reason, and did not belong to the individual but to God.[24]

French spiritualism, as McGrath suggests, was committed "to the existence of a subjective dimension of reality accessed via inner experience." He traced its roots to the seventeenth century, with renewals of this tradition throughout the nineteenth century. It became old-fashioned after the materialism of Condillac dominated everything because of its "dualist commitment to the existence of material and immaterial realities," but it was reshaped by Cousin and Maine de Biran into a project of selfhood based on an individual's "reflexive relationship with experience," or what was called interiority.[25] The duality of exteriority and interiority became a political and scientific source of inquiry and the foundation of a masculine identity sparked by men's different subjectivity. This is what defined Argentine men engaged in Romanticism and eclecticism, as will be analyzed below.

It was at this juncture in European history that the young poet Esteban Echeverría (1805–51) arrived in Paris from his native Buenos Aires in 1825. Analyzing this arrival ten years later, he concluded that before traveling his life was all "external: absorbed by sensations, love affairs, ramblings, blood passions, and sometimes reflection." He spent his time "ignoring where I was going, who I was, how I lived." Satiety "devoured" him, and he "devoured time." This description clearly explained his personal transition to Cousin's and Maine de Biran's philosophies, and for this reason he decided to learn everything from this new spiritualist perspective: "Philosophy, history, geography, mathematics, physics, and chemistry," occupied his time, but he also read "Shakespeare, Schiller, Goethe, and especially Byron," who revealed to him "a new world."[26] After a visit to London in 1829 he returned to Paris to study political economy and law, but he was recalled to Buenos Aires a year later.[27] He arrived home transformed by what he had experienced, and immediately began promoting a project focused on the revival of the May Revolution's ideals, a return to the original revolutionary source as had been done in France after Cousin.

Echeverría's stay in Paris was significant because there he witnessed the rebirth of revolutionary ideas that would lead to the 1830 Revolution in France, and the renewal of French spiritualism that now emphasized the inner reality of experience and was associated with cognition, consciousness, or mind. He also saw and participated in the renewal of spiritualism that ended the supremacy of ideology. This had a gendered component in that after the end of Napoleon's empire, French intellectuals and artists had a sense that the ideologues' materialism had condemned men to total passivity. In reaction, interest in spiritualism increased; spirit in this context involved understanding selfhood at the intersection of science and society.[28]

By 1827 Echeverría was a friend of Frédéric Albert Stapfer (Albert) (1802–92), the son of Philipp Albert Stapfer (1766–1840), an ex-representative of the Helvetic Confederation in Paris and an important philosopher among those interested in German literature and philosophy. The father (Philipp) knew well François-Pierre-Gontier de Biran (1766–1824), Joseph-Marie de Gérando (1722–1842), Pierre Paul Royer-Collard (1763–1845), and Victor Cousin; he was also well versed in Kant, being a source in Paris for many who were interested in this philosophy. His son, Albert, was known among the young Romantics; his translation of Goethe's *Faust* to French was so important that in 1828 he would republish it with seventeen lithographs done by Eugène Delacroix. The book was considered "a landmark in the history of graphic art" and French Romanticism.[29]

One letter from Echeverría to his Stapfer friend was included in his complete works. Written in French in 1827, it indicated their friendship and Echeverría's thinking about South American politics. He criticized the despotism that dominated the area at a time in which even Bolívar had turned into a despot and "aspired to be the monarch of the South."[30] Regardless, he affirmed that this present situation could not "paralyze the natural development of things." He also mentioned how much he missed his walks with Stapfer as well as his devouring of Schiller's and Goethe's poetry, which reveals his devotion to German literature.[31] Attached to the letter was a philosophical essay, written in French, which provides us with a view of his ideas at the time.[32] Clearly, the South American poet had discussed the confrontation between spiritualists and materialists with his friend, and probably knew the criticism of Pierre Laromiguière (1756–1837) by Cousin and Maine de Biran, originally published in 1817 and reprinted in 1828, since he made some points similar to those they had made.[33]

Taking what seems a position favorable to Descartes, Echeverría affirmed that it was "only as a result of the sensations that our soul experiences and the attention [the soul] attaches to [these sensations that] ideas are born and the work of intelligence begins." But he questioned what was "this soul that demands an impulse, a motive, some agent to put it into activity?" So, the next question was whether matter had more power than the soul. The answer was negative, since matter was inert. "For matter requires the hand of man or any other agent to set it in motion, to transport it from one place to another, to produce marvelous effects." By itself, it could do nothing. Echeverría rejected the idea of a soul with the "property of matter and bound to sensitivity." Finally, he admitted to his friend, we must accept that "we are only a machine endowed with activity by the spring of sensibility," an idea he further developed as follows.[34]

> As soon as any object strikes us, our organic being is set in motion; this movement is communicated to the brain, puts it into action and makes it produce ideas, intelligent acts, ultimately the greatest combinations according to the intensity of the sensitive impression. Let us deduce that the various circumstances in which we find ourselves, the various changes which our way of being, our habits, our passions undergo can modify our character and even our morals in that it acts only by the sensitivity.
>
> The existence of ideas is subordinated to sensitivity. It is impossible for man to have ideas without feeling. So, the source of all ideas is sensitivity. The existence of thought is subordinate to sensitivity. Man cannot think without feeling, therefore the source of thought is sensitivity.
>
> All bodies are inert, that is to say, all bodies persevere in the state of movement or of rest in which they are. We have given the name of inertia to this lack of capacity that bodies have to produce by themselves the slightest change in the current state. Inertia at rest is demonstrated by constant observation. We have never seen a body at rest put itself in motion.
>
> Sensitivity is to ideas what friction is to electricity.[35]

The context of Echeverría's citation is important to comprehend his understanding of sensitivity. It ended with a reference to André Marie Ampère (1775–1836), a physicist and mathematician educated according to Rousseau's *Emile*, and known for his discovery of the law of force that established the existence of attractive and repulsive forces between two parallel cables carrying an electric current, which ended in his conclusion that "magnetism is electricity in motion."[36] In addition, Cousin's and Maine de Biran's reinterpretation of Descartes attempted to debunk materialist claims affirming that the soul could not function without the body having the sensations. Here, Echeverría agreed with them, concluding that the soul could think pure thoughts without the body but needed input from the body to imagine or feel sensations for itself.

More importantly, this demonstrates how his own brand of spiritualism was being developed away from ideology. The nation, in his view, possessed a soul, which shaped his metaphysical interpretation of Argentina. While men were the political protagonists, the matter of the new republic, its soul, was female and allowed men to feel and retain political sensibility. This soul was represented by the figure of a woman, while materialist politics, lacking a soul, was depicted as masculine by him.

De Biran had published a work on habit (*habitude*) that appears very important for Echeverría's new ideas.[37] Accordingly, will was needed to have resistance and generate effort; if without effort, no knowledge could be produced because there were no perceptions. An individual who did not experience effort did not have an existence; therefore, through movement the individual perceived existence for the first time. Maine de Biran and Echeverría believed that creation was the result of active forces, life itself was the result of an originator that was a force, pure powerful intelligent activity that was present everywhere and in everything. Both men were interested in the relationship that existed between matter, movement, force, and experience.

The relationship between French spiritualism and Echeverría has not been adequately explored, despite his connections with intellectuals pivotal to the 1830s French Revolution and the new currents of philosophy and science that emerged from it. For example, his shift from ideology to spiritualist ideas introduced him to the concept of interiority, which would become crucial for his generation. Additionally, Echeverría developed an interest in a German Romantic science that, as Jürgen Barkhofft explains, was partially influenced by Schelling. This movement sought to uncover the unifying principle of nature, challenge the Cartesian divide between subject and object, and resolve the alienation between humans and their environment brought about by modernity and modern science. Romantic scientific thought offered a synthetic perspective that integrated all human capacities, aiming to foster a different understanding of nature and humans. It also presented an alternative approach to addressing secularization and the rapid accumulation of empirical knowledge that were crucial for making the new republicanism.[38]

The same need to reconcile also existed in France. Eugène Lerminier (1803–57) was another influential philosopher at this time; in the words of Bonnie G. Smith, he "became an academic hero who could synthetize German idealism, French legal rationalism, British liberalism, and centuries of intellectual history in general, with greater effect, so it was said, than Victor Cousin." Lerminier's emphasis on the legal system, and rational law, "articulated a common, youthful faith in the use of jurisprudence to pursue the yet unfulfilled promises of the Revolution." His generation had faith in the power of jurisprudence "to complete the Enlightenment task of creating a rational society."[39] The problem was that these attempts ignored the important differences that existed in the national intellectual traditions that had been developed in Europe.

The mixing of diverse traditions and contexts multiplied uncertainties concerning the power of philosophy to assimilate the constantly changing political and scientific environments. In Europe, the major objective was to stabilize the canon of European thought and its meaning. One example is an essay on Kant, published in 1840, in which Cousin made clear that there was a German philosophy and a European philosophy, and that his project was devoted to the second. According to this interpretation, Kant was not only the father of German philosophy, but "the author or rather the instrument of the greatest revolution that has taken place in Europe since Descartes." Here, revolution meant "the daughter of time and not of man. The world walks, but no one makes it walk, as no one can stop it." Cousin saw in Kant's philosophy two great antecedents, "the general spirit, the universal movement of Europe, and the particular spirit from Germany."[40] The reconciliation of European philosophies was the spirit of the time, which Cousin characterized as a spirit of unity, but understanding the varieties that appeared in local contexts.

Unity also applied to the human races, there was no "race without truth, for the beautiful, for the good," and a commune belonged "to the entire human species in all parts of the globe." However, "if mankind is one," it was no less true that according to the circumstances, the times and customs affected civilization in very different ways.[41] This idea of unity and variation partly explains the intellectual synthesis that regenerated France in the July Revolution of 1830 that made Louis Philippe I the king under a constitutional monarchy, deposing Charles X as the ruler of France. This revolution sealed the ideological shift from empiricist and sensationist ideas to Romantic conceptions of philosophy, science, and politics, and Cousin's work. It also unleashed the work of social reformers who demanded communal property and the emancipation of women, which created more dissent about the true meaning of emancipation.

Echeverría was quite aware that 1830s Romanticism was the project of his generation. Discussing it in an essay, he explained that the Romantic movement had started with Madame Germaine de Staël, who "imported it from Germany," and from her he took the notion of "genius," which he used to explain that while the "classical genius" was pleased by the contemplation of matter and the present, the "Romantic genius" was "reflexive and melancholic, it swayed between the memory of the past and anticipations of the future," like an in-between genius. Melancholic, "it went searching, like the pilgrim of an unknown land, his native country from which, it was believed, he was banned and to which peregrinating" he would arrive someday. Human genius, he concluded, was at once "a reflection and a revelation" of the divine.[42]

Among the English Romantic poets, melancholy fuses consciousness with "the quality of a persistent, albeit negative feeling."[43] Echeverría similarly describes his emotional state when defining his belonging to something that did not yet exist, which over the years he connected with his madness (*locura*) and ambiguous connection with his country.[44] Romanticism was for him "the modern poetry," which did not imitate or copy, but sought "its type and colors, its thoughts and forms inside itself, in its religion, in the world around it and produced beautiful and original work."[45] The emphasis on originality and genius was a common element in the Romantics of Europe and the Americas.

Another unpublished and undated piece included in Echeverría's complete works dealt with the philosophical problems provoked by the ideas of Franz Joseph Gall and Johann Gaspar Spurzheim (1776–1832), the creators of phrenology. They were rejected by the spiritualists who believed that psychology had to remain separate from physiology because the latter should not overreach into the realm of the mind or into moral, pedagogical, and political matters.[46] Faced with these ideas, Echeverría confessed that he was not a supporter of systems and "had, a long time ago, resolved to follow none of them, or to adopt if necessary one that was entirely negative" in relationship to politics and philosophy, which formed "the axis over which almost the whole social machine" rolled.[47]

This interpretation had two systems: sensationism's "a priori," which was systematic, and "a posteriori," which resulted from gathering materials first to analyze them deductively to produce results. The first system produced work for the present, the second for posterity. He recognized that both could be wrong, and that "being eclectic" might be the best position. Both "rationalism and sensualism" were "legitimate children of philosophy, and all systems that get support from both" must, in his understanding, "have some verisimilitude if not truth."[48] His rejection of taking sides and choosing to select from the sources he considered helpful were connected with Cousin's thought, though his own synthesis would be different.

THE MULTIPLE SOURCES OF ROMANTIC THOUGHT

Henri Lefebvre distinguishes three main philosophical threads that were remixed by this intellectual regeneration. Two of them synthetized French, German, and Scottish sources, and defended spiritualism. The third one rejected metaphysics to develop positive philosophy and science in the work of Auguste Comte. Eclectic spiritualists and socialists were politically in opposition; eclectics were viewed as part of conservative politics, and socialists "as the left wing of romantism."[49] While Comte's positivism was opposed to spiritualist positions, he had been a friend of Saint-Simon, even writing some of his work, but had moved his conception of science in a different direction. Doris Goldstein recognizes that Cousin revived the perspective of the previous eclecticism to update French philosophy to provide "those principles which were considered necessary to an orderly, stable society: the existence of God, free will, and objective standards of good and evil." Politically, this new philosophy meant "reconciling the demands of freedom and order, the principles of the French Revolution with those of the old *régime*."[50] The main objective was to avoid the violence and excesses of extreme positions that had destroyed the country before.

The editors of Cousin's *Cours de l'histoire de la philosophie*, published in 1828, wrote in the preface that eclecticism (*l'éclectisme*) presented "to the youth of France" a system that addressed the insufficiencies of both sensationism and German idealism as represented in the work of Kant and Fichte. This was an attempt to reconcile both schools through a system that also considered the problems of the "present social order," and the need for conciliation between "the monarchical element and the popular element." Similarly, in literature, it reunited "the classic legitimacy with the romantic innovation."[51] Interestingly, for all the claims of newness, originality, and looking forward typical of the Romantics, their philosophical approach was a return to the mistakes of the eighteenth century to address the problems they had created in the present.

The second main philosophical thread was French Socialism derived from the work of Henri Saint-Simon, Charles Fourier (1772–1837), Pierre Leroux (1797–1871), and Pierre-Joseph Proudhon (1809–65), and was related to the economic changes experienced in France, particularly regarding industrialization. The number of ideas circulating at the time caused a philosopher in England to call the first three of these philosophers "Socialistic Mystics," because of their interest in spiritualism.[52] In the same way Proudhon was classified as a "skeptic socialist," which some claimed misrepresented his ideas.[53]

During the 1830s and 1840s socialism branched into numerous variants, and disputes among them became volatile, mostly because of its relationship with religion. A recent study by Julian Strube explains that in the early 1830s, socialist identity had strong ties with religion, which was no accident. In his famous *Nouveau christianisme* (*New Christianism*), published in 1825, Saint-Simon openly proclaimed himself a prophet destined to fulfill "true" Christianity. This New Christianity would achieve a "progressive providential evolution into a new, rational, scientific," and positive form of religion.[54]

Keith Taylor observes that social phenomena, according to Saint-Simon, shared significant similarities with biological and physiological phenomena; therefore, the new science should be regarded as a branch of the life sciences, described as a "social physiology." Human society was an "organic" entity "whose development was governed by natural laws which the social physiologist must attempt to discover."[55] History, as the study of civilization's development, reveals the laws of social evolution, offering humanity a new sense of purpose to replace the theological dogma once relied upon.[56] The use of metaphysics to provide teleological support became a common practice of the Romantics. They believed in a universal association that "led by a class of priests, would fulfill the *perfectibilité* of humanity and result in nothing less than the creation of the Kingdom of God on Earth."[57] In some cases this thread led to an interest in the sciences of the occult that had started attracting the attention of some of the elites.

Saint-Simon's challenge was to create a new social organization aligned with positivism, a secular moral code that prioritized earthly happiness as humanity's main goal.[58] To be happy, man had to achieve both material and spiritual well-being, which required the development of an industrial society and a new social system to rule it. Power should lie with the most enlightened individuals—savants, both scientists and artists—who were the natural intellectual elite of the future, destined to serve as apostles of a new humanitarian religion: a Christian socialism that emphasized "social cooperation and brotherly love," following Christ's teachings.[59] Meanwhile, earthly power was in the hands of the "captains of industry" who applied "scientific rationality in the organization of the various branches of material production" and, in this way, promoted "the social welfare of all social groups" including the proletarians (*les proletaires*).

This system of administration put general directors with professional expertise in charge of politics—the "science of production"—establishing a meritocracy that replaced aristocratic privilege based on bloodlines. As a result, peaceful productive activity would supersede military conflict as the foundation of social order, potentially fostering an international community and political integration.[60] This characterization of society brought clear changes in the view of the sexes in the work of Saint-Simon and followers of his school; until then only men changed and developed different roles for themselves, but among socialists and Romantics both men and women evolved, and the evolution of sexuality became for some a crucial sign of the arrival of the final society.

Charles Fourier never met Saint-Simon, who died in 1825, and knew little about his work before 1829 when he made contact with the Saint-Simonians, but he simultaneously developed some similar ideas about association, in his case in the creation of communes, or phalanges, which he believed would help humanity to experience social evolution.[61] Saskia Poldervaart has identified these characteristics among those who were part of the Fourier movement: an emphasis on the significance of sexuality, a rejection of rigid, "natural" concepts of masculinity and femininity, and an acknowledgment that love, sexuality, and sexual relationships fall within the realm of social science. According to Fourier, this made social science an "uncertain science," one that would ultimately uncover the ideal society capable of achieving the harmony necessary to guide humanity toward happiness.[62]

Poldervaart concludes that while clearly hostile to Christianity, Fourier simultaneously advocated for the creation of a scientific religion, seeing himself as the successor of both Isaac Newton and Jesus Christ. He followed Newton's physics to outline eight stages of social order, progressing toward a final harmonious state through the work of the phalanges. He argued that the Enlightenment principle of self-regulation had suppressed human passions. He recognized the interconnectedness of economic and emotional-sexual repression and exposed the restrictiveness and hypocrisy of marriage.[63] Love was both material and spiritual, and sexual needs "differed enormously" among individuals; while some preferred purely spiritual love, others love "the opposite sex," "*la monosexie*," "*sapphisme*," "*péderastie*," "*flagelantisme*," "*l'androgrénité*," "*bissexué*," and "*trissexueté*." All "sexual expressions would be permitted so long as people were not abused," because there was a process of individuation in the march to social integration.[64]

Fourier wrote about his rejection of the Enlightenment of the ideologues, who, in his words, had a "mania for upsetting each morning the systems of the night before," and only knew "how to mystify more and more every subject which they pretend to clear up."[65] Also, he attacked physiologists, who "were prey of a crowd of trading writers," seeking to "fabricate systems by the fathom, and to produce, by mercantile views, all kinds of discoveries that nature puts into our hands."[66] This criticism, according to Hugh Doherty, shows that Fourier's approach was organic instead of systemic, he compared the association of all nations into one that combined the unity of the human race to a "human fetus in the womb, when all the organs are united into one complete organism or body." Once the collective body had evolved beyond the incoherent state of society, humanity would be ready to enter the world of truth, peace, and harmony—its ultimate destiny.[67]

Pierre Leroux, whom we will revisit in chapter 6, popularized the term "socialism" in 1833–34, initially using it as the opposite of individualism. In 1847 he noted that it had come to describe all forms of religious democracy (*démocratie religieuse*).[68] He was a doctrinaire liberal, briefly became a Saint-Simonian, and developed his own philosophy against eclecticism. He pursued a framework for association that sought to harmonize individualism and socialism, and an affiliation of the two extremes that nevertheless preserved their individuality.[69] He synthesized ideas and defined man as a mixture of sensation, sentiment, and intellect, which also explained how to arrive at a state of harmony.

Auguste Comte (1798–1857), the creator of the positive philosophy that competed with Victor Cousin's system, had gender ideas that differed from the more egalitarian views of Prosper Enfantin (1796–1864) and the Saint-Simonians, or the communist ideas that attacked marriage because it was an economic institution designed by men to exploit women.[70] He wrote that "the social mission of woman in the Positive system" was related to "her nature." Women were superior to men in the sense in which they had "the tendency to place social above personal feeling." Morally she deserved men's "loving veneration, as "the purest and simplest impersonation of Humanity, who can never be adequately represented in any masculine form." Regardless, a woman could not have political power, as some were proposing "without their consent."[71] They were inferior to men in attaining social goals because this was a practical activity that required great energy and force, for which men were naturally better prepared like males throughout the animal kingdom. Men were supreme in the realm of the material.

Comte, like the other philosophers, promoted his system among the intellectuals of the Americas, influencing the Methodists in the United States and the governing elites of Latin America during the second half of the nineteenth century when Cousin's philosophy became less popular. Since Comte was not interested in Romanticism and metaphysics, which was at its highest point when he published his main work in 1830, his influence would be felt more strongly later.

RACE, SCIENCE, AND AESTHETICS

In terms of historical narrative, a student of Kant and friend of Goethe, Johann Gottfried von Herder (1744–1803), with his universal interpretation of history, offered a perspective that would be influential in various ways throughout the nineteenth century, both in Europe and the Americas. Modern Europe in his account was the result of natural processes of adaptation over time as a result of the "intermingling" of different races. In Greece it had produced an *artistic expression* in which the "human form ascended Olympus, and clothed itself in divine beauty."[72] But "the people of Europe" were ranked "as barbarians with the negro and american" before acquiring the culture they had in the present (1:478). Each of Europe's nations had "endeavoured rather to retain it's [*sic*] ancient barbarous manners, as long as it possibly could," and educated culture derived "from the alphabets of other nations," sprung "from roman, greek, and Arabic seed," which attributed to foreign influence what was then considered superior culture (1:490).

Finally, it was "a foreign religion" that accomplished a spiritual conquest that finished the shaping of Europe, with the aim of "moulding all nations into one happy people, both in this World, and in the next," a force that had operated nowhere "so powerfully as in Europe" (1:489–90). The same spiritual agent had shaped all humanity, though in different ways, which was possible because there was "the great and good law of human destiny: that, whatever the nation, or a whole race of men, wills for it's [*sic*] own good with firm conviction, and pursues with energy, Nature, who has set up for man's aim neither despots nor traditions, but the best form of humanity, will assuredly grant" (1:441). This law of transformation through a nature that was good and moral inspired many interpretations created by Latin American intellectuals after reading Herder. More importantly, it challenged the idea of the superior European masculinity related to the conquerors since the Europeans themselves had been formed by being conquered by others.

In Europe over time this process had formed a disposition in "this small portion of the Globe" to "a grand *union of nations*," that the Romans had initiated. The truth was that "in no one quarter of the Globe have nations been so intermingled as in Europe; in no one have they so often and so completely changed their abodes, and with them their way of life and manners" (2:360).

To corroborate his view, Herder reminded readers that Europeans were not able "to say of what race of what nation" they were, whether they were "descended from goths, moors, jews [*sic*], Carthaginians, or romans," whether from "gael, cimbri, burgundians, franks, normans, saxons, flavians, fins [*sic*], or illyrians, and what intermixture of blood took place among their ancestors." Over time, "the ancient family stamp of many european [*sic*] nations" had been "softened down and altered by [a] hundred causes," and without this "the *general spirit* of Europe could not easily have been excited" (1:488). In this history of mankind, Europe was the result of mixing and not of purity.

Regarding racial variation in humans, Herder did not give skin color any essentialist qualities, based on Petrus Camper's (1722–89) writings. He probably had read the latter's lecture "On the Origin and Color of Blacks" (1764) in which he established that "Black, tawny and white men are simple varieties"; the different colors did not reflect "essential differences."[73] Miriam Claude Meijer understands that this perspective was designed to challenge the common belief that unique racial traits arose postnatally; while the nineteenth-century anthropologists misused his facial angle theory to support polygenism, which was precisely what Camper opposed, he blamed philosophers who believed that "Negroes and Blacks descended from the mingling in older times of white people with great Apes or Orang-Outangs," when empirical evidence proved that they were the same as white men."[74] Still in the fields of physiognomy and phrenology his measurement was adopted to support the multiple origin of present races.

Physiognomists incorporated Camper's measuring of the angle formed by lines drawn from the forehead to the jaw and from the ear to the upper lip and based on this they compared racial differences; in turn, this would also be assimilated by phrenologists but based on different philosophical grounding. The issue became more critical by the beginning of the nineteenth century when the problems of human origins became entangled with those of the nation and its people, at a time in which colonialism was being renewed, bringing race to the forefront of European preoccupations.

This is important because Camper was known in Spanish America. Francisco Caldas, the naturalist who criticized Humboldt, wrote a fascinating essay in 1808 on the influence of climate, in which, citing Camper, he touched on racial differences. Physiognomy was important for comparative anatomy at the time because the measuring of differences revealed how organs that controlled "intelligence and reason" developed; if the Camper angle was bigger, these faculties were also better. Caldas noted that the European angle was

85 degrees, and the African angle was 70 degrees, which for him explained the historical trajectory of the races; "art, sciences, humanity and the control of the earth" belonged to white Europeans. On the other side, "stupidity, barbarism, and ignorance" characterized those who were Black.

On the other hand, he also explained racial difference by investigating how climate (*clima*) affected it. Caldas defined climate as formed by powerful agents such as atmospheric and weather conditions, geographical and hydrological features, ecological systems, and environmental interactions. All these forces operating together were "powerful agents on living beings," and combined in various ways what he called "the influence of the climate [*influjo del clima*]." Together with food, climate was crucial to understanding variation among species, humans being one of them. Caldas also understood the influence of these elements on both the body and morals. But if climate had an influence, it was not deterministic; it only increased or decreased "the stimuli of the body," while humans would remain "free to choose good or evil." Virtue or vice was always "the result of our choice, in every temperature and at every latitude." Caldas's understanding of human variation as the result of environmental pressures would remain important in Spanish America over the nineteenth century in the context of the battle between polygenists and monogenists.[75]

As it was for Caldas, Herder's framework in writing about race was colonialism, human diversity, and the existence of hierarchies that linked both, which complicated scientific and philosophical thinking. In Germany, researchers such as Kant, Samuel Thomas von Soemmerring, Camper, and Johann Friedrich Blumenbach supported the monogenesis of the human species, while others, like E. A. W. von Zimmermann, focused on morphological features or infertility (Kant, Christopher Girtanner) to define racial boundaries, despite their belief in species unity across racial lines. Racial hierarchy was understood in terms of beauty (Johann Reinhold Forster, Blumenbach), mental disposition (Soemmerring, Kant), and attitudes toward the abolition of the slave trade. These views were grounded in egalitarian (Blumenbach), commercial (Kant), humane (Herder), and moral stewardship (Forster) concerns.[76] This was far from a coherent understanding of human diversity and its meaning.

Besides the more empirical interpretation of human variation, another understanding came from the arts, where the cult of form in neoclassicism had produced the most popular interpretation about race. Johann Joachim Winckelmann (1717–68), the German art historian previously mentioned, introduced a Platonic notion of beauty in his *History of the Art of Antiquity* (1764), where he defined the highest artistic style as a form of beauty that both drew from and enhanced natural forms, yet ultimately pointed the viewer toward higher, more perfect forms generated by an ideal produced by imagination.[77] In Alex Potts's analysis, Winckelmann assumed that the "white Greek ideal was closest to the original type of humanity, to the ideal human being as it emerged from the hands of God, while other racial types were deviations from this model." While Camper, as mentioned, was focused on empiricism, the idealist model was more deterministic and took beauty as an a priori condition that created hierarchies based on its perfectibility.[78] The German Hellenism that Winckelmann originated "equated the study of beauty with the rise of political freedom," which influenced Friedrich Schiller, who created the "aesthetic state," philosophically supported by Kant's views.[79] Among the Romantics who, like Echeverría, were interested in metaphysical notions of ideal beauty, this perspective became very influential, and in some cases led later to racist ideas that promoted European superiority.

Beauty then became political and went beyond the artistic boundaries in which Winckelmann operated. In a chapter about the organization of the Americas, Herder faced a problem in trying to connect the many different climates on this continent with the emergence of the population that lived in them, which led some to doubt Montesquieu's ideas about the influence of climate. The natives had come from Asia, and "the gradual transition, line for line," could be observed. They had passed over "many, even of various races" to get to the present one.[80] In conclusion, Herder proposed that at the time in which the world was not "traversed with the sword and the cross," because "in modern times the laudable spirit of observation" had begun "to be excited toward studying the human species," it would be important to "collect such scattered delineations of the varieties of our species as are authentic" in order to lay the "foundations of a perspicuous *natural philosophy and physiognomy of man*."[81] Herder's historical approach made him a favorite of the Argentine Romantics who used his work to create a narrative of the modern population transformed over time through both cultural and racial mixing.

THE INTELLECTUAL RENEWAL OF THE AMERICAS

Victor Cousin's books were bestsellers in the United States in the same years that saw the emergence of the transcendentalist movement. From 1829 to the mid-1840s he was the most important philosophical figure there and in Latin America among the young generation that emerged in the 1830s. Partly, his reputation was related to his being "both a creator and a beneficiary of the Romantic generation in Restoration France," his role in the secularization of education, his promotion of German and Scottish philosophies, and support of secular religions together with Saint-Simon and Auguste Comte. Materialism was linked to the science that explained everything according to changes in matter without consideration of the soul, but those who opposed this conception began to defend religious principles, even pantheist ideas.

The Church of the New Jerusalem, for example, was part of an international organization that believed in the writings of the Swedish philosopher and scientist Emanuel Swedenborg (1688–1772). Originally founded in 1792 in Massachusetts by believers from England, it became a national organization in 1817, representing a philosophical mysticism that included an interest in the occult. Victor Cousin's writings were popularized partly by this community when, in 1832, Henning Gottfried Linberg's (1784–1836) translation of Cousin's *Introduction to the History of Philosophy* was published in Boston. Linberg lived in St. Croix, where his father was a prominent judge and a follower of the Swedenborgian Church there, a small community connected with United States believers in this religion, as was his son. Linberg traveled to this country and corresponded with intellectuals who knew about his work, which led to the translation. For example, in 1823, Theophilus Parsons, a professor of law at Harvard College, received a letter from Linberg communicating his desire to move to Virginia to live with John Smith (Johnny Appleseed) at Lynchburg because he was attracted by the New Jerusalem doctrines.[82]

In the preface to the book, Linberg explained that Cousin's lectures possessed "an interest seldom to be met with in philosophical publications." This book represented a new generation because its readers were "a numerous class of intelligent and well-informed young men, who may be fairly considered to represent the flower of the rising generation in their respective countries." Against the accusations of materialism and infidelity that characterized the introduction of French philosophy in the United States because of sensationism, Cousin's work was dedicated to placing philosophy "on a new path, and impressing it with a higher character." He expressed confidence that many readers would sympathize "deeply with the genuine and enlightened love of humanity" of Cousin's philosophy, and by the fact that this philosopher avowed "every where distinctly," a firm belief "in the truth of the Christian religion," even in the "mysterious forms in which Catholicism has enveloped it." Moreover, this philosopher considered its religious mysteries "proper subjects of philosophical inquiry and elucidation."[83] With him, Christians had a modern philosophy connected with Descartes's rationalism, which did not clash with faith.

The other source of interest in Cousin came because of his writings on education; in 1831, as a member of the Royal Council of Public Instruction, he traveled to Prussia to report on its advancement in elementary education; this was published two years later.[84] He was part of the team that worked with George Guizot on the French bill of elementary education approved in 1833, which provided a model for many countries. In 1834 the English translator Sarah Austin published the English version of the report to defend the importance of the creation of a system of national education, which became very popular in the United States. two years later J. Orville Taylor, a professor in New York, organized the first edition published in the United States.[85] New York Schools used the book, and the legislatures of New Jersey and Massachusetts distributed this translation in all their schools. As a gesture of gratitude Cousin was appointed a member of the American Institute of Public education based in Boston.[86]

As was the case with Destutt de Tracy, Cousin maintained communication and promoted his ideas in the Americas; in the United States he had prominent correspondents, including Charles Brooks and Henry Tappan from New York, Charles Sumner, George Ripley, and George Bancroft from Boston, K. True, a professor of moral philosophy at Wesleyan University in Connecticut, and Henry Harrisse from Chapel Hill, North Carolina. He also had a close relationship with the brothers Everett—he had met William in Dresden in 1817 and was in contact with his brother Alexander Everett, to whom he sent a copy of his work in Prussia.[87]

In Latin America, Cousin's influence was stronger in the 1840s, due to a political renewal related to Romanticism's influence, and the interest in popular education; a translation of Cousin's *Cours* to Spanish by N. R. de Losada was published in 1847 in Madrid and was sold in Lima, Peru.[88] In Argentina, Echeverría's arrival in Buenos Aires in 1830 sped up the knowledge of the most recent European philosophy as he became the leader of the Romantic youth. José Tomás Guido in 1834 published a partial translation of *Cours*, for example. But the poet's life in his country was not happy; in his diary he described "how many hopes he was bringing!" when he returned to Buenos Aires, but they were all "sterile: the fatherland did not exist anymore."[89] During his absence Dorrego had been assassinated, the party of civilization had lost power, and the enemy of liberals, Juan Manuel de Rosas, had become the governor of Buenos Aires with extraordinary powers (*facultades extraordinarias*).

Two years after Echeverría's return, on November 20, 1832, Charles Darwin arrived in Buenos Aires, though there is no evidence that they crossed paths. Darwin described the outskirts as "quite pretty" and the city itself as "large," and the "most regular in the world" because of the streets' right angles. At the time he calculated that sixty thousand people lived in it, and the "general assemblage of buildings" possessed "considerable architectural beauty."[90] He stayed at the house of Edward Lumb, an English businessman from Leeds, and had contact with the British community living in the province. He knew about the nature of the area because he had read about the discoveries of gigantic fauna and the botanical research of Bonpland.

The assassination of Dorrego seemed to renew the civil war and the conflict in the fragile frontier that existed with the Indigenous population. The latter had begun to attack areas under control of Buenos Aires's government. In 1832 this forced Rosas to organize a campaign to contain the advancement of what was considered a savage enemy at the time. Darwin was exploring the southern border of the province of Buenos Aires and mentioned in his journal that "the wandering tribes of horse Indians, which have always occupied the greater part of this country," were battling Rosas who had organized an army "for the purpose of exterminating them" (78). During these feared incursions they stole not only cattle but also women who were known as "captives" (*cautivas*).

In his analysis, Darwin mentioned that he was not in agreement with those who defined "the primary races of mankind" and separated "these Indians into two classes." Some of the women deserved "to be called even beautiful" (84). The men "fought, hunted, took care of horses, and made riding gear, they were expert horsemen" (84). Their enemy, Rosas, was "a perfect horseman" who conformed "to the dress and habits of the Gauchos [cowboys]," a camaraderie that had given him "unbounded popularity in the country, and in consequence a despotic power" (86). In conversation Rosas was "enthusiastic, sensible, and very grave." His gravity was carried "to a high pitch": Darwin heard stories from "one of his mad buffoons (for he keeps two, like the barons of old)" referring to his sadistic behavior (210).

According to Darwin, the character of the more educated classes in the towns shared some of the positive traits of the gaucho, though perhaps to a lesser extent. Their men were tainted by vices such as sensuality, "mockery of religion, and the grossest corruption" (183). Darwin's first impressions of the area were the polite, dignified manners across all social classes, the women's excellent taste in dress, and "the equality among ranks." At the Rio Colorado "some men who kept the humblest shops used to dine with General Rosas." Many officers in the army were illiterate, "yet all meet in society as equals." Nevertheless, "the absence of gentlemen by profession appears to an Englishman something strange" (184). Ironically, it was this view of

Figure 4.1. "Gauchos on a Rural Estate." Emeric Essex Vidal, "Gauchos en una estancia," 1818. Copy of a watercolor. Courtesy of the Museo Histórico Nacional, Buenos Aires, Argentina.

social egalitarianism, praised by a foreigner, that men like Echeverría deeply disliked about their country.

In this brief description, Darwin pointed out how Rosas was willing to meet and share his time with everybody, including his friends among the Indian chiefs, because he believed in natural divisions that could not be disturbed by social interaction. Liberals, though, wanted to create a new society based on scarcity and limited access to resources, creating social divisions that were enforced by class, culture, and law. In a letter to Darwin in 1834, Edward Lumb mentioned that the "Indians vanquished by Rosas at all points have lately made their appearance on the frontiers of Cordova & San Luis & have made terrible havoc so that they are not so completely destroyed as some anticipated," and the fear that resulted from this threat made Rosas the only man who could control the situation; which led the Buenos Aires legislature to offer him absolute powers as governor to pacify the region in 1835.[91] Darwin wrote in his diary having heard "that Rosas has been elected, with powers, and for a time altogether opposed to the constitutional principles of the republic."[92] While not ideal, he thought that this might be the best decision, considering the existing chaos of the political life.

John Lynch's classic biography of Rosas explains that the governor "was born to property and privilege in a new land and an old society. The family and the frontier were the first influences which formed him. His heritage was colonial, and his people had been established in the Río de la Plata for some generations, patricians by virtue not only of their lineage but also of their positions and properties."[93] This cattleman was able to pacify the country and achieve a period of political stability unknown since independence. He returned to his post as governor in 1835 and remained in power until 1852, when he was ousted by the movement led by Justo José de Urquiza, the caudillo of Entre Ríos. He detested the liberal Unitarians, not because they wanted a united Argentina under one law, but because they believed in the secular values of humanism and progress. He identified them with Freemasons and intellectuals, "'men of enlightenment and principles,' subversives who undermined order and tradition and whom he held responsible for the political assassinations which brutalized Argentine public life from 1828 to 1835."[94] He departed from the previous federalist ideology that fought to create a constitution that reflected its values.

Rosas's replacement of Dorrego as leader of the federalist forces created a political movement that was not ideological, but practical. He was not interested in the constitutional doctrines proposed by Unitarians and federalists; he created a new type of conservatism defined by Rosas's enemies as "Americanism." According to him, the nation existed, it was material and not ideological, and, as occurred in matter, it evolved in time and according to the events that shape it. Dorrego, to the contrary, was an educated man, a follower of the Enlightenment, a journalist, a republican, and a federalist, closer to the ideas of Pazos Kanki in the 1810s.

Dorrego's opposition to the Unitarians was not because he rejected civilization, education, and science, as was published in the newspapers of the opposition, but because of his unwillingness to accept the "political and Masonic sacred league of Buenos Aires" that was authoritarian and did not respect the freedom of other provinces. The Unitarians at the time of Rivadavia "used as an excuse the need to organize the country while they were shaping it with partial laws" without the approval of those who disagreed with them.[95] He acted in the name of the rights of the people who were victims of the plan of *Grande oriente hispano bonaerense* (the Great Eastern Spanish-Buenos Aires).[96] The caudillos were accused of being responsible for the chaos and "*social war*," but Dorrego asked if those who said these things were ignoring "the influence and clientele" held by those who owned property "over the majority of the other men."[97] Philosophy was important for Dorrego; it needed to "humanize" conflict making it less deadly if it was possible, but he ridiculed ideology—"a practical science that had mathematical demonstrations," based on "experience"—even when his followers attempted to do things never done before so no experience existed.[98]

The federalist leader reminded his rivals that "the times of superstitious fanaticism, the spirit of conquest, feudalism and arbitrariness were far away" from being accepted by the population of the country. The dominant taste of the time was "the improvement of the social order."[99] The claims that the caudillos promoted "a war of the poor against the rich, barbarism against civilization, and ignorance against enlightenment" were absurd; travelers' books revealed that these leaders of the provinces "behaved themselves with great organization and circumspection."[100] On the other side, their enemies were "furibund" and "proto-anarchists" who wrote against the federation, fueling the "public passions" through publications that polarized public opinion.[101] Dorrego defined his actions as ruled by "legality and publicity"; and said that he would never accept "despotism, anarchy, immoral passions, and underground maneuvering"; he created *El Tribuno*, a publication to defend these principles.[102] Federation was presented as a more realistic way to work toward the unification of the country under one government and without oppressing the majority to achieve it.

In 1829 Rosas denied affiliation with the federalist party or Dorrego. He ruled as a centralist, advocating for the hegemony of Buenos Aires and maintaining the existing social structure.[103] As Jorge Myers has explained, Rosas believed in the "'natural' existence of social hierarchies," which were defined by three elements: merit, property, and social distinction.[104] In Rosas's words, above the human laws there was "a natural and divine law" that he followed strictly.[105] Unlike his enemy, the liberal Unitarians, he did not believe that rational design was the foundation of society. He was a typical representative of the Littoral region, defined by the rural life on his estancia; and, as a consequence, according to Tulio Halperín-Donghi, he supported "the more masculine character of Littoral society compared to that of the Interior."[106] This gender characterization would become a key point of conflict with those who belonged to the Romantic generation and embraced a new way of understanding what a citizen should be. Once Rosas banned the study of modern science and philosophy, the engine of the conflict became gender, and how each side would characterize the model of man the nation needed.

It was not true, though, that Rosas was deprived of knowledge about the Enlightenment; his administration employed Pedro de Angelis (1784–1859), a Neapolitan journalist and politician who represented the interests of Naples in Paris at the end of the 1820s. While there, he established connections with the liberal establishment, including Destutt de Tracy, Lafayette, Cousin, and Michelet. Destutt de Tracy was one of those who advised Rivadavia to hire the exiled intellectual to promote his educational policies. De Angelis arrived in Buenos Aires in 1827, just as the attempt to develop the ideologues' project was declining. After Rosas became governor of Buenos Aires, de Angelis, a journalist and publisher, became his ally, enraging the Unitarians who considered him a traitor and an opportunist. As a man of another generation, he also rejected the new Romantic ideas and contextualized the Rosas regime using his knowledge; for example, in a book published in 1845, he explained that the social order was "natural," and nothing could stop the action of people. "If in the natural order nothing is able to contain the development of general causes, under which influence the universal movement of beings is maintained, in the order of societies nothing can repress the impetuous action of the people when its will has taken the character of a firm sanction."[107]

By the mid-1830s, besides the conflict between Unitarians and federalists, a new group of young people emerged that was criticized by both; inspired by the new Romantic worldview that was changing their perception of politics, they were known as the Generation of '37, and their lives were shaped by the failures of republicanism and the violence that this caused to society. Instead of focusing on the rural areas, as Rosas did, they based their future in the promotion of urban life, liberalism, and European philosophy and science, against the defense of local tradition led by Rosas.[108] Halperín-Donghi characterized these intellectuals as defenders of the supremacy of the educated classes they represented.[109]

Echeverría's experience in Europe at the crucial time of the debates between spiritualists and materialists had exposed him to the problem of philosophical systems, one of his generation's main uncertainties. Echeverría's friend, Juan B. Alberdi (1810–84), mentioned that Echeverría had introduced him to the work of Victor Hugo, Alexandre Dumas, Alphonse Lamartine, and Lord Byron, "and all of what then was called romanticism, in opposition to the old, classical school" that he knew from his studies in Buenos Aires, where he had already studied "Condillac and Locke" and had "for years absorbed liberal readings from Helvetius, Cabanis, Holbach, Bentham, and Rousseau." Echeverría also brought about "the evolution of [Alberdi's] spirit, which took place following readings from Victor Cousin, Villemain, Chautebriand, Jouffroy, and all the German eclectics in favor of what was called spiritualism."[110]

This was confirmed by another close friend, Juan María Gutiérrez (1809–78), who explained that Echeverría paid little attention to Spanish literature, and knew English poets, though "he was more inclined to the Germans, especially Schiller and Goethe," whom he studied in French translations.[111]

Vicente F. López (1815–1903), another member of the Generation of '37, wrote about the intellectual renewal of this time in similar terms. He also explained that members of this generation called themselves "positivists" due to the influence of Saint-Simon, who aimed to use science to understand society and elevated history to that category to explain the arrival of a transformative future.[112] This would explain the attraction that history had for this generation, particularly the works of Jules Michelet (1798–1874), which will be analyzed in chapter 5.

Echeverría made known the *Revue Encyclopédique*, published by Carnot and Leroux, which expressed the ideology of the Revolution of 1830. He introduced his friends to the liberal ideas of Lerminier, who was the philosopher of choice for many in France.[113] Lerminier was understood at the time as a believer in the law of the human race that was unity, "which in its essence [was] equal to that of the divinity," developing itself through all times and ages with a mode of development that was "logical and impassioned, unequal and necessary." More importantly for those who grew up in new nations that were ex-colonies, the "human intellect" was for this thinker "complete in each century and in all nations: only, amongst its adaptations some may be relatively superior or inferior in a point of time or of space." The existence of a "law of progress and of triumph" meant that all humanity was moving toward a state of glory.[114] One element that was common in most of the readings was the implication that evolution and adaptation to the environment shaped outcomes according to the universal role of progress and spiritual design.

The same can be said of what was happening in Chile, where some of these ideas circulated during the 1830s and 1840s. It is useful to return to Andrés Bello, whose thought at this time was a mixture of sensualist, Romantic, and eclectic sources. Bello had moved to London to do diplomatic work and stayed there for nineteen years before moving to Santiago, Chile, in 1829, where he became the country's intellectual leader until his death in 1865. He was still devoted to science, an interest he had developed since his days with Humboldt, and would have known the French naturalist Claude Gay, a student of Georges Cuvier, who had arrived by the end of 1828 and would stay until 1842 to complete research in Chile; in the years to come Bello promoted this research. Patience Schell shows that for Bello, science provided a way to strengthen the newly independent Latin American nations and impart knowledge to future generations.[115] In 1830 he created a journal, *Araucano*, that over the years published articles on scientific news, keeping the intellectual community updated about the latest important discoveries and publications.

In August 1834, a few years after Bello had settled in Chile, another great scientist arrived in the town where he lived; Darwin stopped in Santiago for a week, though it is not known whether they met. Chile was also experiencing the pains of building a republic, undergoing a civil war from 1829 to 1830 between conservatives and liberals. Certainly, Bello would have recognized the importance of the English naturalist's observations; in 1839 he translated and published in *Araucano* two articles that had appeared in England about the *Beagle* expedition.[116] As a poet, Bello was involved in the Romantic movement; at the same time, he taught philosophy, including an explanation of the problems he had with the new eclectic philosophy.

Bello's philosophical treatise occupies the entire first volume of his complete works; it was written with the intention of providing a basic text for students. The introduction, written by the editor, explained how Bello's *Philosophy of the Understanding* defended "sensualist or skeptical principles," refusing to recognize the idea of "infinitude" and "denaturalizing other notions and metaphysical principles."[117] Having been written in 1880, this determination ignores that during the 1830s some of the intellectuals of Bello's generation developed their own philosophies according to local needs and the different options that appeared in front of them. For example, while in London in the 1820s, Bello recommended a list of seventy-eight books to teach at the University of Caracas, including William Paley's *Natural Theology*; Locke's *Principles of Moral and Political Philosophy*; Thomas Reid's *An Inquiry into the Human Mind* and *Essays on the Powers of the Human Mind*; and George Campbell's *The Philosophy of Rhetoric*.[118] These were the same sources that many members of the intellectual elites, born around 1810, read before Cousin's philosophy became known in Spanish America.

Bello's philosophy had been developed in the first half of the nineteenth century, and it related to the sensationalist and Scottish tradition. He mentioned that Cousin had accused Locke of confusing two different ideas, succession and duration, and explained that he was not in agreement with eclecticism because for him the "perception [of succession] was not a simple conception of the idea of duration." Succession was for him "an element that was part of the duration. A thing lasts because it succeeds itself continuously, without any intermission."[119] Young followers of Cousin in Chile provided their own interpretation in 1845, through the journal *El Entreacto*. This publication analyzed Bello's interpretation, indicating that he maintained his support for "sensualism" and defended "space and nothingness" as "the same thing"; an "ontological question" that was attacked because the "leader of the eclectic school, M. Cousin, resolves satisfactorily this apparent contradiction making a distinction between *logical and chronological priority*."[120] Citing Cousin, the article explained that "the idea of space *logically* preceded that of the body, and the latter preceded that of space *chronologically*," while in the work of Bello there was only consideration for the *chronological* priority. Those collaborating on this journal specified that the reason for its existence was to serve the youth; it was a generational project as was Echeverría's own.[121]

ECHEVERRÍA, IDEOLOGICAL REGENERATION, AND ROMANTIC SENSIBILITY

Rosas's ban on modern science explains why scientific ideas were his generation's central preoccupation. As we have seen, the proliferation of philosophies and the ongoing emergence of scientific fields that were more independent from philosophy made it difficult to return to the universal systems that characterized early republicanism. The science of man of Cousin and the Romantics provided another universal option, and it is in this context that literature emerged as a viable civilized pursuit and a modern activity for the youth in Buenos Aires. As a result, the development of a national literature, shaped by Romantic sensibilities, intensified the polarization of how men regulated their relationship with masculine and feminine ideals, which became a central aspect of the civil conflict unfolding in the 1840s.

According to Echeverría, the word "romántico" arrived from Spain and started circulating and being ridiculed in Buenos Aires as it was on the Iberian Peninsula. The enemies of Romanticism saw in this movement "an exaggeration and extravagance that was evident in everything—clothing, writings, and manners."[122] While he did not mention it, this rejection also expressed Rosas's ridicule of the culture being exhibited by the youth, which seemed foreign and unnatural, and is similar to what happened to members of the Romantic movement in other countries.

Echeverría had dedicated himself to promoting Romantic literature to advance the nation's progress and as an expression of the emancipation achieved in the Americas. According to his close friend, Juan Manuel Gutiérrez, in Echeverría's spirit renewing revolutionary ideas became "a kind of religious mission" and he invested everything "in acquiring the means and instruments to convert his theoretical solutions into realities in this part of America." He had concluded that the emergence of national literature should be the solution to the emergence of a national identity.[123] He learned how to translate, and Gutiérrez explained that he slowly started "to understand that among the empty phrases and mystical aspirations of the old ascetics, he could find expressions and language twists that gave color and energy to modern thought expressed in our language" (5:xix). Translation became one of the key activities of the Romantics both in Europe and in the Americas.

In Gutiérrez's view, Echeverría was "a romantic in good faith," but he made his own synthesis according to his country's needs; he searched for inspiration in "the serious and philosophical schools of North Europe," in Goethe, Schiller, and Byron, who were "very careful creators of form" (5:xxi). Echeverría could turn thought into an expression of the universal spirit because languages "changed over time, in the same way as beliefs, ideas, and needs of which they are the representation." These forms of language were preferred because they expressed the genius of each thinker, and, because they revealed locality in their writing, it was also a patriotic act (5:xxi).

Antoine Berman has analyzed the importance that the experience of the foreign had in German Romantic literature and how translation was "one of the instruments for the constitution of a universality." For these Romantics, translation on a "grand" scale was a crucial step, alongside criticism, in the creation of a universal, progressive poetry—one that affirmed poetry as absolute.[124] This was the programmatic practice in the work of Schlegel and Novalis also important for the Romantics in Argentina. They also saw translation as a metaphysical project in their belief that everything was translatable, particularly poetry. In his *Aesthetics*, "Hegel stated that poetry could be translated from one language into another (and even into prose) without any loss, because spiritual content prevails in it."[125] To translate, or to read in translation, made it possible to experience belonging through the experience of turning the foreign into something that the poet made his own.

Language was no longer empirical because it had a metaphysical purpose; it was not just the translation of word by word, but the capturing of the universal spirit through meaning to reveal it in local language. As discussed, Condillac and the ideologues introduced the idea of words as signs, but the novelty here is the metaphysical aim that articulated a new relationship between the foreign and the local through a process that started with the recognition of difference between the text and the translation, through universality the poet could arrive at locality. This process helped the poet to understand who he was; through deduction, the translator then eliminated the foreign, creating in his own words the spirit of the text. Consequently, the work of the Romantics made possible the translation of Argentina into a modern country; the existence of national literature became evidence of how modernity sounded in the local context.

Politically, the true "foreign" elements in Argentina were Indians and those of African origin because their bodies and experiences were not translatable or representable in modernity. This generation's obsession with the best aesthetic archetypes, those who expressed higher civilized ideals, would assume an idealist morphology that assimilated the racist ideas of the time. In terms of gender, they obsessed about beauty expressed in female form, and how Argentine women represented the nation.

By the end of the 1830s it was clear that there were two types of nation and citizens, which drove the conflict in the years to come. The conflict became polarized along gender lines. On one side, the younger generation focused on using their bodies to express change, transformation, and inner identity. On the other side, the governor of Buenos Aires dismissed the ideas of transformation and cosmopolitanism, showing no interest in the youth's culture of representation, which he viewed as embracing a true enemy: foreign influence. The interest of young liberals in the representation of the feminine manifested a self-reflective, questioning, and shifting form of identity, constantly defined by its own otherness.[126] This was a departure from the past because the conflict between Unitarian and federalist divided men according to ideology and access to power, whereas the new generation was transgressive and understood the meaning of man in a completely different way.

This aspect appeared in Echeverría's work; his writings abound with references to the feminine, so much so that some scholars have noted "considerable grounds for reading *La Cautiva* as a feminist text."[127] It would seem that in its female protagonist "gender differences are blurred."[128] This interest in undermining the "natural" and permanent categories defended by Rosas appears in the identification of civilized principles with the destiny of women. Echeverría described the importance that the feminine had for his generation, and how the female had become a symbol of modernity, something that was not new, but reshaped through the lenses of Romantic ideas. In his essay about poetry he explained that the new literature attempted to repress the military and violent impulses in order to "make love into a kind of cult, and to make women into deities [*divinizar a la mujer*], these fragile creatures whose perfection symbolized the beauty and candor that the imagination enjoys when it recognizes angels."[129] We can see a similar representation of the culture of sensibility in his poem "La Cautiva."

The poem tells the story of María and Brian, who ended up trapped by Indians in the Argentine desert. Outside Buenos Aires there was a place of abjection that Susana Rotker characterizes by the frontier, mestizaje, barbarism, uncontrol, the sexualized body, and the Other. It was where the Indigenous raids happened and where the white women were kidnapped from civilization. The "capture" in this text represents a textual space, "that of the Indians and the whites"; and in the "extratextual space, the 'noble' Creoles versus the rest (blacks, mulattos, Indians, and Rosistas)" who had the real power in the country.[130]

This poem is considered the first example of Romantic literature in Argentina, and starts with a quote from Lord Byron that clearly indicates the gender preferences: "Female hearts are such a genial soil / For kinder feelings, whatsoe'er their nation, / They naturally pour the "wine and oil" / Samaritans in every situation."[131] While Brian is described as having a "virile face" (*rostro varonil*) (1:106), María is characterized as possessing "male strength" (*fortaleza varonil*) together with youth and beauty (1:131). Her existence, though, is only possible because she loves a man, and once he is dead, she sees herself like a tree without roots, an existence that cannot continue. Women are only valuable as an attachment of men, their characteristics naturalized by their destiny to love and to feel.

Echeverría's description of Romantic poetry as an "intimate voice of consciousness, the alive substance of passions, the prophetic stare of fantasy, the meditative spirit of philosophy," a "wonderful instrument" by which sublime "harmonies express the human and the divine" indicates the importance to him of interiority and the sublime, both praised by Cousin.[132] This was not unusual among the Romantic writers, as George L. Mosse indicates; for Karl Wilhelm Schlegel, a poet Echeverría admired, and other Romantics, "only the love of an understanding woman can spur a man on to true manliness."[133] She is the one who made a man conscious of being a man, but women who performed this role were expected to maintain it permanently within society. Sexuality was not the main expression of this type of man; his life was consumed by the search for a sublime universal aesthetic experience, and, unlike women, who were motivated by love and seduction, he preferred to remain chaste.

A protagonist of Echeverría's poem "Elvira," is described as the "candid essence of beauty under the form of a woman, she was promised to the ardent aspirations of that male soul." The female form was not desired by the male body, but by "the soul" because the female form shaped his notions of beauty and morality, and let men access the spiritual, the divine. In his writing about elementary education, Echeverría also exemplified this view; mothers needed to oversee "the education of the child's feelings" because in women "the emotional faculty" predominated. Rational education required more work and "was more manly, more adept to strengthen in the child's consciousness the notions of duty, to get him used to reflection, to ground the beliefs, and lastly, to form useful citizens in a democracy."[134] In terms of inheritance, women passed on permanence, while men were the agents of change—also a common understanding of heredity at the time.

The availability of journals such as *Revue des Deux Mondes*, *Revue de Paris*, *Revue Britannique*, *Revue Encyclopédique*, and the *Edinburgh Review*, was changing reading habits among the educated class throughout the city.[135] New bookstores were established, numbering five in 1830 to double that by 1836. During the early months of 1837 the *Salón Literario* (Literary Salon) opened as a cultural center for those who felt the need to create an intellectual space from which to promote their new ideas. The group gathered in Marcos Sastre's bookstore, Librería Argentina; its members were Echeverría, Alberdi, Juan María Gutiérrez, Manuel Quiroga Rosas, Juan Thompson, Félix Frías, Vicente López, Carlos Tejedor, Enrique de la Fuente, Pastor Obligado, and Andrés Somellera, among others. There were thirty-five young men at the first meeting in 1837; they assembled with the intention of discussing the cultural status of the country, rejecting Spanish influences and asserting modern ones, and, in sum, changing the social norms and prevailing mores in Buenos Aires society, especially among the men in power. The insignificant number of participants confirmed that they were a minority, and, as Rosas mentioned, they did not represent the interests of the whole country.

In an 1837 speech at the Literary Salon, Echeverría mixed socialist and eclectic ideas to explain how different the men of the previous generation were from those of his. The first era that shaped the historical development of his country was that of the sword, "an enthusiastic, noisy, warrior-like, and heroic period that ultimately led to the independence of our political regeneration." The second was led by forces that were "pacific, hardworking, reflexive," and were working to give Argentina its freedom. The era of the sword had a disorganizing effect, "because the sword does not build," but "destroys and emancipates." The second era was destined to fix the havoc created by the first, "a social regeneration" intended "to cure the wounds and provide the foundations of the country. The present time was about "law and reason" in this understanding of the nation. The fatherland demanded "a rational cult," a reign of "cold and carefree reflection." Echeverría's generation had not been awakened by the "martial spirit" because they were "rational and sensible beings" who sought knowledge and emotions for their hearts.[136]

All three inaugural speeches of the Literary Salon, by Marcos Sastre (1808–87), Juan B. Alberdi, and Juan María Gutiérrez, emphasized their country's inevitable progress because of its participation in a universal historical process. They highlighted the need for the educated members of their community to study the most prestigious thinkers of Europe's intellectual tradition as a path toward better inserting their country into that providential system of progress; the writers they singled out for special attention were Herder, Vico, Jouffroy, Condillac, and Cousin. All three speakers mentioned the special role of philosophical or literary pursuits in the improvements of morality, which was, in turn, a key factor in society's advancement. In this regard Sastre and Gutiérrez singled out certain authors of the Romantic canon: Byron, Lamartine, and Chauteaubriand.[137] The emphasis of this generation is on the development of a national literature and social sciences to improve the conditions of the nation. The Literary Salon did not last very long; by October 1837 it was all over. Despite Sastre's public support of Rosas, he was forced to end his meetings and auction off his books.

At the time, Alberdi was working on a new project, a newspaper, with Rafael Corvalán, the son of a man who worked in the Rosas administration. They counted on the collaboration of Gutiérrez, Vicente F. López, Carlos Tejedor, the brothers Demetrio and Jacinto Rodríguez Peña, Carlos Eguía, Manuel Quiroga Rosas, and José Barros Pazos.[138] On November 18, 1837, *La Moda* (Fashion) appeared. It was inspired by France's highly popular genres "that evolved in the publishing industry in this period . . . the journals of *modes* and *moeurs*," such as *La Mode* (1829–54) and *Psyche* (1835–78), which provided detailed accounts "of the latest trends in fashion and toilette, the magazine."[139] The purpose of these publications was "the cult of the female consumerism that reached its full maturity during the Second Empire and cultivated interest among middle-class women in a new preoccupation: obsessive self-adornment."[140] The target of *La Moda*, though, was not female consumers, it was part of a political project to regenerate society and teach both sexes to embrace a new morality.

According to Doris Sommer, this publication "was correctly suspected of fronting for the unmanly Europeanized 'fops'; it was a coy screen in both senses of hiding and showing, a womanly voice as the men's public organ." Moreover,

"Alberdi did not hesitate to describe himself as feminized, although the suggestion of homosexuality would have been an outrage."[141] Gender categories became a tool to distinguish the two parties fighting to impose their own model of nation and man; deprived of the production of philosophy and science as markers of progress, a new revolutionary ideology was related to a moral regeneration that created a new form of association among men to transform society.

The appearance of this publication was especially important if we consider the conditions under which the press was operating at the time. Due to a law enacted by Rosas in 1834 there was no freedom of expression; beginning in 1835 control actually became even stricter—only four newspapers were being published in Buenos Aires: *Gaceta Mercantil*, *Archivo Americano*, *El Diario de la Tarde*, and the *British Packet*.[142] Jorge Myers estimates that from 1835 to 1840 only three to six newspapers were published in Buenos Aires, and these were almost always in the service of the government.[143] Regardless, behind their descriptions of the latest fashion, the editors of *La Moda* introduced the signs of the latest politics; for example, one article presented Mazzini, "the leader of young Europa," who was fighting to give Italy a "representative republic" like the one that was now governing the Americas; he was the "Apostle of the European Republic and must count on the support of the American Republic."[144] On the same page was an article about modern literature that praised Romanticism and lauded poets who were politically involved, such as Leroux, Quinet, and Mazzini.[145]

Rafaél Corvalán analyzed the ideology of this generation in an article about literature, arguing that they were not harsh critics of Spanish literature but held "the deepest conviction" about an art that "must be in very close harmonic intimacy with the purpose of a society." The purpose of society, he asserted, was "progress, development, the continuous emancipation [*emancipación*] of society and humanity." Therefore, art and society were "modified constantly," as indicated in the writings of "Fortoul, Leroux, and Mazzini," men associated with "progress and freedom." [146] The emancipatory mandate of this project caused a clash with the governor of Buenos Aires who was aligned with the protection of the Spanish tradition that was culturally against any change.

The generational confrontation between fathers and sons appeared frequently in this publication. One article criticized Spanish tradition for forcing sons to copy and imitate their fathers from the beginning of time. Adam was described as the first "innovator and experimenter," which explained why Spanish men reacted with "horror to the tree of science," a reference to the ban on sciences implemented by Rosas at the time. However, it was a crucial obligation of a father to educate his son according to "the current ideas"

Figure 4.2. Sociability in Buenos Aires, ca. 1830. Carlos Enrique Pellegrini, "Bailando el minué en casa de Escalada" [Dancing the minuet in Escalada's house]. Lithograph. Courtesy of Museo Histórico Cornelio Saavedra, Buenos Aires, Argentina. Catalog no. MHS020953.

because intellectual progress was continuous. A son needed an education grounded in the teachings of Germany, France, and England to be useful to a "representative Republic." In contrast, Spanish studies were deemed useful only for "forming Spanish men," which would diminish the influence of Enlightenment culture in the Americas.[147]

La Moda elucidated that the rejection of "Spanishism" (*españolismo*) was based on the idea that despotism and freedom "resided in the people's customs, and not in what was written." By customs, the author referred to "ideas, character, beliefs, [and] habits." Despotism, he argued, was instilled in the population by the Spaniards, and "under the synthesis of Spanishism," everything retrograde in the country was Spanish. Even the Spaniards sought to move away from being Spanish, as demonstrated by the Young Spain movement, based on Mazzini's ideas, which the author described as "our sister" movement. The failure of past constitutions, both federalist and Unitarian, indicated in this article that the only solution was to change "ideas, beliefs, and habits," indicating the use Lamarckian ideas, which supported belief in soft inheritance. In this context, it was imagined that an education emphasizing habits would, over time, organically transform the culture. For the younger generation, democracy was a "lighthouse" (*faro*) illuminating "legislation, morality, education, science, art, and also fashion." Fashion was significant because it fostered both a habit and a sign of novelty.[148]

In June 1838, Echeverría urged his colleagues to create profound change in Buenos Aires's society, which led to the creation of "Young Argentina" following the guidelines of "Young Italy," founded by Giuseppe Mazzini (1805–72) in 1831.[149] Mazzini's program contained a regeneration of revolutionary republicanism to form an "independent sovereign nation of free men as equals," which led to the creation of similar "Young" organizations not only in Buenos Aires but also in Ireland, Switzerland, Poland, Ukraine, France, and Germany. Mazzini warned that revolutionary societies needed to avoid admitting "heterogenous elements into their ranks," and should possess a "defined program" to avoid "discord" and have a "unity of action and aim."[150] He understood Young Italy as "Republican and Unitarian," and believed that "the progressive transformation" of Europe was directed toward "the enthronement of the republican principle" that had failed during the previous revolutionary wave.[151] Unitarianism was needed over federalism because without unity there was no "true" nation or strength.

Mazzini's had significant revolutionary influence in the 1830s; in 1834, while in exile in Switzerland, he created "Young Europe" with other political refugees from Poland and Germany; its principles were declared to be "an association of men believing in a future of liberty, equality, and fraternity, for all mankind." The members desired to consecrate "their thoughts and actions to the realization of that future." Humanity would only be constituted "when all peoples" of which the nation was composed had acquired "the free exercise of its sovereignty," and would be associated "in a Republican Confederation" to be governed and directed by "a common Declaration of Principles and a common Pact, towards the common aim—the discovery and fulfillment of the Universal Moral Law."[152] As we will see, this universalism coexisted with the scientific racism that emerged in the 1840s.

5

Racial Man

In 1850 Domingo F. Sarmiento (1811–88), another member of the Generation of '37, was among the leaders of the opposition to Juan Manuel de Rosas. He was living in exile in Chile, where he published *Argirópolis*, a proposal for creating a new capital of Argentina to be constructed after the governor's defeat.[1] His country, he wrote, would be called the Confederate States of the River Plate, and the city to be created would eliminate the problems generated by the interests of Buenos Aires every time there were plans to unify the nation. Distinctive in this narrative, and in most of those written by Sarmiento during this period, is the weight given to racial ideas in his descriptions of the world order, which included the renewal of colonialism, and a modern nation that was now understood as the establishment of a race, or a homogeneous group with a glorious history of reproduction and permanence.

As noted, race was an important consideration during colonial times, but early revolutionary leaders rejected colonial castes and lineages as the foundation of the new nations, even though they did not consider themselves equals of enslaved people or the Indigenous population. The emancipatory will of the leaders of the independence movements was based on erasing colonial society to build one with no links to the past. However, by the 1840s, things were different; European invasions of the Americas and other continents had been revived, bringing with them violence and efforts to exterminate native populations. Once race became associated with the modern nation and the formation of a modern national population, masculinity started being understood differently. Reproducing the "right" people, while exterminating bodies that were unable to represent modernity, became part of the civilizing mission by mid-century.

The early revolutionary process in the Americas was motivated by the desire to create new republics and to end colonialism, which was not deemed a modern system by many intellectuals who followed the Enlightenment such as Simón Bolívar or José de San Martín. But radical revolutionary politics were ideologically less successful in Europe; the anti-Jacobin conservatism that challenged egalitarianism and universality in England began a process that accommodated ideas to more limited goals, which was also linked to the development of imperialistic politics over time. Thus, the emancipation of the American colonies did not end European colonialism or its accelerated exploitation of other parts of the planet by nations invested in the growth of free trade and the traffic of commodities.

Sarmiento had noticed in 1841 that "almost all Asia" was British property. The "English race" also occupied most of Africa; meanwhile, in Buenos Aires, Rosas was carefully killing "citizens of the Spanish race" every day to terrorize the local population.[2] In geopolitical terms, "the dignity and future position of the Spanish race in the Atlantic" demanded the creation "of the body of a nation" that could challenge in the future "the Saxon race of the North in power and progress." Unlike other Spanish American nations, the Argentine confederation could be that future power, bearing in mind its wealth if it removed the "tyrant and his rabble" from power.[3] In exile, Sarmiento imagined that even if England were to disappear as a nation later, "the calculation of its population reproduction and the increase of the population of this race" meant that it would leave behind "so many nations in all the terrestrial globe" that its continuity would be assured.

It was not clear how South America would be saved "from this universal invasion" that was happening everywhere.[4] As Sarmiento's narrative described, after the initial wave of European support and enthusiasm, a feeling of disappointment set in about the new Spanish American nations. The invasion of foreign territories everywhere was inspired by "*the pity caused in Europe to contemplate that such a rich soil is in the hands of peoples who do not know how to govern themselves, who do not have enough industry or population.*"[5] In his view colonialism was the result of "pity" and not of greed and exploitation. But not all Europeans were able to take over the American territories that were badly

Figure 5.1. "The Slave Women of Buenos Ayres Demonstrate They Are Free Thanks to Their Noble Liberator." D. de Plot, May 7, 1841. Courtesy of the Museo Histórico Nacional, Buenos Aires, Argentina. Text: "Not a single woman slave will wail in chains anymore on the River Plate. Their bitter crying ended when the humane Rosas became governor. Proud of their liberty, compassionate and generous. Rosas lavished this precious gift on the unhappy African."

administered by locals. Sarmiento believed England was the only country that had all the characteristics needed for modern colonization.

Due to the wars of independence, many new nations, such as Mexico, rejected foreigners, but "in Montevideo and Buenos Aires" European immigration had increased the number of inhabitants, which demonstrated that these cities had already started the process of diluting their colonial population.[6] The diverging path of patriotism created by those in exile started gradually; Romanticism had created historical narratives praising the formation of a people/race that by this time characterized the modern nation, which facilitated the racialization of peoples everywhere. But in Argentina it was also related to the political frustration felt by Rosas's enemies when confronted with his continuity and popularity. The exile community attributed its failure to govern to the support received by the governor of Buenos Aires from a racialized mob of Indigenous and mixed-race individuals predisposed to chaos and violence. The Black population, many still slaves, viewed him as their liberator and remained fiercely loyal.

The Rosist system, Sarmiento declared, consisted "in making the Argentine Republic a bag held only by him [Rosas]," and he would one day present himself as "the gravedigger of the Spanish race in South America, laughing at the latter's degeneration and ignorance." Sarmiento implied that the peoples of South America were "witnessing in an impassible way the slow agony of the populations that did not even ask for freedom anymore, being satisfied only by eating in peace."[7] But if their immediate interests did not move them to recover their rights, this "decrepit race condemned to disappear," should at least prevent the dismemberment of the republic.[8] The replacement of the local population so as to outnumber those who opposed civilization in Argentina was based on the belief that "the emigration of the excess of population from old nations to new ones," would "increase strength and produce in one day the work of a century." This was the way in which the United States had developed itself, "and this is how we will grow ourselves." Those in North America had the benefit of being the descendants of "manufacturing England," in its "national traditions, education, and their racial propensities to elements of development, wealth, and civilization," and they did not require external aid to advance.[9] Nationalizing foreigners was a quick process to ensure that the transformation would be successful. "Let us make the Argentine Republic the fatherland of all the men who came from Europe; let us give them the freedom to mix with our population, taking our jobs, enjoying all our advantages. This is what happened in North America, which had three million inhabitants when it became independent and now has twenty-five [million]; which had only thirteen states and now has twenty eight, many of them populated only by emigrants."[10] This liberalism was accepting the idea that the country's women "mix with the population of countries more advanced than us, so they can communicate to us their arts, industries, activity, and aptitude for work," believing that these traits were inheritable. The role of leaders like Sarmiento was to "make things easier for them, attract them to our soil, help them to become established, and make them love our country, so others would hear about their prosperity" and would also move to Argentina. Men needed "to solicit, seduce, and offer advantages" to European men to make them decide to establish themselves in the country.[11] This narrative proposes a racialized masculinity that gave up the most basic privileges of traditional conceptions of manhood that were based on the control of women and protection of men's rights to be able to reproduce and continue.

HISTORY, SCIENCE, AND THE MODERN NATION

The 1830s represented a decade of generational renewal for a young cohort born around 1810, but, at the same time, this renewal was the result of the failures of the Enlightenment to build a permanent model for the modern nation. The problems, creating institutions, morals, and manners were still the same, but now these traits were seen as attached to the organic transformation of peoples over time—they were the result of time and history, as explained by Herder. In part, this explains how the concept of "people" started to change from a revolutionary and disruptive notion based on the emergence of the patriot/citizen, to a masculine concept related to biological continuity and reproduction, competition, and war. While the earlier model of a citizen praised the emergence of the individual, now the idea of people as a race emphasized a collective identity that was inherited, shared, and protected by its members. Rights were important not only to protect the individual but also to preserve a group's bloodlines.

The sources that explained human diversity at this time were many and included comparative historical linguistics. In 1788 Sir William Jones had proposed a connection between Sanskrit, Greek, Latin, Persian, Celtic, and German languages, and in 1813 Thomas Young used the term "Indo-European" to describe the populations linked by these languages. In 1831 John Cowles Prichard introduced the term "Indo-Germanic," and, with the development of comparative grammar and the beginning of what would become modern linguistics, the idea of language families, such as Aryan, started being used among antiquarians, archaeologists, and those interested in literature.[12]

The conception of a humanity divided by families was more chaotic in the natural sciences. At this time, there was no clarity about human origins and variations, and the abundance of different speculations in this area of research gradually created confusion. Carl Linnaeus, Georges-Louis Leclerc (Comte de Buffon), Johann Friedrich Blumenbach, Georges Cuvier, and Conrad Malte-Brun were frequent sources when the topic was addressed, but they all had very different interpretations regarding how modern humans had emerged. Linnaeus defended the existence of four races: American, European, Asiatic, and African. Buffon believed instead that there were six divisions: Laplander (the polar nations), Tatar (eastern and central Asia), Southern Asiatic, European, Ethiopian, and American; later he reduced the number to five because the Laplanders and the Tatars were considered part of the same group. Blumenbach adopted Buffon's classification but changed the names and the geographical distribution; he used the term "variety" instead of race and analyzed the existence of five groups: Mongolian, Malayan, Caucasian, Ethiopian, and American. Cuvier proposed only three races: Caucasian, Mongolian, and Ethiopian, while the geographer Malte-Brun identified sixteen races, one of which was American.

There was no agreement even on the unity of the human species; Linnaeus, Blumenbach, Cuvier, and many others believed that all races had a common origin, but many others rejected this view. In addition, an understanding of variation in nature was influenced by Alexander von Humboldt's system of organic unity that linked all living organisms, and in 1809 in his *Philosophy of Zoology*, Jean-Baptiste Lamarck argued in favor of the gradual transformation of all living beings in his conception of progressive evolution.[13] Geoffroy Saint-Hilaire adopted this belief in the mutability of species proposing that cosmic influences could trigger sudden changes in the embryo, while he also defended the unity of all organic beings.[14]

Another serious problem was a lack of clarity as to how heredity worked. Blumenbach, for example, thought that the different properties of both parents were "on the whole pretty well blended in the offspring;" it was possible "by breeding successively from offspring and one of the original parents" to produce "an offspring exactly resembling this parent." It was said that some dissolute Europeans had begun "with a black woman and copulated with their offspring till they made her the great grandmother of a white." National features and character arose "from a nation marrying among themselves," and their characteristics would be more marked "in proportion to the rarity of connection with foreigners," which was exemplified by "the amazing peculiarity" of the Jewish race. The "advantage of crossing breeds" was well known and could be explained "by the transmission of the parent's qualities." If "an unfavorable deviation in structure or constitution" was transmitted, and "the descendants who receive it hereditarily intermarried," the deviation was doubly enforced in their offspring; but if a connection was made "with another family or breed" it was, "on the contrary diluted." Blumenbach indicated that "the soundness of breeds would require crosses."[15] In addition, physiology was developing different methodologies and practices to explain racial changes among humans.

Physiologists' research on the brain led to a new discipline that had great importance in the development of racial ideas in the Americas. As mentioned, phrenology was created by Franz Joseph Gall (1758–1828) and his assistant Johann Spurzheim (1776–1832) during the first quarter of the nineteenth century, and was based on three principles: the brain was "the organ of the mind"; it was made up "of a number of separate organs, each related to a distinct mental faculty"; and the size of each organ was "a measure of the power of its associate faculty."[16] It spread around Europe, but the center of the opposition was in Edinburgh where in 1815 phrenology received a strong rejection that reduced its popularity among those teaching at the university because it challenged the traditional philosophy of common sense that dominated in higher education.[17] Regardless, it became an important source of popular science.

Spurzheim understood phrenology's leading principles to be observation and induction. These principles posed a challenge to the prevailing view of perfectibility because they focused on "what man is, and not what, according to prejudiced opinions of some philosophers, he ought to be."[18] Unlike other Enlightenment philosophies that were popular in Edinburgh, phrenology denied the idea that education produced faculties, believing that it only worked among students who already possessed the conditions of their manifestation. Education was "nothing but exercise, cultivation and direction," which contradicted the view of children's minds as tabulae rasae. From this doctrine it followed that "material organs produced the difference of the manifestation of the mind; man did not have any power upon his faculties," the latter were conditioned by material organs.[19]

According to this interpretation, man was submitted "to the general law of living nature" and like other organized beings he participated "in the properties of his parents so far as their organization is propagated," which made marriage extremely important because of its influence on heredity. For Spurzheim it was a pity that there was greater care "of the races of our sheep, pigs, dogs, and horses, than of our own offspring." To perfect mankind, the same method should be used, first taking care that "the germ be good" and then educating the child both morally and physically because man had two natures that battled each other: "man as animal, and man as man; or man endowed with faculties common to man and animals, and with others proper to himself."[20] By this time phrenologists were discussing human race without implying racial divisions; individuals of all classes, races, and genders expressed superior or inferior qualities, and because of this phrenologists needed to analyze every individual case.

From the 1820s to the 1830s the progress of phrenology was due mainly to popular public talks and debates in which a significant number of women participated. The discussions started connecting with political and social concerns that attracted those who were not devoted to medicine. For example, the Scotsman George Combe (1788–1858) became the main phrenologist in Great Britain, albeit without completing a degree at the university. In 1819 he published his defense of phrenological studies explaining that human faculties and functions were "radically the same in all the human race"; but they differed in their combinations "and in their degrees of energy and activity"; genius was "nothing but eminent energy and activity of the faculties which form ideas," unlike Romantic belief.[21]

In 1835 Combe repeated the established belief that heredity worked in a way in which features and constitution followed "more generally the father." Breeding of horses and oxen indicated that "great importance was given to the male," and among humans, "the complexion chiefly follows that of the father;" the offspring of a Black father and white mother was "*much darker* than the progeny of a *white father* and a dark mother." This curious observation, which should easily have been disproven by observation, was one of the tenets of phrenology. It was based on "a general fact" related to the "stock," whereas the mother was not impressed "by her own color, because she does not look on herself; the father's complexion must strikingly attract her attention, and may, in this way, give the darker tinge to the offspring."[22] The importance of males attracting female attention through color would resurface years later when Darwin introduced sexual selection.

Slavery was the greatest immorality affecting both Great Britain and the United States, according to Combe. Defending this institution because "Negroes are incapable of civilisation and freedom" was not relevant because their heads presented "great varieties of moral and in theory intellectual devel opment," which could be seen and "appeared fully equal to the discharge of the ordinary duties of civilised men." But the race had never received justice from its European and American masters, "and until its treatment shall have become moral[,] its capabilities cannot be fairly estimated," and the judgment against it was premature.[23] The white race was "exclusively to blame for the origin of the evil and for all its consequences"; it was urgent for its members to recognize that "the natural laws never relax in their operation; and the existing evils" would increase until a remedy was adopted. If nothing were done it was probable that a "war of extermination between the black and the white population" would take place, or that enslaved people would create a kingdom for themselves within an existing state.[24]

It is important to note that by the 1830s the concern driving the European interest in race was related to the pressures to form the modern nation in territories that were ethnically diverse and had populations with different historical trajectories, which called into question the association between national borders and the ethnic diversity of European countries. In the case of Combe, himself Scottish, the Celtic race was his preoccupation; Colin Kidd recognizes that in Scotland "the obsession with racial classification led to other demarcations being drawn between the different categories of white peoples," as started to be the case all over Europe by this time.[25]By the 1830s phrenologists' assumptions gained them even more enemies in Europe. In a letter sent by Charles Darwin to a cousin, William Darwin Fox, he mentioned that he had dined with Sir James Mackintosh and talked about phrenology. The latter had "entirely battered down the very little belief of it" he had, and it became clear that only supposing that education had "any effect in decreasing the power of any organ of the brain," could one believe in the truthfulness of this discipline.[26] Heidelberg physiologist Friedrich Tiedemann (1781–1861), who had studied with Blumenbach, Schiller, and Cuvier, gave phrenology the hardest blow. He was a Foreign Fellow of the Royal Society of London and of Edinburgh and in 1836 had concluded that there was no perceptible difference between the brains of the Europeans and "the Negro." The internal structure was the same, and the "Negro brain" did not exhibit "any greater resemblance to that of the ourang-outang than the brain of the European, excepting, perhaps, in the more symmetrical disposition of its convulsions."[27] Regardless, Combe rejected this evidence because it assumed that animal and moral feelings have no seat in the brain, an idea that diverged from phrenological principles.

Human diversity presented a real problem for the development of the nation because it demonstrated that the unity existing in parts of Europe was the result of a violent process that had shaped nations and continued to do so. A new historical narrative explained this process, providing an explanation of national origins over time. In the British Isles the most influential figure was Sharon Turner (1768–1847), who wrote a history of the Anglo-Saxons in volumes that appeared from 1799 to 1805. The introduction explained that by then philosophy had rejected "the vulgar error of antiquity" affirming that men had "sprung fortuitously from the earth." Instead, at this time, the modern population was understood as "the result of emigration from some primeval residence," and individuals such as himself could trace from "historical documents the colonization of many parts of the world." If that failed, evidence could also be found in "analogous manners and language." The most intelligent scholars had "discredited the unnecessary fable of various original races, as well as of spontaneous animal vegetation."[28] Nations were the result of branching off "from preceding nations sometimes by intentional emigration, and sometimes by accidental separation."[29] Turner located the origin of human population in Asia, from which there had been a migration to Europe, and from the continent to the British Islands. In 1832, Turner published the first volume of *Sacred History of the World*, in which he argued that the British Empire's future depended on maintaining "its reasoning mind firmly attached to the great Newtonian principle of Divine causation of all things," ensuring its scientists remained "in the foremost ranks of intellect, honor, and celebrity."[30] Science and nation needed to remain interconnected.

Augustin Thierry (1795–1856), a French historian who had started his career as the secretary and collaborator of Henri de Saint-Simon, was an admirer of Turner and used his work when writing the history of the Norman invasion of England that appeared in 1825.[31] He belonged to the Romantic school of French historians that, according to Lionel Gossman, "renewed the writing of history in the early nineteenth century" and was "thoroughly committed politically and openly active in contemporary politics."[32] Thierry's liberalism and his war against the Bourbon restoration were the results of his own personal history; he saw historical research as deeply tied to contemporary experiences and concerns, asserting that a historian's grasp of the past arose from personal experience, which shaped what questions to ask the past.[33] His book about the Norman Conquest emphasized that the Anglo-Saxons had been invaded and defeated by those who lived in what was now French territory, pointing out that the nation was not the result of unification but of conflict. As Krzysztof Pomian indicates, François Guizot and Thierry introduced a novel idea; for them, the nation was not "a unified, homogenous entity personified by its king and court but a composite of various groups in conflict," it was the result of conflict that over time led to progress.[34]

Philippe Roger estimates that "it was not until Augustin Thierry that the Anglo-Saxons would regain their strength and majesty. And even then, it would be in the role of the conquered." But this ethnic Romantic style "remained impervious to English or American attempts to transform Anglo-Saxon studies into a laboratory of ethnic hierarchies." Thierry used previous historical sources to write, but he did not transfer "the myth of superiority that had become the heart and soul of English and American Anglo-Saxon legend." In fact, he followed Turner, who believed in humankind's racial unity, and had "quarreled with racist Anglo-Saxon scholars such as John Pinkerton."[35] Thierry believed that modern nations were the result of racial conflict that had led to the formation of modern populations.

Reginald Horsman's analysis reveals that the British narration of a past related to the invasion of several groups over many centuries was transformed between 1830 and 1880, when "the dark period in the national history is gradually transformed into one of apotheosis."[36] By the 1830s the emergence of a narrative that emphasized national unity led "to describe the people living within the bounds of England," but, at times, it was also used to describe "a vague brotherhood of English-speaking peoples throughout the British Isles and the world."[37] This view placed Anglo-Saxons as the most modern, in the sense of updated, of all human races.

In France, Michelet addressed similar issues of national and human origins in the 1830s. His *Histoire de France* began publication in 1833 and was completed over eleven years in nineteen volumes. The history of his country is the result of a blend of "chemical principles" akin to oil and sugar. All the races that had been part of this blend were known, but a unique and distinctive quality had nevertheless persisted, which required explanation. This quality became even more significant when the focus was the origin of a living, dynamic entity like a nation, which had a union capable of "internal development and self-modification!" The process that explained how a country continually evolved, was the core subject of French history.[38]

Michelet also redefined materialism and spiritualism; for him, matter, inherently divisible, fostered disunion and discord, which denoted that material unity was a contradiction and, in governance, meant tyranny. The spiritual realm had the capacity to forge unity because it bound, encompassed, and, in essence, loved. Love and spirit forged unity; only through spiritual means could a nation be united (148). Isolation and "rejecting all foreign ideas" led to being "incomplete and weak" (433). France owed much "to foreign influence; all the races of the world had contributed to endow this Pandora" (129). This allusion to Greek mythology portrays France as a vessel filled with unique gifts—its people—that formed it.

By the 1870s, Michelet's historical interpretation would converge with the new evolutionary science. In a letter he sent to Darwin in 1872, he wrote that "a submarine passage" would be created, allowing France and England to "finally recognize their proximity, or rather, their ancient unity." The English naturalist and his followers were contributing to renewing this union, reminding both nations of their proximity and "kinship of race and spirit." Both countries shared "an observation more precise than that of the Germans, but in a vigorous simplicity that frees the great English and French minds from the scholasticism that delays and often misleads Germany in countless ways." Darwin reciprocated this respect, contributing to the funds for Michelet's tomb at his grave in 1877.[39]

National concerns and the problem of unity in the context of human diversity were expressed in a second wave of European colonialism that began after the end of the Napoleonic wars and became prominent by the 1830s. In this background, we need to understand the impact of the Slavery Abolition Act of 1833, and, in 1837, the creation in London of the Aborigines' Protection Society mostly by Quaker leaders such as Thomas Hodgkin (1798–1866). The society's purpose was to protect the rights and promote the civilization of native populations that were threatened by the advancement of white men over their territories due to colonialism. Hodgkin traveled to Paris to try to open a branch of the society there, which led him to work with a member of the Royal Academy of Medicine, William Edwards (1777–1842), who, in 1839, created the Société Ethnologique de Paris (Ethnological Society of Paris) for "the study of the human races according to historical traditions, languages, and the physical and moral traits of each people," expanding the study of human origins to other fields besides natural history.[40]

Unlike phrenology, the Ethnological Society attracted members of the French scientific establishment, such as Michelet, Marcellin Berthelot, Alcide d'Orbigny, and Saint-Hilaire. The elite of the European scientists and scientific institutions were also represented, including Hodgkin and Hermann Schlegel, and those of the Americas: Pedro de Angelis from Buenos Aires; Ramón de la Sagra, director of Havana's Botanical Gardens; Teodoro Vilardebó, director of Montevideo's Natural Science Museum; William Brown Hodgson, ex-consul of the United States in Algiers; Peter Claussen from Brazil; Martin de Moussy, a correspondent member from Uruguay, and Agostino Codazzy from Venezuela; Peter Du Ponceau, president of Boston's American Philosophical Society; George Ticknor, the author of a history of Spanish literature, from Boston; the Brazilian ministers of Foreign Affairs and Navy; the secretary of the Brazilian emperor, and the president of the Institute of History and Geography of Rio de Janeiro.[41] Its honorary members were equally respected, including among them Alexander von Humboldt and the presidents of London's Royal and Asian Societies.[42]

Edwards was born in Jamaica of British parents, studied in Bruges, moved to Paris in 1808, and became a French citizen twenty years later. The first issue of the Ethnological Society's memoirs appeared in 1841 and included a reprint of his book on race published in 1829.[43] It was originally written as a reply to Augustin Thierry's brother, Amédee, who had written *History of the Gauls*, which linked the present population of France to the country's past. This historically oriented approach to studying human races stood in stark contrast to the taxonomic frameworks of Linnaeus (1707–78) and Blumenbach (1752–1840).[44]

Edwards's writings recognized "the confusion of times and authors" that prevailed in the understandings of human races and how the French had established historical ones that could be unrelated to those recognized by natural history.[45] Blumenbach's classification was not helpful in clarifying the development of a nation because each one of the varieties/races "embraces and confuses too many nations." Languages, though, were helpful in tracing the "unbroken connection between the old inhabitants and the new" (4). In terms of physical development, nature sometimes confused, sometimes separated, racial types. There was little consistency in the way each progressed; while nature was "always busy producing," its efforts could lead to both conserving and destroying racial types (22).

This progression of racial development was a process of adjustment to nature and historical developments without a clear plan; "human nature must have a great force of resistance to know how to triumph" over the challenges of historical cycles. But "not all peoples were equally able to resist in the same way." While climate in nature tended to conserve the same type, the "crossing of the races" was more powerful because acting on "intimate organization," it presided "over the first formation of being" and seemed always "having to alter the forms" (18). As a result, like Herder, Edwards supported the idea that modern European nations were of mixed race.

In 1832 London's popular *Fraser's Magazine* published an analysis of Edwards's book. The magazine, founded by Hugh Fraser, had appeared two years earlier and was a Tory (conservative) venue, having as the main editor William Maginn (1794–1842). This anonymous article conveys that Edwards helped to dispel "much of the confusion encountered" when reading historical narratives about European nations. Physiological science could separate the races "more effectually than the best historical data, if there be any truth in the assumed permanency of varieties among human beings," as was known to be the case "in other species of animals, and in the vegetable creation." Humans were affected by the same laws applied to other species.[46] Form here was "assuredly hereditary, when naturally acquired." Among civilized ladies, "the barbarous custom of altering the natural form," using a corset that compressed the chest was an example of unnatural bodies (675).

The article explained that hereditary characters of form were "those which originally" belonged to "the foetus," and grew with its growth, as Camper had proposed. It was in the "*figure and general proportions of the body*" that one could find "some remarkable variations among different nations." Many believed that the natural diversity of human form was the result of "deviations from *one primitive type*" and this included skin color, texture of the hair, stature, and the relations of parts (675). But it was clear by then that the idea of "distinct species among mankind" was contradicted "by every appeal to physiology and zoology." The physical and moral characteristics demonstrated "so close an approximation of individual varieties to each other" that the whole human species had resulted from one common stock "spreading over the earth, and separating into distinct families" over a very long period (676). Edwards permitted an understanding that the modern nation consisted of "many stocks or families, and their different physical characters" were insufficient for "the purpose of tracing out their origin"; it was in "language and moral and physical attributes" that human races seemed "to possess a greater degree of variety than is expressed in Blumenbach's arrangement." It was for this reason that the connection between "ancient and modern nations" had been greatly interrupted (676).

The European colonies in the Americas offered an example of this disruption, in that "the aboriginal inhabitants" were "disappearing and giving place to new languages and races." Europeans might be forming castes "by a medley of interconnexions," and "the white population" had not lost "in its genuine characters," because each nation recognized its proper offspring (677). People established in foreign climates "preserved their types indefinitely, notwithstanding the modifications of temperature and climate, and social intercourse," which put to rest the concerns about Europeans turning black in the tropics. Edwards's application of physiological principles to history filled the voids that existed, confirming and supplying data that validated historical accounts and offering insights beyond what could be gleaned from simply categorizing heads into five types (679). As Efram Sera-Shiar indicates, by the late 1840s ethnologists were reliant on travel narratives to understand the state of the races of the world after centuries of European colonialism.[47]

In 1839 a transformative scientific travel chronicle was published—Charles Darwin's journal of the *Beagle* expedition appeared after great anticipation. He provided examples of concerns about the extinction of native populations, and different timescales of racial human development. He noted that the presence of "civilized people" among "savages" indicated to him that far from a future of unity due to the influence of civilized ideas, the separation among humans had accelerated. The presence of "white men" such as himself in many parts of the world suggested that the "savages" would not exist in the future. "The thoughtless aboriginal," was blinded by the advantages that came from "the approach of the white man," but they failed to notice that the latter seemed "predestined to inherit the country of his children."[48] The description of the different outcomes of human populations due to colonialism appeared in several passages of the journal.

Another note mentioned that in Australia "some even think that [an Indigenous] race will soon become extinct" after hearing about what had happened in Hobart Town.[49] But while he recognized that there were "several evident causes of destruction," there appeared to be "some more mysterious agency generally at work. Wherever the European has trod, death seems to pursue the aboriginal." Whenever one looked, "Americas, Polynesia, the Cape of Good Hope, and Australia," the same result was found, as if an "invisible and powerful force" had a plan to repopulate the world.[50] Far from the Romantic association with beauty and harmony, men are here surrounded by death, greed, competition, and fighting. His first experience with this phenomenon was Rosas's campaign of extermination against the Indians in the desert outside Buenos Aires.

Also in 1839, Combe explained how the existence of "wide differences of opinion" about the origin of human diversity had led some to reject all classification in anthropology, "but the same objections would apply with equal force to the whole range of natural science, which, divested of arrangement, presents an uninviting chaos."[51] This was an accurate observation highlighting the challenges of establishing a precise human origin—challenges that complicated efforts to ground modern nations in the legitimacy of scientific evidence.

AMERICAN RACE SCIENCE

Cameron Strang indicates that in the United States "a more biologically deterministic explanation for human difference did take hold among many Anglo-Americans between [the] 1780s and 1840s," which, by the 1850s, culminated in the creation of a new national science based on racial beliefs.[52] This process can be traced to 1822, when a group of physicians formed the Philadelphia Phrenological Society, and the first American edition of George Combe's *Essays on Phrenology* was published.[53] In 1825 a Washington newspaper published two lectures of Johann Spurzheim taken from the London *Times*, but the editors commented with skepticism that "the rage for examining the human cranium" had "extended itself" in England, and was not infrequently "found associated with the rouge pot and Cologne water in the toilet tables of our belles," indicating that in the United States women were also an audience.[54]

In 1829 a review published in Philadelphia by its Medical Society, reported on Edwards's *Physiological Characters* relating that its author did not lay "much stress on the modifying influence of climate." Instead, he believed that "the physical peculiarities of race may be preserved for a long series of ages, and that even after a people are dead in history, they may yet be shown to be in fact numerous," which proved "their consanguinity or common origin with other and occasionally remote nations" by the possession "of common physiognomical traits." This constancy, exemplified, as usual, by the Jews, indicated the simultaneous continuities and discontinuities that resulted in the existing modern populations. For this reason, it was "very obvious that history may derive useful aid from this department of physiology," as it had already done "from a careful comparison of the languages and dialects of the different nations scattered over our globe."[55] The persistence of traits over a long time was a concern in societies that were racially divided.

It was in the 1830s that phrenology achieved success among the educated elites through the efforts of Spurzheim, who introduced craniology as a methodology useful for determining human variation. Its phenomenal success in the United States was important for the survival of a discipline that was largely rejected by established scientific fields. In 1832, this scientist arrived to teach in Boston, but he died after a few months of living there. Phrenological societies were organized nevertheless, and the discipline was discussed "at the lyceums, in the social and debating clubs, and by the public press."[56] On the day of Spurzheim's funeral, the Phrenological Society of Boston was formed, and in a letter to Blumenbach, the society's corresponding secretary, Samuel G. Howe (1801–76), wrote that the new organization had as its objective "the examination of the principles of science of Phrenology directly; and indirectly all of the physical that has a bearing upon the social, moral, and intellectual conditions of man" (122). The society had 150 members, which made it the "leading society of the United States," consulted by men from the country and abroad (124).

The writer and publisher Nahum Capen (1804–86), a friend, fellow student, and a phrenologist himself, described Howe's indifference "to the metaphysics of Locke, Stewart, and others of the old schools." Spurzheim's work opened Howe to "a new world of mental order and activity such as he had never imagined or realized. He first began to see man in his relations of power and duty, and to comprehend the value of self-knowledge in the cause of education" (154). While many phrenologists were committed to abolition, they did not see other races' children as possessing brains that could be filled with knowledge and culture that they did not possess by nature.

Susan Branson indicates that at the time, opinions about race were shifting in the United States "toward a belief that difference was immutable and unchanging" but wars and external events could drastically affect human populations.[57] The concerns about reproductive rates among races was well established in the United States, as the work of the American Colonization Society, founded in 1816, demonstrates. This association was created to organize the transportation of those who could not become part of the "People" and needed to return to their ancestral African home. In a speech given in 1827 to the society, Henry Clay (1777–1852), a member who was to lose the presidential election against Andrew Jackson the next year, explained the importance of relocating Black people over time.

Clay did not believe in the possibility of a multiracial society; in his view, transporting free Blacks and enslaved people to Africa, with consent of the states, might "lead this country to rid itself entirely of that which is justly admitted to be its greatest curse."[58] He proposed sending six thousand individuals each year for a total of thirty-three years; as a result, a proportion of the "European Anglo-Saxon" population "of twenty million to two million of blacks could be achieved," and in sixty-six years, a proportion of forty million to two million. Through this operation "the black population" would "be kept down" completely. While justice and morality were mentioned in the presentation, it was clear that the foundation of a new colony in Africa served the purpose of separating the Black man "from that race with which he can never associate."[59] Anglo-Saxons needed to avoid the dangers of being invaded, conquered, and destroyed numerically by those who could change their European form, obtained in this continent over a long period of time.

In 1835 the *Annals of Phrenology*, a journal published in Boston, edited by Capen, and created to disseminate the research produced in Europe and the United States, published a review, which had originally appeared in Edinburgh, written by George Combe of William Edwards's book,.[60] It praised Edwards for "discovering among modern nations, the descendants and representatives of various ancient races" that had been considered lost "in the mixture of tribes which followed the various conquests and settlements" that had taken place in Europe.[61] This review also explored what happened when societies had no restrictions on the mixing of races, and discussed the possibility that a race would disappear if its members were in the minority.[62]

Combe concluded by explaining that after reading the book everyone would regret the absence of phrenology in it, "which would have doubled the interest and importance of his discoveries." He implied that by using measurements of the skull and the brain to determine the existence of types, pure types would appear, and then phrenologists would find "likewise the prevailing cerebral development of that race," which led to a plea to scientists to mix Edwards's work with phrenology.[63] This was a stretch, as Edwards did not consider phrenology a science and did not believe in "pure" races. Still, this desire was fulfilled a few years later in the work of Samuel Morton.

In 1840 a review of Morton's *Crania Americana* appeared in an important scientific journal published in the United States, and was also written by Combe, though it was unsigned. He praised the book as "the most extensive and valuable contribution to the natural history of man" that had appeared on the American continent.[64] According to Combe, the reader of the book was left "strongly impressed with the conviction that this branch of science is still only in its infancy." He noted that it was not pure phrenology, but a local interpretation that had some weaknesses. For example, the descriptions of the mental qualities of the different human families were "vague and entirely popular" and there was "scarcely an instance of the specification of well-defined mental faculties, present or absent in the races."[65] But Morton's book confirmed the perception that existed at the time about the future of those who were not white.

> One of the most singular features in the history of the continent [America], is, that the aboriginal races, with few exceptions, have perished, or constantly receded, before the Anglo-Saxon race, and have in no instance either mingled with them as equals, or adopted their manners and civilization. These phenomena must have a cause; and can any inquiry be at once more interesting and philosophical than that which endeavors to ascertain whether that cause be connected with a difference in the brain between the native race and their conquering invaders? Farther, some few of the American families, the Araucanian, for instance, have successfully resisted the Europeans; and the question is important, whether in them, the brain be in any respect superior to what it is in the tribes which have unsuccessfully resisted?[66]

Spanish American colonies' racial culture called attention to the results of racial mixing that had occurred for centuries, which provided evidence of the fluidity of color and features according to ancestry. For example, Conrad Malte-Brun (1775–1826), a Danish French geographer and a founder of the Geographical Society of Paris, authored *Précis de géographie universelle*, published from 1810 to 1829, and at the time one of the most complete studies of the world's geography and populations. As an example of human diversity, he used the system created in the Spanish colonies, based on castes, to control and regulate racial variation. Spaniards were classified as white, and there were also castes of mixed blood that resulted from intermixture with the pure race. A "fresh alliance with the white race so completely obliterates all remaining traces of colour, that the children of a white and a female Quinteron [mixture of a 'quateronne' (mulatta and a European) with a European or a Creole]" were "also white."[67] Whiteness was, as a result, something that could be lost or gained through mixing, which explained the need for a system of castes to maintain social and racial differentiation.

The decisive considerations of Spanish American colonial societies indicated that "more or less European blood, and the more or less fair skin" decided "the consideration which a man should enjoy in society, and the opinion he has of himself."[68] Climate had an important influence on the changes of different races, and next to physiological characteristics, language was the "most indisputable proof of the common origin of different nations," and it was these elements that supported the assumption that the population of the American continent was derived from the emigration of people from Asia.

SCIENCE, RACE, IMPERIALISM, AND MASCULINITY

The relationship among philosophy, science, masculinity, and politics acquired a new dimension by the mid-1830s when expansionism in the United States began to be associated with the country's racial difference from the rest of the American continent and the men it had created. While politically there was a continental agreement by the last quarter of the eighteenth century about creating new nations removed from colonialism and European institutions, by this time the relationship had shifted to one defined by confrontation. The understanding of human diversity and origins now focused on embodiment as a sign of social and political power, precisely something that the independence movements in Spanish America had wanted to end with the aid of science.

By this time, however, phrenological science—a modern field arising from an interest in linking the interiority and exteriority of the body—was being used to promote very different political agendas. This science formed one of the roots of imperialism in the United States, alongside the triumph of materialism over the philosophical principles of the nation—the opposite of what occurred in France with the renewal of spiritualism. This process led to the rise of an American science, ethnology, which marked a new stage in US history. The vision of the United States that emerged in Spanish America connected this science to the image of colonists expanding the nation's territory, and to the defense of slavery, at a time when Europe, led by the British, was working to end the slave trade. More importantly, this national science contrasted with the local philosophy that gave rise to transcendentalism, highlighting the growing ideological split between science and philosophy in this country.

The relationship between phrenology and race in the United States was spread in the popular press, as it had in Europe. For example, an extract of *Phrenology Vindicated* written by the polygenist physician Charles Caldwell (1772–1853) emphasized the differences that existed between African brains and those of Caucasians. The former had "far more of the animal and less of the man" than the latter. If a colony "of the most cultivated Africans in the United States or the West Indies" was planted and "only let [to] be full-blooded," and entirely apart from Caucasian influence, "instead of advancing cultivation and improvement" it would "retrograde and degenerate." In a few generations it would "return to barbarism." Caldwell expressed fears that the Liberian colony in Africa was not improving because "*the condition of man is never stationary*." He affirmed that "no form or degree of education" that man could bestow, "aided by all other *earthly* causes," could "ever raise the African to a level with the Caucasian mind."[69] This article was placed below one that reproduced a poem that was a Texas call to arms published in the *Lousiana Advertiser*.

In 1836 US colonists established the Republic of Texas in Mexican territory, an action that was criticized in strong terms by those who embraced spiritualist positions in the country. For example, the most famous Unitarian preacher, William E. Channing (1780–1842), denounced in 1837 the seizure of Mexican territory by men he called "barbarians" because they had "proclaimed that the Anglo-Saxon race" was "destined to the sway of this magnificent realm; that the rude form of society which Spain established there is to yield and vanish before a higher civilization."[70] Channing found it curious that this development had happened in "a civilized age, and amidst refinement and manners," linked to "the lights of science and the teachings of Christianity, amidst expositions of the law of nations and enforcements of the law of universal love." It was from a "free, well-ordered, enlightened Christian country" that "hordes have gone forth in open day [*sic*] to perpetrate this mighty wrong."[71] He emphatically rejected this view of civilization that supported warfare among different races of men divided between conquered and conquerors; for him this kind of thought was against the faith of civilized Christians and was materialist and based on some dubious philosophical and scientific conceptions.

> It is sometimes said that nations are swayed by laws as unfailing as those which govern matter; that they have their destinies; that their character and position carry them forward irresistibly to their goal; that the stationary Turk must sink under the progressive civilization of Russia, as inevitably as the crumbling edifice falls to the earth; that by a like necessity, the Indians have melted before the white man, and the mixed, degraded race of Mexico must melt before the Anglo Saxon. [. . .] We boast of the progress of society, and this progress consists in the substitution of reason and moral principle for the sway of brute force. It is true that the more civilized must always exert a great power over less civilized communities in their neighbourhood. But it may and should be a power to enlighten and improve, not to crush and destroy.[72]

An article published in London in 1838 about the Romantic writer James Fenimore Cooper (1789–1851), the author of *The Last of the Mohicans*, mentioned that "Americans are proud of being thought of [as] that race [Anglo-Saxon], and no greater affront could be offered to a Yankee than to question his title to an Anglo-Saxon origin."[73] A year later, the same newspaper published an article about Canadian immigration and explained this preference more clearly. The "Americans of the United States apply the term Anglo-Saxon to distinguish the various families of the various branches of the Teutonic family" from the "French and Spaniards who are of another family of nations." While this term might not be correct, it was used because the Anglo-Saxons had given "their name to the English," the settlements founded by them were "all powerful in North America as they seem destined to be in many other parts of the globe," and, as a consequence, it was not "unnatural that the people of the United States should use the more distinguished name of the branch rather than the term Teutonic."[74] In the United States the press reported the linkage of the modern nation to a specific race locally and globally.

In 1842, while in New York, the American historian and linguist John Russell Bartlett (1805–86) founded the American Ethnological Society following Albert Gallatin (1761–1849), an ethnologist and linguist originally from Geneva who was by then a prominent figure after serving as adviser to Thomas Jefferson. This society took up the example of the association established in France. In a speech published by the *National Intelligencer*, Bartlett defined ethnology as related to the origin and history of nations, including an analysis of the physical character of races, their conformation, their languages and national peculiarities, their artistic knowledge, types of government and laws, and manners and customs.[75] Clearly, this definition of the field differed from the French and British ones, which were related to the "protection" of Indigenous societies understood as disappearing because of the intensification of colonialism but not to the nation's formation.

In 1845 a Vermont newspaper noted the popularity of this declared national science. Due to the increased conflicts among nations, the writer asked: "Shall philology and ethnology overturn the nations?" explaining the expansion of pan-Slavism in Europe, and the dangers that it presented for Germany, which announced the possibility of future war if the political independence of the "Slavonians" was established on the basis of race.[76]

In 1842 the Ethnological Society of London was founded to study the natural history of man. A notice about its creation was published a year later mentioning that its purpose was to investigate "the Natural History of Civilized as well as Uncivilized Man." At a meeting at Hodgkin's house, Ernst Dieffenbach (1792–1847), a surgeon from Giessen who had completed research in New Zealand, presented an essay in which he mentioned that among those who were civilized the "distinction of races, as well as their admixture," had been exhibited with some success in maps presented in Germany and France. It should now be the intention of the Ethnological Society "to extend the plan to the whole habitable globe." England had the greatest responsibility for producing this research because its empire extended "over a greater diversity of races than any nation upon earth," which allowed scientists to be brought into "contact with nearly every race" that connected imperialism with science.[77]

In a meeting in 1844, the Arctic explorer and physician Richard King (1810–76), the secretary of the society, emphasized that the London association was following the example of the United States and France and that its intention was to focus on the fact that "whole nations . . . have disappeared" because of the "rapid progress of commerce and colonization, which so eminently characterized the present century." This new historical process had directly affected the development of humans, and a science was needed to understand the changes and transformations.[78] It was then necessary to preserve "a record of the living, and of that which remains of the dead." But King acknowledged that the "means of preserving existing nations" created disagreements, although he affirmed that everyone could agree that "in a philosophical point of view it is necessary to obtain much more massive information" than what was possessed at the time of the "physical and moral characters" of colonial and primitive populations that were in the process of vanishing, which indicated an urgency that forced them to lose no time.[79]

French writers became a good source of ideas that contradicted this thinking. For example, those who defended the role of religion for developing the morality of the nation, such as Victor Cousin, also used the work of Felicité Robert de Lamennais (1782–1854) who established another form of Christian socialism that was the foundation of France's moral stance. His newspaper, *L'Avenir*, whose motto was "God and Liberty," became the voice of liberal Catholics in the 1830s, and became known in the Americas. An 1834 translation of this author's *Paroles d'un croyant* (A Believer's Words), which had been banned by the Catholic Church for attacking its orthodoxy, was published in New York.[80] In the preface, the book's translator explained that *L'Avenir*'s principles demanded that liberty of religion, education, press, association, and the right to vote be extended to the mass of people, and political centralization be rejected. A Spanish translation was published the same year, taken from the eighth French edition, which contained an appendix on liberty and absolutism.[81]

The book's importance was partly related to its fight against materialism and racial egotism. According to Lamennais, individuals and nations were united because they were all God's creations, and those who separated family from family, and nation from nation, dividing what the creator had put together, were doing the work of "Satan."[82] In this type of thinking the interest was not in the law of nature, but the law of God, justice, and love; He recommended avoiding the statement that "he is of one nation, and I of another people," because "all nations have had on earth the same father, who is Adam, and have in heaven the same father, who is God."[83] The defense of human species' unity was reinforced by the scientific idol of the Romantic generation, Alexander von Humboldt. In 1845 the first volume of Humboldt's *Cosmos* was published, and in the same year English translations appeared in London and New York; a year letter several editions appeared in French, and in 1851 a Spanish translation was released.[84] In the edition published in New York, the translator's name is not included, perhaps on purpose because it was not good; the same happened with the English edition, which Darwin deemed written in "wretched English" that he barely understood.[85]

The intention of *Cosmos*, in Humboldt's translated words, was to "comprehend the phenomena or corporeal things in their general connection; to embrace Nature as a whole, actuated, animated by internal forces."[86] Curiously, this is reminiscent of what Caldas had written about climate in 1808. But, at a time in which scientific fields were claiming independence to observe nature within specific interests, this wholistic view of nature seemed outdated and speculative. The connection between poetry, philosophy, and the sciences remains in Humboldt's work, but in the 1840s it was not typical of scientists' writing. Darwin himself was disappointed after reading the first two volumes, even though he read them with such intensity that the book was returned damaged. He wrote that overall, he was "rather disappointed with it; though some parts" he considered admirable, but he found that there was too much "repetition of the Personal Narrative," and "no new views," though he was pleased by the citations of his *Journal*.[87]

This German approach to science was much more appreciated by those interested in radical politics. In 1845 the American journalist Charles A. Dana wrote a very positive review of *Cosmos* that appeared in the journal of the Book Farm Phalanx, a utopian organization for communal living that existed in Massachusetts from 1841 to 1847. This book was "great" because it was "an eloquent and profound expression" of the "great idea" that had inspired the era of progress: "the idea of Unity." Since this community embraced Charles Fourier's socialism, the reviewer found Humboldt's political and social vision closer to his beliefs. Therefore, the book's combination of the physical sciences and poetry to show "that harmony reigns in the material creation" was highly praised. It was the opposite position of "the war, the falseness, and the misery" that prevailed even in "so-called Christian" countries.[88] At a time of racial division in the United States because of slavery and the imperialistic politics being pursued, Humboldt's writing and embrace of the unity of living things inspired those who were aligned with a spiritualist perspective.

The discipline of ethnology in the United States completely opposed the conception of racial unity in the origins of species, and by mid-century this field had become a science in service to the country's racial politics. The publication in 1854 of unedited papers written by Morton included additional contributions from celebrated scientists such as a disciple of Alexander von Humboldt, Louis Agassiz (1807–73), who had moved to teach in the United States in 1847, but, unlike his mentor, was a defender of polygenism.[89] In the preface to Morton's book, George Robins Gliddon called ethnology the "cis-Atlantic school of Anthropology," an alignment that recognized the compatibility of US theorizing of race with European imperialist politics if not with views about slavery that predominated in some European countries.[90]

In another section, Henry S. Patterson (1815–54) explained the reasons that ethnology should be "eminently a science for American culture." It was in the United States where "three of the five races, into which Blumenbach divided mankind," had been brought together "to determine the problem of their destiny as they best may." Meanwhile, the immigration of Chinese laborers to California was threatening to bring the country "into equally intimate contact with a fourth." This multiracial presence required a way to manage people that depended "in great measure, upon their intrinsic race-character." It was commonly repeated that "the contact of the white man seems fatal to the Red American," whose tribes were fading and threatened "by ultimate extinction." The Black population was described as "thriving" under "the shadow of his white master," falling "readily into the position" assigned to them, which accounted for their existence, reproduction, and physical well-being.[91]

Ethnology was a distinctive "American science" that controlled the outcomes of population growth and racial mixing, which explains why the introduction to *Types of Mankind* described the new scientific field as divided into "two principal departments, the *Scientific* and the *Historic*."[92] In another chapter, Josiah Nott boasted that there did "not exist, in any language, an attempt, based on the highest scientific lights of the day, at a systematic treatise on Ethnology in its extended sense." Morton had been the first to conceive "the proper plan," but his death had crippled the further development of the field; the publication of *Types of Mankind* was an aid "in suggesting the right direction to future investigators" (50). This introduction also made a case for this science as a defender of the country against the foreign abolitionist influence.

Nott explained how John C. Calhoun (1782–1850), in charge of foreign relations from 1844 to 1845, complained that "England pertinaciously continued to interfere with our inherited Institution of Negro Slavery," an act to which he responded using the works given to him by Samuel Morton since they supported "his personal observations in America, on the Anglo-Saxon, Celtic, Teutonic, French, Spanish, Negro, and Indian races," now corroborated by "modern science" (51). According to his version of the events, Calhoun had sent writings about ethnology in the diplomatic correspondence directed to the United Kingdom's Foreign Office, which created a controversy in the English press.

The conflict was resolved with an announcement from the government of Great Britain stating that the country "had no intention of intermingling with the domestic institutions of other nations." But the alleged influence of American science in England was strong perhaps because "no class of men" understood better "the practical importance of Ethnology than the statesmen" of this country, "yet from motives of policy, they keep its agitation studiously out of sight" (52). Although I did not find references to this incident in the British press, many articles opposed the annexation of Texas by the

United States, primarily due to concerns about the expansion of slavery in the Southwest, which most European countries opposed. Mexico had abolished slavery in 1829, with exceptions for Tehuantepec and Texas, the latter due to pressure from Anglo-colonizers who owned enslaved people. In 1837, slavery was fully abolished, which contributed to the Texas Revolution. A British news article reported that the conservative French newspaper *La Presse* had covered an agitated conversation between the British minister in Washington and Calhoun, triggered by the Texas situation. Allegedly, the former claimed that if Texas were annexed, England would fulfill its "designs" over Cuba that had been refrained from solely at the US request. Such articles fueled anger among Spanish American leaders, sparking localist sentiments and strong opposition to this renewed imperialism, which, shockingly, now also involved the United States, a pioneer in fighting colonial power in the Americas.[93]

RACE SCIENCE IN SPANISH AMERICA

The Argentine Confederation also experienced the aggressiveness of the United States in trade and the control of territory. As discussed in chapter 2, the report of Juan de la Piedra recommended colonizing the Malvinas Islands (Falkland Islands in English), a strategic point of navigation between the Atlantic and Pacific that had been used in the past to hold prisoners. In 1826 Luis Vernet, who was born in Hamburg, finally obtained a grant to start colonization. In January 1828, the legislature of Buenos Aires named him governor of the Malvinas and allowed him to initiate the process. Frédéric Lacroix, a French traveler, used a letter written to Captain King that was in his possession to understand the state of the colony around 1830; the guest was surprised to encounter in this distant territory a house in which he enjoyed "music, singing, and dance" and a "good library with works in Spanish, German, and English." Vernet had brought to the island and enslaved fifteen people whom "he had purchased himself" with the condition that "he would teach them some useful trade and grant them their liberty after a few years of servitude." They were between fifteen and twenty years old, and "seemed to consider themselves happy," according to this account.[94]

About one hundred people were living on the island, including twenty-five gauchos and five Indians. There were also two families from Holland, two or three from England, and one German; Spanish and Portuguese merchants made up the rest of the population. The gauchos were mixed-race individuals from Buenos Aires, whose boss was French.[95] G. T. Whitington was one of the Englishmen who had received permission from Vernet to obtain land,

and in a book he authored he mentioned that in 1831 three sealers from the United States resisted the authority of the governor, "in persisting wantonly to destroy indiscriminately the rookeries of seal." The men were taken to Buenos Aires, "where they were condemned as lawful prize, in December."[96] In reprisal, the authorities in Washington decided to send the Lexington corvette, "from which a considerable force was landed on the Island, who literally destroyed the whole of Don Louis Vernet's establishment," shocking the leaders of Buenos Aires.

According to Whitington, it was rumored that "the American government actually contemplated taking possession of the Falklands" (Malvinas), with the objective of "establishing there a naval depot to protect their fisheries and promote their trade with the South American Republics."[97] He believed the rumors to be true because during this period he had received an offer to purchase his grant, which he communicated to London. This possibility "excited immediate attention, and once more England began to take an active interest in the control of the island."[98] The incident also excited the press in the United States that covered the incident.

In April 1832, the *Boston Courier* reported that this act had caused "some excitement in Buenos Ayres," and that the government had suspended "the functions of the United States Consul." In a speech before Congress, given in December 1831, President Andrew Jackson explained that the incident had been the result of "acts injurious to our commerce, and to the property and liberty of our fellow-citizens." He also emphasized that regarding the inspection of the world, "we give no advantage to other nations, and lay ourselves open to no injury."[99] This response made it clear that the action had been legal, not a violation of established international law, and was officially supported by the president.

The reaction was also explosive in Buenos Aires. In February 1832 *Gaceta Mercantil* denounced what had happened as a "scandalous violation of international law" (*derecho de gentes*) and "a cowardly attack" on a place that was peaceful, which demanded "a reparation of the offense that had [been] inflicted on Argentine honor."[100] On February 14, the governor of Buenos Aires, Juan Ramón Balcarce, who was finishing Rosas's term while he was fighting on the southern border, informed the other provinces that the government had confirmed "the truth of the scandalous acts said to have been committed in the [Malvinas] Islands." The colonists had been "assaulted under [a] friendly flag;" and some of the country's citizens were conducted "as prisoners of the U. States, with the apparent object of being tried there." The "unanimous indignation which this odious outrage" had produced was fully justified; and "no doubt the same sentiment" would be felt by "men of honor" in every part of the world. But, Balcarce insisted, it was "impossible that the Government of Washington" had approved these acts since it always behaved up "to the principles of moderation and justice," which led him to expect that the United States would give the government proper satisfaction.

Three days later, Balcarce sent another letter to the governors to notify them that he was "far from believing that the authorities of the United States" could approve of "a transaction as contrary to the laws of nations," and as adverse to "the relations of friendship and good understanding maintained by the two Republics."[101] In December 1833 President Jackson reported to congress that "negotiations commenced with the Argentine Republic relative to the outrages committed on our vessels engaged in the fisheries at the Falkland Island" together with "other matters of controversy." He announced that they were expecting the arrival of a representative of Argentina to continue negotiations.[102] This ended up being unnecessary because in this same year the English took control of the territory, a fact reported by Darwin in his journal entry of March 1834 when the *Beagle* stopped at what he called the Falklands Islands.[103] The government of Buenos Aires protested the action of the British, but the political uncertainty of the time relegated this conflict to a low position on the list of priorities.

Internationally, this incident was not as important as the creation of the Republic of Texas, which sent shockwaves throughout the Spanish American nations, making members of the regional elites understand the aggressive approach of the United States, and contradicting the previous belief that the American continent was associated with creating a new type of government and society. US expansionism was based on men's aggressive and brutal power, as expressed in racial destiny. In 1837 the Mexican intellectual and politician José María Tornel y Mendivil (1795–1853), considered by his contemporaries "the most important and influential figure of the first half of the nineteenth century," wrote a criticism of the United States affirming that since the country's creation, its predominant intention had been "the occupation of a great part of the Spanish and [then] Mexican territory." This aggressive stand was not based on civilized ideas of higher moral conduct, but on sheer material desires and greed.

This "proud Anglo-Saxon race" belonged to "the barbarians of another North and another time," unlike the modern civilization created in eighteenth-century culture that had replaced brutish men who only knew about violence and conquering.[104] In no other part of the globe were "racial antipathies against people of color [*gente de color*] more perceptible than in the United States," which included the Indigenous population that they had killed; and they sent those who survived to Mexico together with Black people.[105] The mistake of the Mexicans had been to believe that "the neighboring power was our best friend, and that a system exclusively American that contradicted the European should be formed," led by the United States and allied republics.[106]

In 1847 Morton's *Crania Americana* was analyzed by the Cuban José María Calvo y O'Farrill (1789–1848). In his view, ethnography was "the science that teaches how to study and generalize the human races," which was an "entirely modern" discipline, but the translator makes it clear that if Morton wanted to prove to Mexicans that "the American continent was populated without any communication with Asia," this idea was "in contradiction with the Old Testament, immemorial tradition, and the current doctrines of natural philosophy," which were defended by "the most eminent thinkers of the time."[107] A year later, in 1848, José Antonio Saco expressed his concern about the possibility of the incorporation of Cuba, still a Spanish colony, into the United States. He was in exile fighting colonial "despotism" and bothered by the proposal of Cuba's annexation because of the possible disappearance of Cubans. He wanted not only that the island be "rich, enlightened, moral and powerful" but also that "Cuba be *Cuban* and not *Anglo-American*." His position was that the idea of "immortality was sublime; and nationality was the immortality of peoples."[108] The existence of the modern nation depended on the continuity of specific bodies over time.

In Buenos Aires, the greatest concern was not the incident with the United States but to assure the continuity of Rosas in power and to answer his request to acquire absolute power as a condition to becoming governor of Buenos Aires again. A law approved by the Buenos Aires legislature in March 1835 gave him all the powers of government (*suma del poder publico*) over the province with only these restrictions: to conserve, defend, and protect the Catholic religion, and to defend and support the cause of the National Federation proclaimed by "all the peoples of the Republic." This extraordinary measure could last if Rosas believed it was necessary.[109] The other provinces also approved his representation of the country with respect to foreign nations, which was very important considering he was against European intervention and influence in the new Spanish republics.

The French became more interested in extending their colonial empire than in making war in Europe and competed with the British and the United

States for influence in the territory that was formerly part of Spain, placing the liberals who defended cosmopolitanism in a difficult position. The French and British triggered conflicts, and the defense of "Americanism" reignited the independence patriotism supported by Rosas, leaving his enemies in disarray in the 1830s and 1840s. In May 1837 Rosas signed a declaration of war against Bolivia, formerly Upper Peru. Andrés de Santa Cruz (1792–1865) was the leader of Bolivia at the time, and he united Peru to create the Peru–Bolivian Confederation, putting himself in place as its leader.

Protected by the British, he gave them preferred commercial status against the interests of France, thus hurting the commercial interests of Chile and the Argentine Confederation. In addition, Santa Cruz had consolidated enormous resources in the new nation, which made it one of the wealthiest areas of South America populated mostly by Indigenous peoples. He had allies among Rosas's enemies and wanted to incorporate territory that belonged to the Argentine Confederation into the territory he controlled.

Chile began to plot against Santa Cruz and asked Rosas to join its efforts. Rosas released a manifesto with many accusations about Bolivia's support of his enemies now living in exile, denouncing Santa Cruz's claims over territory that belonged to Argentina, and the protection of the assassins of Facundo Quiroga, the leader of La Rioja allied with Rosas.[110] The war declaration mentioned the despotic rule of Santa Cruz's administration, including his making those he governed inferior "to the foreign Europeans" and noting that locals were being mistreated by the despotic and authoritarian government, which was ironic considering that Rosas's enemies accused him of despotism.[111]

Santa Cruz replied to Rosas's accusations by echoing many of the denunciations made by the exile community, and he published the declaration in English and French to make an international appeal as the conflict was connected with the more aggressive geopolitical design of the British and French over the American continent. He separated the men of reason from the madness of Rosas who was willing to array "innocent men in arms for a *holocausto*" (holocaust) at the impulse of his "blind passion, or yielding to still baser motives," trampling without remorse "on the sacred rights of humanity, and the sacred laws of justice."[112] The conflict was amplified on April 28, 1838, when France declared a blockade of Buenos Aires and its coast following a dispute over the obligation to serve in the National Guard, which also applied to foreigners, and the threat to execute a Swiss citizen who was accused of spying on behalf of Santa Cruz. The cause was mainly that the

English had preferred commercial status due to a treaty signed in 1825 that gave them special protections, which was not the case for the French. When Rosas refused to change course, France asked Brazil to unite forces.[113]

The war with Santa Cruz lasted from 1837 to 1839, involving the northern part of Argentina and the southern portion of Bolivia. The conflict would be settled through diplomacy after Santa Cruz lost the battle of Yungay against Chile and was forced to resign. Rosas's dealings with foreign powers had deteriorated since his declaration of war on Bolivia, and by this time he had denounced foreign intervention as a significant problem facing the new Spanish American nations. The escalation of the conflict with France in 1838 was part of the aggressiveness that also involved an intervention in Mexico, which was occupied by France from 1838 to 1839, the year in which the English newspaper *Era* reported about the situation in Spanish America. After analyzing newspapers from Buenos Aires, Brazil, Chile, Peru, Bolivia, and Ecuador, it was clear that "a great ferment was working in the public mind" of the region's nations "against Europe in general, and France in particular. The barbarous outrages of the French government against Mexico and Buenos Aires are warmly animadverted upon;" but the British were also accused of favoring Santa Cruz, thus breaking a promise of neutrality.[114] The promise of a transnational alliance among the followers of the Enlightenment was shattered in this environment. The Argentine liberals living in exile believed that the European intervention could help them get rid of Rosas and they aligned themselves with Europe, which did not help to discredit the accusations of treason made by the governor of Buenos Aires.

ROSAS'S AMERICANISM

Liberals attacked Rosas for his nativism, his ignorance and neglect of science, and the ways in which he rejected the cosmopolitanism of his enemies' culture. His political ideology was characterized under the name of "Americanism," which meant his defense of the local culture was formed in the past. But what they presented as a representation of ignorance and barbaric government hid the fact that the governor of Buenos Aires had what they lacked, a militia to fight foreign powers at the beginning of a new wave of colonialist intervention in Spanish America. Following Tulio Halperín Donghi's interpretation, we can understand how the militarization of Argentina's Littoral was at this time continued by Rosas's army at the highest point of conflict against European powers since independence, which reignited a local patriotism that was despised by the youth living in exile. This attitude made the Romantic young liberals appear to be lacking virility and honor, cowards who chose to support the enemy instead.

In December 1839, Pedro de Angelis argued that the governor of Buenos Aires was resisting a new wave of global colonialism, recognizing that French aggression marked a departure from Enlightenment ideals of emancipation, for which Rosas's response should be understood in the context of defense of the virility and honor of the men he ruled. It was important to remember, according to de Angelis, that Rosas led a "young people" and that "Spanish blood" circulated in "the veins of the sons of the country, who, like their ancestors, strongly resented offenses to their dignity and self-esteem [*amor propio*]."[115] The use of the word "barbarians" and the comparison of "an American with an Algerian" by the French admiral Leblanc would not be forgotten and made it difficult to negotiate peace. It was an "*honor*" to be compared with the Algerians, but the French seemed to ignore that the "foreign yoke" had only made Rosas more powerful. While the governor did not intend to be hostile to foreigners, he had also "always wanted to prevent them [foreigners] from becoming more important than the sons of the country." These feelings "had been birthed from the depths of an enlightened patriotism" that explained his actions (30–31).

Moreover, the explanation continued, Rosas's house "was the most cosmopolitan in the city," but the mission of the Americans was "to reject en masse the domination of foreigners" (30–31). Their future depended on this, "and France was just trying to recolonize the New World!" (37). This was a fair remark considering that after the 1830 revolution, the monarchy led by King Louis Philippe I (1773–1850) came to power and directed France's return to colonialism, but de Angelis insisted that this "old chimera of the universal monarchy" that had been "impossible in Europe was absurd in America," which explained why Rosas was fighting (38). Rosas's *Americanismo*, or what the governor called "The Great American System of Confederation," consisted in the defense of the continent's autonomy from the influence of the Europeans who had been responsible for the injustices inherited from the colonial past. He was called the "Defender of American Independence" precisely for his commitment to anti-colonialism.[116]

Those who had left the country to live in exile during the previous years gathered mainly in Montevideo, where they immediately started publishing against the regime, supporting France against the Argentine Confederation. The continuation of *La Moda* was *El Iniciador* (the Initiator), founded by Andrés Lamas (1817–91) and Miguel Cané (1812–63). Most of those who were part of the Generation of '37 moved to Uruguay and collaborated in organizing an international campaign in the press to discredit the governor of Buenos Aires. In addition, Gian Battista Cuneo (1809–75), a member of "Young Italy" and follower of Giuseppe Mazzini, was also part of the team, with Garibaldi arriving later. Cuneo, too, was exiled in Montevideo and shared the ideology of his generation: both "Young Argentina" and "Young Italy" transformed this city's cultural scene.[117]

In a new article written under his pen name Figarillo, Juan Bautista Alberdi turned Rosas's Americanism into a satire through a made-up letter allegedly received from Buenos Aires. In it, a conservative victim of the French blockade explained what a happy life he enjoyed in the city now that they could devote themselves only to social pleasures. He explained that the country had not yet experienced an American revolution because that should be "the triumph of *Americanismo*," which meant "of the elements we own, of the American civilization over the *Spanishism* [*españolismo*] that was the old and exotic in Spanish civilization."[118] Locality, authenticity, hypermasculism, and divisiveness were criticized as the basis of Rosas's barbarism.

Meanwhile, following the British desire to end slavery, Rosas signed a special treaty with Great Britain "to cooperate in the complete extinction of the infamous and piratical slave trade," which also contradicted the image of the governor as a keeper of colonial institutions.[119] The Mackau–Arana Treaty, signed in 1841, put an end to the faceoff between France and Buenos Aires, making it clear that the exiles had overestimated France's solidarity with civilization. As far as French interests were concerned, the agreement was "equitable enough, but it completely abandoned the local allies of France." The proscribed Argentines "who would lay down their arms within a month" would be granted amnesty, but the exile leaders or any whose presence in Argentina was "incompatible with law and order" were not included.[120] Sarmiento, noted in 1845 France was no longer the "ideal and beautiful, generous and cosmopolitan [nation] that has shed so much blood in the name of freedom" given its renewed interest in colonialism; the country that Argentines had learned to love since 1810 through its books, philosophers, and journals, did not exist anymore.[121]

After the peace was signed, Rosas refused to recognize the legitimacy of the Rivera administration ruling Uruguay because he had deposed General Oribe, Rosas' ally. Rosas also needed to recover the trade that Montevideo had gained during the blockade of Buenos Aires, so he ordered the siege of the city. The British representatives attempted to convince him to end his intervention in Uruguayan politics, but he did not accept the request. In 1845 he refused again to end the blockade of Montevideo and the British and the French united started another blockade against Buenos Aires. France attempted to protect its citizens in Uruguay, who, according to David Rock, "were mainly artisans and peasants from the Basque country, living principally in or near Montevideo."[122] Initially the confrontation consisted in a naval war because the European powers did not want to land troops in Buenos Aires. Regardless, a British official decided to wreck the obstruction Rosas had placed in Vuelta de Obligado to stop the navigation of the Paraná River.

A battle followed and 250 Argentines died, which turned the support that the exiles from Montevideo had offered to the Europeans into treason, and even Unitarians criticized the community in Montevideo for their open support of foreign enemies. In 1838 the Unitarian general Juan Lavalle, called those in Montevideo "Saintsimonians" (*Sansimonianos*), individuals driven "by a very misguided self-interest, who wanted "*to disrupt the eternal laws of patriotism, honor, and common sense*," after they supported the foreigners in the conflict between their country and France, which indicated the divisions that existed among Rosas's opponents.[123] In 1842, Sarmiento attempted to justify the exiles' support for the Europeans, and mentioned that unable to defeat the tyrant, "the people of Argentina at least wanted to support the army that was part of civilization," insisting that there were local, "selfish" interests, but his generation sided with universal ideas and interests that belonged to humanity.[124]

Andrés Lamas began to write a series of articles critical of Buenos Aires's government in 1845; they provided a good perspective on the racial and scientific ideas previously described and how they were merging with conflicted masculinities in the context of the liberals' support for foreigners. He wrote that "the exclusive and extreme love of what comes from the land, of what originates from the land," was a "deep and virile trait of the moral physiognomy of the Spanish race." That should never be criticized because it was related to the "purest love and patriotic virtue upon which robust nationalities are built."[125] The problem was that this feeling could be twisted in a way that could make societies return to a savage state, as had happened under Rosas.

Religion and virility had sustained "Spanish nationality" during seven hundred years of struggle; but it ended up damaging the empire. Lack of openness due to a closed virile culture and distrust of the foreigner was the main problem continued by the Rosist government. In promoting hatred of foreigners, Rosas had turned it into an "*American principle*," which had confused how nationality was felt. It was a powerful feeling among the "virile, warrior, and agricultural population" like that of Buenos Aires, whose material needs were satisfied. Such sentiments "exaggerated nationality to a degree much higher" than what their ancestors had felt.[126]

Liberals like Lamas understood the present civilization as "an embryo" whose growth would be "the result of civilization, of industry, of the foreign population that by mixing itself with us, acclimating to our soil, exploiting it," would produce "a civilization, and American industry" through population growth "in the desert" and "uniform education." Lamas used the example of the United States, a country that had "duplicated its population and production in less than a century" because of immigrants from Europe. Another example for him was the city of Montevideo, because foreigners had touched it with their liberty and turned it into "a splendid city."[127] It was for this reason that immigration needed to be planned by the government. Foreigners were the solution, not the problem.

SARMIENTO, ETHNOLOGY, AND MASCULINITY

Domingo Sarmiento was born in 1811 in San Juan, a province of the Cuyo region that belonged to the River Plate Viceroyalty at the foothills of the Andes. Though his family was poor, he belonged to an old colonial lineage, a product of the union of illustrious families on both sides. A religious uncle took charge of his tutelage, seeing that he received the knowledge necessary to acquire an advanced education. By remaining in San Juan, very far from Buenos Aires, Sarmiento was kept outside the sphere of influence created by those in Montevideo. In 1841 he moved to live in exile in Chile, where he started his own campaign against Rosas, which was intense and won him a reputation among those forced to live outside their country. He attacked Rosas's ideology, and in one publication he mentioned that "50,000 souls" had left the country because of the violence of the army controlled by the governor. He complained that liberals fought "for civilization and American institutions" but were ignored because of their "lack of nationalism!" a quality that was very important in Argentina after the conflicts with foreigners.[128]

Sarmiento's fight against nationalism continued, and in Chile he clashed with Santiago's elite because of his social class and ideological differences. He wrote that there was "no other country in South America that had a more hairy, fetid, and monstrously augmented national spirit" than Chile.[129] After his talent for journalism was recognized, he was made editor of the newspaper *Mercurio*, but many rejected him and his kind of liberalism. At this point he had been devoted to Cousin, the French Romantics' project of social regeneration, and French socialists' ideas, which conflicted with Andrés Bello's generation. A group of well-educated young Chilean men created a journal, *Semanario*, with Bello's assistance. It was headed by José V. Lastarria (1817–88), a disciple of Bello and a student of geography and law.

In an article written in 1842, Sarmiento sarcastically criticized what had been published in the new journal against the Romantics in favor of classicism, as written by Bello. Sarmiento reminded the author and those at *Semanario* that "after ten years since Europe's Romantic school had been buried next to its literary ancestor," this article was outdated, and seemed to ignore "the socialist or progressive school" that was by then leading European thought.[130]

Sarmiento declared himself a socialist, understanding the term as "the concurrence of art, science, and politics that only intended to improve the fortune of the peoples, favor liberal tendencies, fight reactionary preoccupations, and rehabilitate the people, the mulatto, and all those who suffered."[131] Philosophy and science worked together in creating "the science of life," through the study of history, humanity, and the march of civilization that formed opinion and directed politics.[132]

His criticisms stirred indignation, particularly because the foreigner felt entitled to challenge those who were behind the creation of a national literature. Regardless, the editor of *Mercurio* replied again, this time using the words of Lamennais to reject "exclusive patriotism, which is the selfishness of the peoples" and was behind the resistances he believed prevalent in Chile.[133] The French writer rejected local patriotism "because it divided the inhabitants of various countries, inciting them to harm themselves instead of helping each other; it was the father of this horrible monster called war."[134] His words supported the international solidarity of the young generation with the advancement of civilization and progress that made them take the side of European powers against Rosas's Americanism.

Sarmiento's poverty and lack of social status created many conflicts for him, including criticism of his manners and clothing. In reply to his critics, he published a satire, using some old themes to ridicule the superficiality of the wealthy. He also added science and race to explain the evolution of ideas of manhood among the wealthy. In 1842 "Physiology of the Packet" was published to introduce a new male character as a product of the times, "the packet" (*paquete*). He was the "*dandy* in France and England," the product of an alliance between these two countries. In the past he had been called "*coxcomb,* rudely in rude England, *petitmaitre* in Paris, and *mountebank* [*saltimbaqui*] in Madrid." After time passed and their culture was adopted by all classes "they agreed to call themselves *fashionable*, *elegant*, and *lechugino* [another word for *currutaco*]." After the French 1830 Revolution they emerged as dandies. In America, "where everything was late," they were just called "packet," and their physiology and evolution were explained in the essay.[135]

The "packet" could exist anyplace but required "the perfumed air of civilization to breathe," and lived "where there was laziness, luxury, coquetry, and above all cologne, a *mark* [*marca*] of Paquete." The word "paquetería" was related to this noun and meant elegance; it also implied that the person followed a fashion appropriate for high class. Following physiological science—physiognomy as well as phrenology—it was explained that the parents of the packet needed to have "both moral and physical qualities," the mother needed to have "a nervous constitution, demanding character, a whimsical will, and a capricious nature." If she added to these traits "a pinkish complexion, soft curves, and light chestnut hair, it could be said that she had the game half won. Blonde hair was extremely advantageous." The father needed "first to be truly the child's father, and from a distinguished ancestry and lineage" (10). Once born the baby brought with him "the organization and the brain organ that indicates the vocation for packetism." The first distinguishing signs were his being "a sucker, annoying, a crier, and slimy;" it was a given that he needed to be "white, blonde, and fat," and that the nanny could not deny him anything. If the child were balked for anything when he was still an infant, there was a risk "that the germ" might be wasted (11).

Once his education ended, and having learned very little, the packet discovered fashion, and he passed "from tails to the frockcoat, to trousers, to a tie, to a hat, to gloves, and to that white glove that he had looked at with indifference until then" (13–14). A process of *dandification* (*dandynarse*) followed, which consisted of learning how to pose in very specific positions; once the process had finished, he enjoyed all-day partying that suddenly ended when he began to "turn around a beautiful planet," a woman. This encounter made him crash against "a boa constrictor that devoured him alive. He married!" (18). The defense of male autonomy and the references to fashion only repeated old themes; at the same time, Sarmiento also embraced the cosmopolitanism that Rosas rejected.

In 1842 Sarmiento wrote that nationalism was opposed to civilization because his generation was living at a time in which "the spirit of civilization amalgamated and gathered everything that is good, whatever its origin," a clear reference to Cousin's eclecticism.[136] In another piece printed the following year, he insisted that those like him had nothing that was "their own [*propio*], nothing original, nothing national," which emphasized the nature of the conflict against Rosas.[137]

Sarmiento wrote against the view held by Chilean nationalists about their past, using narratives of assimilation and extinction that revealed his knowledge of the emerging scientific literature. Analyzing a lecture given by Lastarria about the Spanish colonial system, he criticized its author as not having "emancipated himself from the ideas fashionable during the independence revolutions to incite emotions against Spanish domination."[138] The Chilean Lastarria had lied about "a pretended fraternity with the Indians" to "put us against our fathers whom we wanted to expel from America." For this reason, he had praised the virtues of the Indigenous chiefs Colocolo, Caupolican, and Lautaro, "as if these savage men belonged to our American history," as if they were not "a nation foreign to Chile" who needed to "absorb, destroy, and enslave them as the Spaniard would have done." According to Sarmiento, honesty about the role of the Spaniards was needed—they "had simply done what all civilized people did with the savages when they exterminated them to occupy their territory." The colonial era had effectively done, deliberately or not, what needed to be done with those who were inferior (214).

It was because of this "injustice" that instead of remaining in the hands of "savages incapable of progress," by then the territory was occupied by the "Caucasian race, the most perfect, most intelligent, beautiful, and progressive of all the races that populated the earth." War and conquest were among "the ways in which providence" had given weapons to "those more powerful and advanced to replace those who due to their organic weakness or lateness in civilization's race, could not achieve man's greatest destinies on earth." It was thanks to this injustice that the American continent was not controlled by "savages incapable of progress." This process was possible because the world's population "was subjected to revolutions that recognized immutable laws; the strong races eliminate the weaker; civilized peoples replace the savages in the possession of land." This process was "providential and useful, sublime, and great. In five hundred years the European race with its arts, sciences, progresses, and civilization" would occupy "the majority, and the best, portion of the Earth under the same principle by which Spain occupied the majority of the new world three hundred years ago" (214). Obviously, this argument directly contradicted the socialism that he was preaching two years before, which reflects how difficult it was by then to develop a coherent thought.

Sarmiento opposed Lastarria because, following Herder, he recognized that the racial mixing that had prevailed in Europe was different. While it would be correct to mention the influence of the "savages of Germania" on the contemporary population because of mixing, in Chile this was not the case; the savages had been "exterminated or mixed into the poor popular class." In the United States the condition of the Indigenous population was much worse than in Spanish America: "Two hundred indigenous nations had disappeared in three centuries," and only one had mixed its blood with the Europeans because "there was an antipathy caused by race and civilization" and "there was no possible amalgam between a savage population and a civilized one. Wherever the latter set foot, intentionally or not, the former had to abandon the territory and existence, because sooner or later they had to disappear from the earth's surface." This was the same conclusion Darwin had arrived at in his *Journal*, which might indicate that Sarmiento had read it—the presence of Bello in Chile ensured the circulation of the latest books on natural sciences and travel narratives (214).

Sarmiento was a reader of the history derived from Romanticism and understood that a narrative of historical continuity required linking the ex-colonies to Spanish conquerors to explain the evolution of civilized people in the new nations; besides, by the 1840s, these historical figures were more sympathetic because their masculine power indicated domination and superiority, which were traits that favored male reproduction. Racial concerns put masculinity at the center of the narrative because it was understood that if race was about reproduction, genealogy, lineage, conquest, war, and extermination, men were the ones who defined these categories. The modern nation's narrative had changed and was by this time grounded in the past and not in the future, as it had been before, which required a new national narrative.

Jules Michelet was the historian who probably had more influence over the Romantic generation, together with Herder whose writings had been translated into French by Edgar Quinet, one of Sarmiento's idols. Ironically, Rosas's ally, de Angelis, was the only one who knew Michelet personally; he had even introduced this French historian to Gianbattista Vico's *Scienza nova*; while it was influential among the Spanish Jesuits exiled in Italy, this publication was unknown in France before de Angelis moved to Buenos Aires from Paris.[139]

The influences of the new French history and the racial literature emerging in the 1830s appear clearly in Sarmiento's book *Facundo: Civilización y barbarie* (*Facundo: Civilization and Barbarism*), published in 1845.[140] He affirmed that South America and Argentina needed an Alexis de Tocqueville, who "knowing social theories, like the scientific travel[er] possessed a barometer, would come to penetrate our political life to reveal it to Europe, and France," which were eager to know about the "lives of the diverse portions of humanity, these new ways of being that did not have well defined and known antecedents."[141] Clearly, this is the role that Sarmiento assigns to himself, to translate to the civilized what his country was and the place it had in the war between civilization and barbarism, a process that was common to all societies.

He understood that at this point liberals like himself oscillated between an idea of civilization that was universal and based on human solidarity, and one that was national and racial; also between a civilized manhood that was marked by men's manners, ideas, and brotherhood, and one that was based on race, origins, and reproductive needs. *Facundo* attempted to balance all these elements. Multiple sources were used in writing the book, particularly the accounts of scientific travelers to describe regions of the country its author did not know, such as Buenos Aires, or to narrate events that had taken place without his being a witness. For example, the already mentioned writings of Lacordaire, which Sarmiento cited and plagiarized in *Facundo*, were helpful because the French scientist had lived in Buenos Aires and described the competing political dynamics of the Argentine Confederation as "the struggle of the old stationary mores of the country, of the local habits transmitted from generation to generation, against the modern civilization that seeks to enter America with its inflexible doctrines and its lack of pity for the affected."[142] Civilization was recognized by its cruelty unlike the past view dominated by optimism and positivity.

We can see all these elements of ethnology present in Bartlett's definition of the term in Sarmiento's *Facundo*. Though it is impossible to establish how much he knew about what was happening in the United States, he clearly knew about the existence of ethnology. International masonic societies spread and debated scientific ideas, and Sarmiento was definitely a Mason; he was listed as a Grand Commander of the Scottish Rite lodge in Buenos Aires in 1858. In addition, being in Chile, he had access to the latest ideological novelties that Bello introduced through his writings and in publications he edited.[143] French literature was available to him in Chile, and by the 1840s, it was clear that the modern nation was understood very differently—in terms of gender the main problem was how to represent civilized masculinity in its racial dimension.

In *Facundo* Sarmiento explained how *Americanismo* (Americanism) was manipulated by Rosas to promote opposition to the culture of civilization.

According to Sarmiento, civilized clothes were banned and "these European exteriorities" (*exterioridades*) were replaced by "national clothes that were eminently American."[144] In *Facundo* the civil war was also about the defense of two types of men and masculinities, and Sarmiento was on the side of the "man of the city who dresses in European clothes, lives a civilized life as we know it everyplace where there are laws, ideas of progress, means to educate, some municipal organization, regular government, etc." Moving away from the city, "everything changes in its aspect; the men dress differently, in what I call American dress because it is common to all those living there; the habits of life are diverse, their needs peculiar and limited; it seems as if there are two different societies, two peoples strange to each other." Those who lived in rural areas did not "aspire" to live like those in urban ones, they rejected "with disdain their luxury and cordial manners; and the dress of the citizen, the tailcoat, cape, saddle, no European sign" could be present without triggering punishment in nonurban places. If a man dared to show up wearing a frockcoat in an English saddle he was risking "brutal aggression from the peasants."[145] The rejection of men who followed European fashions was also a rejection of the forces of progress.

This dismissal of the transformative force that led to civilization was already affecting the development of race; and in this context it is not curious that Sarmiento mentioned James Fenimore Cooper's *The Last of the Mohicans* when addressing the battles against the "Indians" who inhabited the desert in Buenos Aires's frontier.[146] He noted that this writer had transported the scene of his descriptions "to the borderline between the barbaric and civilized lives." The stage for many of his books was the war in which the Indigenous races and the Saxon race were "battling for the possession of the land." This book made him understand that it was true that the modifications of the environment brought analogous changes in customs, which explained why Cooper's descriptions, in Sarmiento's view, seemed to "plagiarize" what happened in the pampas, which was an interesting comment because he had never been in either place, and only knew what was mentioned by travelers.[147]

Facundo referred to mixed races, Indigenous races, the pure Spanish race, and the "black race, which was almost extinct already except in Buenos Aires, where it had left its zambos [mixed African and Indian] and mulattos, the inhabitants of the city." While the latter were the link connecting "the civilized man with the coarse one; a race inclined to civilization, endowed with talent and the most beautiful instincts of progress," the opposite was the case of mixing the Native Americans. The mixture of the Spanish, Indigenous, and

Black races was "a homogeneous whole distinguished by its love for laziness, incapable of industrial work," produced by "the disgraceful result of the incorporation of the indigenous people during colonization."[148] This repeated the ideas noted in his criticism of Lastarria, but in *Facundo* the solution consisted of opening the country to the European immigration "to populate our deserts" giving the foreigners lands for free.[149]

It is in this idea that the example of the United States is clearer in the book. Rosas was the worst enemy of foreigners, but "a government sympathetic to the Europeans that protected individual security would have populated the borders of our rivers in the past twenty years, realizing the same wonders that had been consummated on the Mississippi's riverbanks."[150] The source of this information was a popular book published in 1847, *The Progress of America*, written by the Scottish economist and statistician John MacGregor (1797–1857) who, as a merchant, had lived for several years in Nova Scotia.

MacGregor argued that all the statistical and historical information analyzed gave proof that the United States, "the great Anglo-Saxon Republic," had advanced "steadily along with the march of civilization, in civil liberty, and religious freedom." On the contrary, the moral and physical state of the Spanish race in America was in serious decline. It was "an extraordinary fact in the history of a people once so formidable" that in the years 1846–47 wherever Spanish was spoken there was no "civil liberty or religious freedom," the "spirit of anarchy" dominated, and "confidence, or security in the government" was lacking.[151] This racial problem was also related to masculinity; the "Mexican character" had all "the public and private vices bequeathed by the Spaniards to his colonial descendants," which included men who were "exceedingly effeminate," and had an expression "stamped upon them by low vices and sensuality."[152] In Lima, "the sons of the brave and stately Spaniard dwindle away into effeminacy."[153] While Sarmiento cited the racial evaluations made in the book, he ignored gendered comments. Moreover, by the end of the 1840s it was clear that racial experiences in the Americas contributed to the split of European races in two, as was explained in a review of MacGregor's book, published in London's prestigious *Athenaeum* in 1847.

This article observed that after reading *The Progress of America* it was "impossible to glance at the progress of America without feeling impressed by high destinies of the Anglo-Saxon race—and the contrast which they afford to the fate of French and Spaniards in the New World." Europe may be said generally to exhibit two types of civilization, "the Latin and the Teutonic;" the "Latin type," represented by the Spaniards, "conquered the natives by the sword, while the Teutonic, represented by the English, subdued nature by the industrial arts." The achievements of the Latin civilization were "the more brilliant—those of the latter [Teutonic], the more enduring."[154] A few decades after the independence movements had emancipated the Spanish colonies, the Spanish men who were so despised in the past were now the only ones respected in the European literature so admired by liberals.

6

Sodomized Man

As Jorge Myers demonstrates, Rosas was not ideologically representative of the past's continuity but an expression of a republican tradition that the Generation of '37 claimed to embody.[1] In 1849 the conflict with the British and French ended with the signing of the Arana–Southern Treaty and, a year later, the Arana–Lepredour Treaty. European powers recognized the sovereignty of the Argentine Confederation over its rivers and ended the negotiations to break up territory of the Confederation to create a country to be ruled by Rosas's enemies supported by foreign powers. His defiance of Europe enhanced his prestige and the patriotism of Argentines and Spanish Americans who saw him as the true defender of the independence's ideology.

General San Martín, the liberator of Argentina, Chile, and Peru, praised Rosas for his actions in a letter he sent on May 6, 1850. The purpose of this note was to express his "true pride, seeing the prosperity, internal peace, order, and honor restored in our beloved Homeland" by the governor of Buenos Aires. This achievement "amid such challenging circumstances faced by few nations," led San Martín "to sincerely congratulate" Rosas and the entire Argentine Confederation. He wished the governor "full health and, upon concluding your public life, the just recognition of every Argentine," and ended the message calling Rosas his "passionate friend and compatriot," promising that he would always support him. Such recognition overshadowed those in Montevideo who had supported their country's enemy.[2]

While the governor of Buenos Aires had recognized that the British and the French had returned to the past, treating the new republics as colonies, the young generation was unable to support their country at the cost of abandoning universalism by embracing locality. Science formed the foundation of civilization, and, as the English noted, commerce—its sister—was the engine of progress. The British and French attacks on Buenos Aires to secure free navigation and increased trade was an opportunity for them to bring their country and themselves up to date, which meant to be modern. Their preference for foreign victories and reluctance to join the army further tarnished their image during the years of the conflict.

In an 1843 letter from the writer Luis L. Domínguez (1819–98) to Félix Frías (1816–81), Domínguez described how Juan B. Alberdi, who claimed to be "the leader of the young generation," had deserted like a coward when the siege of Montevideo (1843–51), backed by Rosas, began targeting the exile community in the city and the commercial interests supported by foreigners. Although Alberdi had urged others to enlist to support the Unitarians, he joined the reserve militia to avoid direct combat, staying at home and preparing for a yearlong trip to Europe. In addition, he was also described as envious, vain, and vengeful in protecting his sense of superiority.[3] Domínguez labeled him a "deserter" and declared the "young generation" a disappointment. During this time, Echeverría was in Montevideo and largely inactive, leaving a leadership vacuum.[4] While some men of his generation advocated restoring the 1824 Unitarian constitution, this poet sought to regenerate the spirit of the 1810 revolution, which originated tensions among Rosas's opposition.

This situation diminished the young generation's influence on national politics. However, their survival was ensured as Montevideo became a hub of revolutionary activity, attracting members of Young Italy, who were also exiled in Uruguay. Their association with Italian leaders provided Young Argentina with a network to organize an effective international campaign against Rosas with an organization they already knew and followed.

Alessandro Bonvini highlighted the "global scale" of Young Italy, describing it as a "multipolar network of associations across Europe and the Americas, governed by a single leadership." This organization educated revolutionaries and conducted "an intense propaganda" that significantly influenced the exiles in Montevideo. Prominent Italian leaders in Uruguay and the United States turned the Americas into a political platform for Italian patriots. They launched newspapers, established branches, forged alliances, and developed innovative programs for transatlantic cooperation. These activities, carried out in collaboration with exiled members of Young Argentina, connected Rosas's opponents to a broader transnational movement based on universalism.[5]

Bonvini argues that the political strategy and propagandistic style of Young Italy contributed to "formalizing a definition of an alternative Argentine nationalism" in opposition to Rosas, but this seems to be an exaggeration. The Generation of '37 was deeply frustrated by their inability to address the complexities of nation-building, particularly after the evident failures of republicanism in Europe and the Americas and the rise of race science. As mentioned in chapter 4, along with the emergence of new spiritualism, Giuseppe Mazzini (1805–72) offered a spiritual conception of "dynamic activity connecting the individual mind with the larger world of which it is a part."[6] Mazzini's followers aided a small group with little knowledge of large-scale politics in developing propagandistic journalism that expressed its idiosyncratic interests.[7]

Echeverría and his followers were not an empty ideological vessel. By the 1830s, they believed the country's divisions stemmed from a lack of healthy associations among men, which hindered the establishment of a civilized sensibility that fostered the study of science, literature, and the arts. This led to disputes over how to represent the spirit of the times: Rosas's vision was materialistic and masculine, whereas that of the young generation was understood as spiritual and feminine, the attributes of modern civilization. By 1837, this opposition was expressed in clothing, manners, education, sensibility, and masculinity. While these contrasts were not new, the intensity of the violence they provoked was.

Rosas's opponents fled into exile to escape the violence inflicted upon them, including the physical punishments imposed by Rosas's police, the *Mazorca*. The terror attributed to this organization by the propaganda printed by the exile community added a new dimension to the goal of weakening the Rosist regime. They portrayed society as divided into masculine and feminine roles, presenting themselves as victims of dishonor and degradation for their support of enlightened ideas. In doing so, they appealed to European countries to end Buenos Aires's dominance and protect civilized men. Their prolonged stay in Montevideo, surrounded by an international male community of revolutionaries, amplified their cause. Renowned European writers and intellectuals, critical of Buenos Aires's despotism, brought their struggle into global focus. Gender became their most effective tool to promote their struggle.

Second only to gender, the destruction of education and science in the Argentine Confederation featured prominently in their exile writings. Rosas's association with racialized groups, perceived as foreign to civilization, underscored how far the country had declined since Bernardino Rivadavia's reforms. In their view, restoring the republic of science, whose universalism remained intact, was the only path to progress that Argentina could take.

This marked the final stage of the civil war, which ended in 1852 with Rosas's defeat, thanks to the leader of Entre Ríos who had suffered the consequences of civil war in Uruguay and broke the alliance with Buenos Aires to remove Rosas with the aid of foreign-composed armies. Those who favored foreign ideas and culture took power, realizing their long-conceived project, including passing the 1853 constitution, written by Alberdi, without opposition.

ROSAS AND THE REPRESSION OF SCIENCE

Rosas's stand against science was exacerbated by the Anglo-French intervention in the River Plate (1845–49) and his suspicions that foreign scientists and artists were spies working for their countries. In fact, the conflict with Buenos Aires's British community began because of the treatment of European scientists and artists who were visiting the city. Those allowed to pursue their interests had to avoid promoting foreign ideologies that questioned national sovereignty. For example, the French botanist Aimé Bonpland who had a ranch in the Littoral province of Corrientes, was allowed to pursue his research. His proximity to Uruguay meant that the area was often occupied by troops engaged in the civil war dividing Montevideo, with one faction backed by Juan Manuel de Rosas and the other supported by Unitarians, Young Italy members, France, and England. In 1839 Bonpland had established a good relationship with Governor Genaro Berón de Astrada, an enemy of Rosas, who was defeated and killed in a bloody battle in which soldiers from Buenos Aires massacred hundreds of prisoners.

The winner of the battle allowed Bonpland to continue his research in his estate, but the naturalist complained about the loss of his pedigree sheep and horses. In 1840 he went to Montevideo, a place he knew already, to reside with the priest Dámaso Larrañaga (1771–1848), with whom he had corresponded since 1818, when Bonpland expressed admiration for Larrañaga's work, which exceeded even his highest expectations.[8] This priest was the founder of the country's public library and a supporter of Lancasterian schools and liberal education, besides being a naturalist who had important collections and was known internationally for his work. At this time, he was in communication with de Angelis, who wanted to purchase some of his possessions.

Stephen Bell documents details of Bonpland's trip showing that he sympathized with those in Montevideo; he also admired the city's progress due to the influx of foreigners from France, Italy, and Buenos Aires, and wrote in a letter that he was "astonished" by the enormous number of houses in Montevideo, many of them new. Moreover, 159 ships were in the port, and, in the crowd, he discovered many people he knew, who were surprised that he was not wearing civilized clothes.[9]

Once in this city, Bonpland wrote to his peers in France about his research; the letters reveal that even when he had been in the middle of a civil war, the scientific networks still worked for him. The correspondence analyzed by Bell lets us know how effectively knowledge traveled even in the most difficult conditions. One of his correspondents was Charles-François Brisseau de Mirbel (1776–1854), who had sent him seeds previously, including for Australian pine, which he planted successfully in Corrientes. Considered the founder of plant physiology, Mirbel provided information to Bonpland about current research in Europe and access to his own international network of researchers. For example, Bonpland knew about Matthias Jakob Schleiden (1804–81) who was classifying plants through the observation of their cells under the microscope, which Mirbel was also doing.

Schleiden was one of the founders of cell theory, a follower of Friedrich Wilhelm von Schelling and *Naturphilosophie*, and an important figure to those interested in vitalism. While recognizing the importance of this new approach to botany, Bonpland expressed his preference for the observation of the relationships that humans established with plants, which had been extremely helpful to him in his travels. He reminded his correspondent that those in Paraguay had deduced the usefulness of yerba maté without any cellular observation, which provided a more holistic understanding of the function and usefulness of plants.[10]

The French presence due to the war against Rosas was also important in promoting the sciences in Montevideo; for example, in August 1843, Montevideo's newspaper *Le Patriote français*, printed a letter sent by Arsène Isabelle (1806–79) to the government of Uruguay to advise the creation of a museum of national history. In it, the French naturalist asked the minister of interior to follow the example of North America where "incredible prosperity" had been obtained quickly "through natural resources alone," and where "miracles" took place every day, thanks to laws, "public order, internal peace, and a patriotism" that was "ardent and enlightened!" The museum had been created in 1837 due to the efforts of Larrañaga to place his collections but was organized by Teodoro Vilardebó (1803–56), who was trained in Europe as a naturalist and had recently returned to Uruguay.[11]

In the international liberal model, science, economics, and politics were connected with the presence of a type of local and foreign man who was industrious and promoted policies that placed "mercantile, industrial, and scientific speculations on truly solid ground." This man was defined in the epigraph of the newspaper's title as championing "honor and fatherland!" Therefore, the first step of his action was the exploration of the territory "to classify all zoological, botanical, and mineral" species, depositing them "in an ad hoc museum, so national and foreign savants had the means to describe and analyze these resources conveniently," to compose a complete body of work describing what was valuable in the national soil.[12] But this was hardly a novelty. The mission coincided with Larrañaga's own ideas. In 1816 he had recommended the same to the young people of Montevideo. He told them that if they wished "to give new impetus to these two riches of nations, that is, to Agriculture and Commerce," they must study "the great book of Nature, that fertile and ever-new mother." Their discoveries would "honor your homeland and increase the lines of its trade and cultivation.[13]" This idea was foundational in the River Plate area.

Meanwhile, in Buenos Aires, self-taught naturalists continued to pursue their interests. For example, Francisco Muñiz (1795–1871), researched megafaunal fossils in Buenos Aires, vaccination, and geological formations, and his work gained recognition outside the River Plate region. He was a medical graduate from the first faculty of the School of Medicine established in 1822, and corresponded with European scientists, Charles Darwin among them, to whom he offered his valuable bone collections, which Rosas ended up gifting to France in 1842, an act that fueled criticism from exiled opponents who saw it as proof of the government's disregard for scientific advancement and civilization.

Muñiz's inaugural lecture as a professor of medicine in 1826 showed the influence of ideology in his thought, so it is not a coincidence that it was about gender and science and the role of women in the conservation of society. Drawing on ideology, he emphasized women's role in preserving society, stating that men and women were naturally different: "Man is active; woman is passive." He described men as "warm and dry" and women as "humid and cold," adding that while men commanded, women nurtured and consoled.[14] Despite this perceived inferiority, Muñiz argued that women, through love and care, ultimately governed men. He further posited that women's restraint and coquetry were vital for reproduction and societal preservation. In a materialist interpretation, he likened the uterus to "a living animal" that controlled a woman's body and emotions, making it the foundation of her existence and her societal role. By the 1840s he was serving as an army doctor appointed by Rosas, at a time when few institutions devoted to science remained. The Tribunal de medicina (the Medical Tribunal) that dealt with health at the national level was one of them.

Also in 1845, José Rivera Indarte (1814–45), to whom we will return, wrote a report on the conditions of the war against tyranny with a section on public education, in which he criticized the government's lack of support for teachers, noting that only parents with means could afford classes for their children. Those wishing to teach sciences were required to obtain a license, which demanded documentation proving their support for Rosas's party. The requirements included being a citizen, as foreigners were deemed unfit to educate the youth; demonstrating proficiency in a "Spanish" style of writing; and, above all, showing "worship of Rosas," devotion to the dictator's "American system" (*sistema americano*), hatred of the "savage Unitarians and foreigners," and participation in Rosas's societal festivities, the *Mas-horca* (more gallows), dressed in a red vest and the *divisa punzó* (red ribbon), the symbol of the ruling party.[15]

Registered teachers could teach physical sciences, theology, political, economic, and moral sciences as long as they did not clash with the official system. The youth had freedom to deliver presentations on all subjects, but the priority was to praise Rosas even if the subject was medicine or astronomy; moreover, the presentations had to be checked by officials in advance, several days before being presented. There was no freedom of the press, and newspapers could only print whatever the governor approved beforehand.

Such accusations against Rosas were part of the propaganda of those in exile and included exaggerations. While by this time the system of public education favored by liberals was not in place, several schools in Buenos Aires were directed by enlightened people such as the Frenchman Carlos Clarmont; Florencio García; Mariano Martínez; Antonio Oliú; Pedro Sánchez; and Gervasio Sueldo. There were also schools for girls directed by Agustín Bailón and Melanie Dayet, the wife of Pedro de Angelis. This couple was superbly trained for education; he had been the tutor of the children of Joachim Murat and Carolina Bonaparte, the sister of Napoleon; she had been a "lady-in-waiting to Countess Orloff," who was the hostess of a famous literary salon in Paris.[16] Besides, the presence of de Angelis ensured that a person who knew and cared about science would be in charge of the scientific collections that existed in Buenos Aires, something that was not that common in South America. He also published important documents that were related to materials important for naturalists produced during colonial times, which means that in terms of science Buenos Aires was well represented internationally.

On the other hand, in 1846 stricter measures to control the materials students could access were approved partly as a reaction against the conflict in Montevideo that was supported by those in exile. A commission, chaired by a priest and composed of four men closely aligned with Rosas, was appointed to oversee teaching programs. This commission decided which books would be used and set curricular standards, ensuring that both adhered to "the orthodox doctrine of the Holy Catholic Apostolic Roman Church, morality, order, the political system of the state, and the progress of the sciences and belles lettres."[17] Books used for teaching foreign languages, preparatory studies, and sciences at the university also required the commission's approval. Additionally, the university's provost participated in the review process, alongside two citizens appointed by the government.[18]

SODOMY'S BLOOD STIGMA

Those in exile associated the repression and tight control over society that was obvious by the 1840s with the hypermasculinity of the society's culture, the kind of male sociability it promoted, and a subversion of the social order that affected the population of the city that was white and civilized. At a time during which science was establishing clear boundaries between the races, Rosas was doing the opposite; his most solid support in the city was with the Black population and those of mixed races. In a book of memoirs, the writer and doctor José Antonio Wilde (1813–85) told an idyllic story about the peaceful and even loving relation that existed between enslaved people and their owners, only interrupted during the years Rosas ruled when everything "was demoralizing and corrupted" leading Black women to revolt against "their protectors and best friends." They spied for the tyrant, accusing many families of being "SAVAGE UNITARIANS"; they became arrogant and insolent and were feared as much as the secret police.[19]

Félix Frías, an enemy of the governor of Buenos Aires, noted disdainfully in 1838 that May 25, Independence Day, was not observed with "official demonstrations," but with "bands of black Africans roaming the streets of Buenos Aires, ending up in the Plaza de la Victoria to perform shameless dances, scaring decent people away from the scene."[20] The propaganda against Rosas published in Montevideo repeated the same stories. Andrés Lamas reproduced a notice published in *La Gaceta* in 1839, describing a celebration attended by Rosas and his daughter and highlighting that "nobody was left out of the dancing, because with everyone *mixed* [*entreverados*]" distinction was not known.[21]

This interpretation of Buenos Aires's sociability was not accurate. according to Halperín-Donghi's analysis, the governor of Buenos Aires followed a tradition that "sharply distinguished [the broad masses] from the *gente decente*, and this dividing-line was more important than that between those of European and those of Indian or African descent."[22] Mark Szuchman agrees that to defend his natural hierarchies, Rosas employed obsessive social control, which was the common practice of the ruling classes of Buenos Aires following the Spanish example.[23] Argentina's colonial heritage mandated control of groups considered "*chusma* [rabble]" by the "decent people," but the constant military conflicts had challenged this discipline, although Rosas was firm in maintaining it.[24] While he did not believe all people were equal, he freely associated with Black, Indigenous, or mixed-race groups, viewing these distinctions as natural rather than artificial, unlike those based on clothing or symbolic representation.

According to George Reid Andrews, the Argentines of African origin made up more than a fourth of the population of Buenos Aires in 1836, but they had contributed more than half of the men participating in the war for independence.[25] The vision of the city controlled by those who had the weapons threatened the tranquility of many prominent families and partly explained their support for Rosas, who kept the southern border and the urban population under control. Andrews argues that Rosas courted the African nations, but "he was also capable of abandoning them when it suited his interests."[26] The governor "made genuine concessions to the Afro-Argentines"; in 1836 he eliminated the mandatory drafting of those who had been emancipated after the age of fifteen, and three years later he finally abolished the slave trade.[27] But he had reinstituted the trade in 1831 and allowed it to function openly, which was used against him in the propaganda campaign led from exile.

In brief, as Halperín-Donghi has concluded, it is impossible to claim that Indians, mestizos, or Blacks constituted the new Argentine aristocracy, since the independence movement had "refrained from introducing innovations which might affect the most significant of the social differences inherited from the past. It still, in a manner analogous to that of the *ancien régime*, incorporated them into the image it projected of the body politic."[28] Rosas's federalism sustained this incorporation; in Rosas's words, "The ideal of good government would be a paternal autocracy, intelligent, disinterested, and indefatigable."[29] It was precisely this paternalistic authority that was rejected by a new generation that wanted to get rid of traditional patriarchy to impose a modern one.

The split among men about the best way to represent their sex was reflected in the journalistic war that involved both sides. By 1840 those living in other countries included Domingo Faustino Sarmiento in Chile and Félix Frías, Florencio and Juan Cruz Varela, Esteban Echeverría, Juan Alberdi, and José Rivera Indarte in Uruguay. They were the most prominent writers of propaganda published in the press in the 1840s. Exaggerations and horrific images of the savagery of the troops supporting Buenos Aires's authorities were common in publications, and the institution that became the focus of the criticism was the *Mazorca*, presented as a hypermasculine organization responsible for the torture and killing of society's best men, though the origins of this society were not completely clear.

Lucio Mansilla (1831–1913), a nephew of Rosas and frequent visitor to his house in Palermo as a child, would clarify years later how the origin of this word was related to the social manners that existed during Rosas's government. As a liberal and civilized man, he pictured a society lacking a clearly established racial and gender order; there were "abandonments and perfidies on all sides, and people of criminal nature [rubbed up against] good and respectable folk; blacks, mulattos, Indians, a Babylon, complete promiscuity."[30] The fear of being assaulted by these dangerous groups caused the "better citizens, the most inoffensive, the most timid men" to join the *Mazorca* out of "the terror" that overcame them (108). According to Mansilla, this was the product of the fear of falling into the hands of the secret police whose audacity rose to the level of "limitless abandon." Their practices had the advantage that no one would report them: what man would dare confess his own violation? (108). "They take an ear of toasted corn and say: we'll be inserting this into such and such part; that's where mazorquero, cornhusker, comes from. What happens then is only known through rumors: Who would ever report it?" (108).

The lack of formal accusations and the atmosphere of secrecy contributed to doubts about the real nature of these secret police. Nobody knew who the victims were, no man was able to go public after this crime, and suspicions and gossip were the order of the day, favoring the government's repressive activities. There was no freedom to express ideas, and there was "a sense of foreboding." Terror resounded like "a dull explosion," and everybody slept with a prayer on their lips. "Who? Everybody. No one is safe!" The only ones who were able to tell this story were the "émigrés with nothing at stake." The danger originated in the very authority watching over the population, and during the nights people fell asleep listening to the watchman sing: "Twelve o'clock and all is cloudy, long live the federation, death to the Unitarian savages, long live representation!" In this narrative, people had become used to living in fear, "just as those who live near slaughterhouses become accustomed to its foul odors" (108–9). The extent of the violence explained why the youth were forced to leave the country to protect their honor.

The uniqueness of Argentina's conflict was that it was gendered in an uncommon way: one side defended the female sign to indicate civilization while the conservative federation exercised disciplinary practices against those who did not "naturally" behave like men. In this context, acts of sodomy were used to punish men seen as unmanly and unnatural, a risk to the healthy development of a new paternalistic system. This action was not sexual but an act of discipline and forced self-identification, the "blood stigma" mentioned by Echeverría; men who were accused of acting like women in the public space were penetrated with a product coming from the national soil to naturalize them, to root them to the country that they refused to belong to. Despite the shame involved, exile propaganda openly publicized this practice in the press.

As we have seen, this was not new, the same had been done in the eighteenth century with the *currutaco*, for example. Also, in the United States, Richard Godbeer has shown how the colonial newspapers linked sodomy to foreign corruption and urban vice in the English colonies of North America before the creation of the United States.[31] The difference in Argentina was that the split was not between the duality local/foreigner but the result of a split among competitive projects to be the citizen who would represent the nation, as Jorge Salessi has suggested.[32] Ricardo Rodríguez Molas has noted the frequency of sexual allusions tied to violence "in the Rosist literature of a popular nature," and this is visible in the division between masculine federalists and feminine Unitarians that gendered the conflict being waged.[33] According to the anti-Rosist campaign, the panic felt in the city was exacerbated by continual threats made against men in the popular press, through songs, and in popular poetry. In 1835 *El Gaucho* published a poem that exemplifies this culture.

Long Live the Mazorca

To the Unitarian who stops to look:
That corncob you see
Dressed in blonde husks . . .
It has sunk to Hell

The Unitarian faction!
And thus with great devotion
You should say to yourself:
Save me from that bind
O blessed Federation!
And be careful
When you go
To see that this saint
Doesn't come at you from behind![34]

In 1830 Luis Pérez, the editor of *El Gaucho*, described the enemies of the governor in the following way: "Hey diddle-diddle / Make the queers wear britches / In a second any tailor / Can sew a fine Unitarian / Who prides himself on his *levita* [frock-coat] and glasses."[35] Many poems repeated the characteristics of the Unitarians, all of them related to the culture of modernity depicted as superficial, inauthentic, and feminizing. Modernity changed the core of masculine identity because its culture was representational, which was attacked as unnatural by the defenders of Argentine tradition that supported Rosas, who emphasized their belonging to the soil and not to external things such as clothing, books, and a new sociability. The question was whether the May Revolution had originated in the desire to start something radically different, or were its origins in the defense of the Americas' form of life against the Spanish authority.

The Unitarian

Federales, pay attention
And look at a faithful portrait
of a unitarian who is
as pedantic as he is fatuous.
A Unitarian was looking at himself
in the mirror during morning
and conceited
with his figure
this is what he exclaimed:
"I am the happiest young man
for fortune flatters me;

to enjoy in the world
what I like and pleases me
if any girl looks at me
she falls in love with me just by
Observing my graceful little person.
If to have a good time
I dedicate myself to visit them
they go crazy for me
and fall in love,
because my elegant figure
and very sweet words
are arrows that penetrate
their hearts and souls.
Comforts I have plenty,
I spend money without rule
because the credit supplies me
with everything I lack.
If any creditor tries to charge me,
as usual in the morning,
I tell him: "come back at one, that the debt will be paid."
But immediately after I perfume myself,
tighten my tie
And prettier than an Adonis
I go out to hit another trap. . . . . . . "

This is how his speech ended
his young man or ghost,
and his servant, who had heard him,
told him these words:
"There is no doubt that you are happy
if this life does not end,
and if one day
they don't send you to jail for being lazy
which is what I hope will happen
if you don't change in your gear
And pay obedience to the government commanding
us today."[36]

The Unitarian

the Unitarian is a ghost
in whose physical being
it cannot be understood
if the soul is national.
Betrayal, intrigue, and plot
he exercises skillfully;
his head is full of air
Immortality he proclaims;
And that's how he calls decent
the one that professes his vices.[37]

Sarmiento commented that even the Unitarian leader José C. Paz, was not exempt from the terror that threatened those who opposed Rosas. According to him "real panic [*pavor*]" was visible in Paz's face when the *Mazorca* came close to his house, "announcing its fatal proximity with military music, fireworks, and the happiness" that accompanied these gatherings. The brave general always lost his "cold blood," and could not relax until these men were gone. Sarmiento explained that even men whose spirit "was indifferent to the blast of a cannon that could kill them" could be turn into cowards when they confronted the possibility of personal abuse of their honor (*vejación*) at the hands of a mob working for the government. Paz was afraid to be drawn through the streets to be hit by sticks "or other violence, common at the time, that shame for humanity does not allow [one] to mention."[38] The humiliations that Rosas's enemies suffered were described as crimes against civilization itself by his enemies, and thus against the European countries that had originated it. This helps explain the requests for an invasion of Argentina made in the press. "Europe has a right to demand these conditions from the American Governments, but it is also her duty to protect amongst these new States, by the weight of her civilizing influence, the Governments which show themselves to be interested in promoting the civilization and improvement of these countries."[39] The shame caused by the acts of sodomy was both individual and collective since the real intention of the penetration was to symbolize the complete submission of the foreign threat.

José Rivera Indarte (1814–45), the most effective propagandist against Rosas, requested foreign intervention to protect those men who defended proper civilized ideas and were endangered by barbaric practices.[40] Born in Córdoba, Argentina, he was a temporary supporter of the governor of Buenos

Aires, a poet, and an official journalist of the regime responsible for writing threatening verses charged with sexual reference in 1835.[41] In this same year, his *El voto de América* was published in Madrid, proposing that each nation send representatives to Spain to gather in a congress to negotiate the recognition of the ex-colonies' independence and their future diplomatic engagement, and to end a situation that had not been resolved after more than twenty years of conflict.[42] Many considered the proposal treacherous and accused Rivera Indarte of being a Spanish agent. In 1836 he published another version of this proposal, expressing the need to establish a new era of coexistence with Spain but without paying reparations.[43]

This kind of position did not fit well with Rosas's patriotism, which, together with other disagreements, led to Rivera Indarte's imprisonment because of allegations that he had stolen from a church.[44] Once freed, he spent time traveling through the United States, and then Brazil, but in 1839 he returned to Uruguay and, as the editor of the newspaper *El Nacional*, he carried on the propaganda against Rosas.[45] Despite his dedication, Rivera Indarte was not well liked by many liberals because of his alleged political opportunism.[46]

Rivera Indarte's first attack on the governor of Buenos Aires was *Rosas y sus opositores* (*Rosas and His Opponents*), published in 1843.[47] This book, which originally appeared as a series of newspaper articles, was popular both in the Americas and Europe. Rivera Indarte criticized Rosas for questioning the masculinity of exiles due to their alliances with foreign men, even though Rosas himself maintained a close relationship with the French consul in Buenos Aires, Washington Mendeville, with whom he shared "those mysterious *soirées* of wine and love, [taking] place at the Palermo estate, in the darkness of the night" (7). The patriotism of the young generation was not characterized by tradition, they wanted to accelerate the arrival of the future, which required radically changing the social order of their country.

This social disorder obliged "youths, the elderly, merchants, priests, attorneys, the well-read, all belonging to the society's first class" to drag the "heavy chains" of the enslavement to which they were subjected by Rosas and his rabble, society's lowest classes, in the Rosist prisons (225); in this narrative, the Popular Restoration Society was a club of drunken Rosists that later developed into the essential organ of Buenos Aires life, and it was this organization that inverted men's sociability. In Rivera Indarte's text, Rosist values sprang to political life when Rosas himself armed his followers with what would be their weapons of choice to injure and dishonor his rivals, the ear of corn, the knife, and the cudgel made with the bull's penis.

> As proof of his appreciation, Rosas sent the society a mysterious gift: a huge head of maize, harvested at his estate in Azul. So valuable a present, adorned with light blue ribbons, was a filthy use of this symbol and a gesture of contempt for the color. It was delivered by his daughter Manuela to the Franciscan Ravelo, with these instructions: "Someone most interested in the society sends you this head of maize so that you can insert it in the Unitarians." The society received this splendid emblem with applause and paraded it through the streets in triumph; so, in addition to mustaches, the red jacket, the dagger, and the stick [*verga*, bull's penis], it became one of the insignias of those who were then called *maz-horqueros*. (211)

The tyrant possessed "ferocious vices"; he dirtied "his own home with scandal and repugnant incest; the jails and barracks, with the blood of hundreds of noble citizens" (84). In Rivera Indarte's account, Manuela Rosas, the dictator's daughter, had been a model of "modesty and sweetness," before the unceasing sexual assaults by her father began. But Rosas would win out and, "to make her lose the coyness of her sex," exposed her to "dangers and to making her ride high-spirited colts," and confronted her at numerous meetings where he would overwhelm her with "orgies [that] exploited even her piety." Once he had "dirtied [her], and she was covered in mud, he raised her up to the highest social ranks," where he ordered the other women to kneel before her as "the most beautiful, the most perfect," the best of all (287). The "innocent virgin" was transformed into "a bloodthirsty *marimacho* [tomboy], wearing across her face the filthy stain of ruin" (287). The hypermasculinity of her father's society did not allow for femininity, not even in his own daughter.

In Rivera Indarte's *Las tablas de la sangre*, published in 1843, the same charges against Rosas were repeated, though with greater intensity, enjoying widespread success if we judge by the spread of Rivera Indarte's stories throughout Europe and the Americas. Every important publication on these continents received a copy in translation of the main stories facilitated by networks of British and French agents that favored resistance to Rosas's regime for its role in opposing the intervention of Europeans in the River Plate. This explains why Rivera Indarte's book opens with a long list in alphabetical order of the names of the men sacrificed by the *Mazorca*. In response to the indignation aroused by this listing of the dead, Rivera Indarte repeated the point he had been making since his days at the Montevideo newspaper *El Nacional*: "*Killing Rosas is a holy cause.*" The entire book is a call to tyrannicide and a justification of it by reference to the tyrant's crimes—theft, incest, torture, homicide, madness, and rape. But of all the crimes listed, the most serious remained the destruction of social order by arousing the "rabble" against the "decent people." The weapons used in the attacks were depicted as sexual, and Rosas continued to be seen as a "pervert" whose main pleasure consisted in debauchery and objectionable sexual practices.

> Those familiar with Rosas's life know that his recreations consist in more offenses to the decency of men and women. He rests from his bloody chores by tormenting the humanity prostrate and degraded before him; he amuses himself with outrage and fondling the naked part that even savages conceal; he laughs with the grotesque contortions that wretched, idiot madmen make between sharp blows. The blows, whippings, offenses, dirty and painful insertions are the harmonies with which he distracts himself during meals and at leisure. He keeps madmen to exploit and offend their modesty in the most oafish manner, and he has reduced some of them to the piteous condition of the eunuch, by a painful castration like that practiced on sheep.[48]

The depiction of these excesses had a political purpose. According to Rivera Indarte, Rosas's sodomite authority justified his assassination, and in support of this verdict he cited the most important specialist on natural law, Samuel von Pufendorf, who had justified the killing of those who attempted sodomy. In *Of the Law of Nature and Nations*, this scholar recalled that general Marius had crowned the soldier who "had killed Marius's nephew, a Tribune, or Colonel in the Army, for soliciting him to an unnatural debauch."[49] Using the same reasoning, Rivera Indarte justifies killing Rosas and the destruction of his political system. "The judgment of Marius has been much praised; not content with absolving a soldier who had killed one of his own nephews who had violently threatened to offend the soldier's modesty, he also crowned him with the wreath given those distinguishing themselves with extraordinary acts of valor."[50] This civilized approach to punishing Rosas for his barbarism was the result championed by this propaganda campaign. The publication proved most effective in fostering a negative perception of Rosas beyond the River Plate conflict zone in 1845, when the British and French navies declared the blockade of Buenos Aires, alarming foreigners who had interests in the region.

The prestigious newspaper the *Times* of London spread some of the accusations against Rosas, publishing a letter in October 1845 from Buenos Aires warning that the conflict might cause a long war and become the "utter ruin of English trade and English influence in this part of the world for generations." Rosas would never accept the British and French demands, and he had become "more popular than ever among the natives. The whole province of Buenos Ayres was under arms." The French were leading the attack and the British government "seemed to be totally misinformed of the affairs of the country," which alarmed those in London.[51] The circulation of *Tablas* was extensive because Rivera Indarte operated on behalf of Great Britain and France when negative information was needed to increase anger in these countries to support interventions in the River Plate, which were unpopular. The horrific actions described in a sensationist tone helped popularize the stories and their circulation, but the rapid escalation of the conflict also caused the emergence of controversial sources that defended the governor of Buenos Aires.

The *Newcastle Journal* reported in 1845 on a pamphlet titled "Rosas and His Calumniators," which exposed the author of the renowned "Tablas de Sangre"—a work that once stirred profound sensation across Europe"—as a "monstrous dealer in tales of horror." This author had recently sent a letter to the emperor of Brazil requesting "his armed intervention to save himself and [his] company" while pretending to be an innocent member of the exile community in Montevideo. Citing Pedro de Angelis's claims published in the *Archivo Americano*, the article portrayed Rivera Indarte as a fraud, a man devoid of honor, and someone willing to do anything for money, including stealing objects from a church.[52]

In the same year, a newspaper in Columbus, Ohio, printed a story on the same subject but condemning British and French aggression in the River Plate region. Holding the British accountable, based on the same pamphlet defending Rosas mentioned in the *Newcastle Journal*, the writer argued that the conflict stemmed from rival political parties in Montevideo, amplified by foreigners "arming themselves under European colors" and funded by "British loans secured on vast land grants in Uruguay." The accusations against the governor of Buenos Aires, "actively disseminated in France and England," were deemed "the most atrocious calumnies" spread by "one Indarte," described as "a political renegade and traitor of the blackest die [*sic*]," a "thief, a forger, a sacrilegious wretch," and, above all, a "foreign spirit" who betrayed his own country. The Monroe Doctrine established principles limiting European interference in the Americas; consequently, these European actions had to be opposed unless authorities were willing to sacrifice "our just national pride, and at the peril of our high national position." Clearly, this conflict extended beyond the Argentine Confederation and Uruguay, challenging the

Enlightenment's philosophies of emancipation and the universal principles governing such ideals that seemed forgotten until the foreign intervention took place.[53]

Rivera Indarte's accusations resurfaced in 1849 when the *Dublin Review* printed a provocative article that reviewed several works published on the River Plate conflict and the siege of Montevideo. It explained that the social struggle of Buenos Aires was caused by the same ills created by modern culture. The city's population was "almost exclusively Spanish," representing "Europe and the habits of civilization." On the other side, the rural population represented "the original inhabitants, that is barbarism, with all the customs and habits of primitive life." The population of the Pampas had "a peculiar physiognomy, such as is to be found in no other part of the world." The desert had shaped them, and no sign of social organization could be found among its inhabitants.[54] The last paragraph seems inspired by Sarmiento's *Facundo.*

In the anarchy generated by a society split in this way, the *Mazorca* was mentioned using the words of General O'Brien, an Irishman who was a British representative in Montevideo. He had sent a letter to Lord Aberdeen in 1844, and parts of the letter were included in the article to explain that the name "Mazorca" was derived "from the inward stalk of the maize when deprived of its grain," used as "an instrument of torture, of which your Lordship may have some idea, when calling to mind the agonizing death inflicted upon Edward II."[55] If we remember that this king died because a steel rod had been inserted into his rectum, the connection between politics and sexual violence becomes obvious.

Knowing the effects that this propaganda was having in Europe, in 1850 the Rosist *Archivo Americano* published a strong condemnation of the *Dublin Review* article and the lies and distortions promulgated by Rosas's opponents; this article was an "abominable fabrication of insults" and indignities "of every kind and degree, that it may fall under the anathema of enlightened opinion."[56] The claims were so absurd, de Angelis contended, that no person in their right mind would believe them. A list of books written by Europeans who praised Rosas followed to emphasize that accounts in the Unitarian press were not based on facts, but on political ambition. The "impure presses of Montevideo" lied about the Restoration Society being a "secret club" and about the existence of torture. De Angelis based his defense on the fact that a close inspection of the list of members of this society, published in *La Gaceta Mercantil* on April 16, 1842, revealed the names of members of the best families, and not the "rabble" described by the exiles.

During times of upheaval the "families who were more exposed to public hate" asked the protection of the "Sociedad Popular, called by the Montevideo press *Mazorca*, even when many Unitarian savages were alive thanks to the organization." This society was not "a club or a lodge": on the contrary its "virtuous citizens" were the most decided enemies of secret societies, and just hearing the words "lodges or lodge brothers" (*logistas*) filled them with "indignation and horror." Its members were "federal citizens, neighbors and owners of property, who loved liberty, honor, and the fatherland's dignity." Rosas despised the introduction of tribalism (*espíritu de secta*) in the country because this only happened during weak or factious governments. He only promoted unity. The society's members were not Jacobins, as depicted in France; ideas like these were created by "degraded beings outlawed by the society of decent men." Everyone was insulted "with an uncommon obscenity even in this class of evil men." The writer living in exile was the true "unheard and barbaric *savage*" who lied to those who did not know the country.[57] The clarification that the society was not a masonic lodge was probably related to the existence from 1835 to 1840 of two secret organizations, one operating in Buenos Aires and the other in Montevideo. Rosas had banned this type of organization from the country.

But despite de Angelis's efforts, by 1849 the stories about sexual violence had spread even in Spanish America. For example, in 1842 the influential *El Museo de Ambas Americas*, edited by the Colombian Juan García del Río in Chile, published an article about the violence in Argentina. With regard to sources, the article mentions that the authors consulted "the work published in 1840 in this same city [Valparaíso] with the title 'El Tirano de los pueblos argentinos,' *El Nacional de Montevideo* [which Rivera Indarte edited]; many papers and documents from Brazil, Buenos Aires, and other places, and the testimony of various trustworthy and respectable people."[58] The article repeats a long list of charges against "the tyrant," some of them referring to his lack of proper boundaries as expressed in his indifference to divisions among the races and sexes: "He has gradually corrupted the rabble, flattering them to the point of allowing his own daughter and his most distinguished partisans to dance, both in public and in private, with black slaves, and thus releasing all brakes on the brutal passions of the very dregs of the people. . . . [From the *Mazorca*] have emanated scenes so inhuman and disgusting that the pen resists putting their account to paper."[59]

In 1842 Pedro C. Avila, claiming to be an ex-member of the *Mazorca* published a book in Lima, Peru, denouncing the governor of Buenos Aires, and requesting foreign intervention to end a system of oppression that was at the service of the most barbaric forces. He began his story by showing how Manuela Rosas was a corrupted figure, used by her father to seduce enemies and maintain tight control over what was happening in the city, where Rosas's revenge over those who opposed him was in the hands of members of his secret police . Confused in the alcohol-infused stupor that accompanied the torture sessions, the "mazorqueros" heard every night such desperate and scary screaming that "not only made the house tremble, but also the entire neighborhood," a scene so disturbing, it led to the terror that dominated the whole city.[60]

Female suffering is an important part of Avila's book. "The painful moans of the beautiful sex reached heaven itself," but they could not stop the carnage. The narrative then turns to detail the extent to which the primitivism of Rosas's supporters degraded the nation. One day a father was beheaded while embracing his six daughters, and once he was dead the secret police started to "kiss them on the head" ending the abuse by "raping them all" and destroying their home so they did not have even a place to sleep (22). The popular rabble making up the mob that participated in such acts took pleasure in their sexual violence against women of the best families.

In another incident, this time in the city of Cordoba, three small children were kidnapped, and a ransom request sent to the families. After gathering all the money they had, the three mothers, who belonged to the best of society, showed up begging for the freedom of their sons. Since they had not been able to get the sum requested, the Rosist supporters forced them to have sex to make the full compensation. After the rape, they brought to the mothers the heads of their children "dripping blood and still moving," a spectacle so shocking that it killed them on the spot. The bosses then started laughing and "tied up the heads [of the children] in their mother's braids" (25). Avila explains that the dishonor of women who belonged to Unitarian families was part of a political strategy to terrorize society. Rosas gave his supporters permission to "rape the women who were known to be Unitarians" and the description of one of these sessions clearly aims to destroy Rosas's reputation as a defender of the laws (25).

Taking place at the San Miguel church, men were being tortured at the same time that a group of young women were being punished, after which "they were stretched" to be covered with asphalt (*alquitrán*), "their clothes were torn off, and they were thrown into the streets naked; other men feeling more pity dragged them around the neighborhood, and contented themselves just with raping them," says Avila ironically (28). He makes it clear that he does not want to mention "the considerable number of people that had been raped[,] because many of them are still alive," and also out of consideration for the families of those who had died, "among them Miss Dolores Amaral, one of the most interesting young women of Buenos Aires, an innocent victim sacrificed in the name of the illustrious restorer of the laws and public tranquility" (26). Following other exile writers, Avila explained that these crimes against women were committed to destroy the honor of the men in the family.

In the exile account, the phallus was the symbol that dominated the public space in Buenos Aires, and all weapons were understood as phallic. Together with the *Mazorca*, Rosas's power was represented by "*a knife and a verga*."[61] The "verga" was "the nerve" of the bull, the penis, which, after being twisted and dried, "does not break" and can be used as a bludgeon to hit. Used as a whip in Argentina, this instrument of punishment became a symbol of domination used to assault those men interested in exploring new gender values. The poet Esteban Echeverría had explained that the members of the secret police wore the "crimson ribbon" together with "the moustache," and "looked, with the verga in hand, for victims or knaves to stigmatize."[62] In his mind, the symbolism of being struck with a bull's penis was not lost, as the slaughterhouse where cattle were killed was one of the most productive businesses of those, like Rosas, who managed large estates.

News about Rosas's abuses against the population of Buenos Aires also spread among scientists. In July 1845, Bartholomew Sullivan (1810–90), a naval officer and lieutenant on the *Beagle*, wrote a letter to Darwin from Montevideo. He reported about the fossils he was collecting and his work in the Malvinas Islands. He also mentioned crossing paths with the German landscape artist Johann Rugendas (1802–58), who was leaving for London; he reminded Darwin that he had met him in Chile years earlier. Finally, he noted the violence unleashed by Rosas among British families in Buenos Aires, writing, for example, about "a Scotch family of nine, Father Mother & Children have all had their throats cut near Buenos Ayres." Sullivan also reported that Montevideo was holding out against the siege organized by Rosas but that the most "dreadful butcheries in the country" had occurred, with troops from Buenos Aires killing "above 500 prisoners in cold blood near Maldonado, principally Indians who had been serving Reveira" (Fructuoso Rivera, Uruguay's leader from 1839 to 1843).[63] Similar reports about the violence were spread in scientific networks because the circulation of information and collections of pieces were disrupted by the war, as had also happened in the case of Bonpland. In 1845 Darwin revised his previous favorable comments about Rosas in *Journal*, writing that they had turned out "entirely and miserably wrong."[64]

In 1852, after the defeat of the Rosist forces, a trial was organized to judge Rosas in absentia because he had fled to live in exile in England. Emilio Agrelo wrote a condemnation of Rosas in the legal case against him. He wrote that the tyrant ordered "the extermination of all that was civilized: he rejected the moderate and progressive element in order to substitute it with ignorance and barbarism." This attempt to destroy civilization was another crime "that increased the catalog of [crimes] committed by Rosas."[65] In the documents presented by those responsible for prosecuting him, he is represented as a figure who used primitive instinctual forces against those who wanted to impose reason in the country. One document introduced in the case quotes Rosas's own justification for the crimes committed during 1840–42. He alleged that the invasion of the "savage Unitarians" was so hateful and atrocious that it "exalted the popular feeling" since the people could not stop a response triggered "under the terrible aspects of a natural revenge." The killings were then understandable from the point of view of Rosas's natural law.[66] Against this logic, his enemies affirmed the undeniable civilized right to freethinking, making the campaign to exterminate European ideas and education a crime against humanity.

Buenos Aires's intervention in Montevideo, and the attempt to remove those in power, happened because the capital of Uruguay had become an international center of political activity financed by French money and unofficially supported by some British agents who did not want France to acquire power in the River Plate. According to Diego Lamas, Rivera had presented a document to Lord Howden explaining that Montevideo was subjected "exclusively to the French influence and the will of [Giuseppe] Garibaldi." These two conspired for a long time and had erased "all Uruguayan [Oriental] influence and elements; thus, there was not a single authority or character that represented the national interests" in the country.[67] The ensuing conflicts continued until 1851 when a peace treaty was signed to end all the hostilities.

Garibaldi wrote in his autobiography about the masculine culture of those from Buenos Aires, who had "the pretension of being the first in elegance." This man had more "imagination" than those from Montevideo, and the main poets "known in America were born at Buenos Aires. Vareta [Varela], Lefimer [Lafinur], Demengue [Domínguez]" were some of the "porteño" poets. Between the men of both cities there was a "rivalry in elegance and courage," the beauty of their women, and the talent of their poets; these men were irritable, "capricious as women, and with all that, sometimes simple as children."[68] This last statement was probably one of the few points on which the Italian revolutionary agreed with Rosas to the embarrassment of his enemies.

In his writings for *El Nacional*, published in Montevideo during 1845, Andrés Lamas recognized the importance of fashion in establishing a sense of otherness and modernity in society. According to him, Rosas violently prosecuted "the dress and manners of civilization," mandating a style of dress assumed to be more proper for its authenticity. Lamas explained that the reason for such measures was that the governor of Buenos Aires "understood through intuition that clothes, manners, way of speaking and purity of word, have more influence than that attributed to them by the frivolous spirits."[69] It was for this reason that any sign of liberal culture raised great alarm in Buenos Aires. Even Rosas's daughter, Manuela, was very careful with what she represented in the public space; he wrote to an English friend thanking him for the dresses he had ordered in England for her, noting that the clothing was not only "very beautiful" but true "federal dresses" that represented the Federal party style and color since they were all red.[70]

In 1843 the exiled Florencio Varela was sent to London as a representative of the interests of Uruguay at the recommendation of the British Commodore Purvis; he was prominent as founder and editor of the *Comercio del Plata*, one of the main newspapers in Montevideo. His mission was to personally request that the authorities send British troops to fight Buenos Aires. According to General José C. Paz, the objective was to propose the creation of a new nation by adding the provinces of Entre Ríos and Corrientes to Uruguay, an idea that Paz opposed and that was scandalous to those in the country. This action reaffirmed the image of those in exile as traitors. Varela did not have any success, and it seemed clear that Lord Aberdeen, the foreign secretary, was using those in exile with no intention of breaking up communications with Rosas.

De Angelis directly responded to Montevideo's newspaper *El Nacional* over the publication of an article written by Rivera Indarte that described the conflict with Rosas as a fight between "*absolutists and free men vs. mashorqueros and patriots*."[71] The response emphasized the lack of morality of its author, citing his definition of the brotherhood that united those in exile as "*the most sweet fraternity that unites those who have born of the same womb*." De Angelis specified that such an understanding was typical of a degraded man lacking in decency

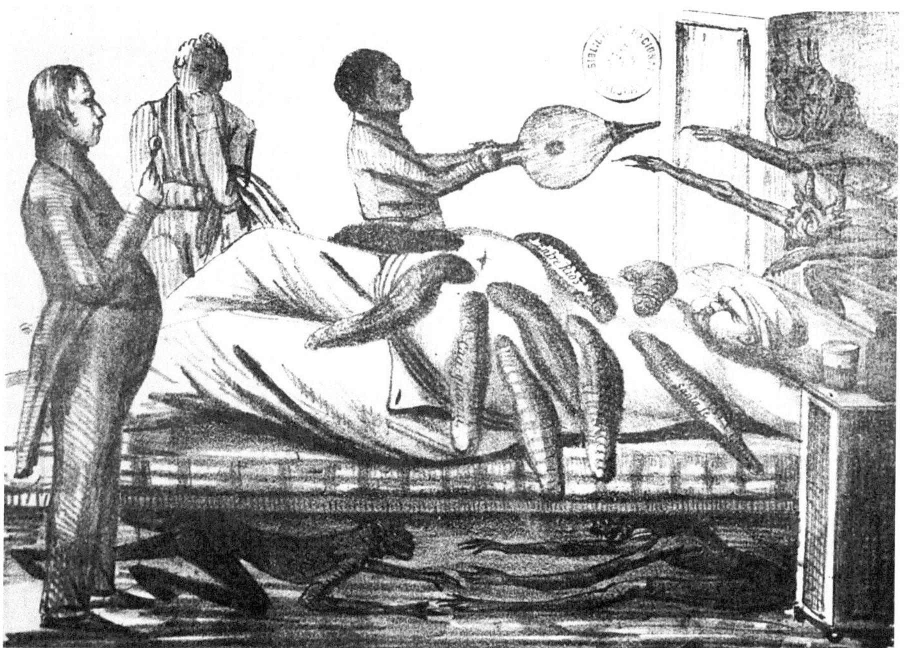

Figure 6.1. Rosas's sickness and science. "Rosas Enfermo," published in *Muera Rosas*, a weekly publication of the exile community in Montevideo. (*Muera Rosas*, no. 3, January 6, 1842). Uruguay's Ministry of Education and Culture, *Anáforas*, Newspapers of Uruguay, https://anaforas.fic.edu.uy/jspui/handle/123456789/4053.

and good habits, who boasted about his treason, "looking in horror at the true defenders of *liberty, civilization, and morals*."[72] While the two parties claimed to be civilized, they did not agree about the meaning of modern civilization.

Varela and others writing from Montevideo were advocating the restoration of everything scientific that Rivadavia had brought to the country. *La revue indépendante,* a journal that was published from 1841 to 1848 and edited by the socialist Pierre Leroux, understood this objective and printed articles about the Argentine Confederation and Uruguay. In one of them, written by the editor of a Montevideo newspaper, it was revealed that the Argentine Confederation missed "these young people trained at the university of Governor Rivadavia, these *doctores* [doctors] with the sweet voice who knew how to talk about love, and inspire it," unlike the federalists who cared very little about these issues, and were "passionate about more massacres and war" that destroyed everything.[73] The Rosist confederation rejected the sciences and preferred superstitious treatments, such as those of the Black population, which was ignorant about medicine and civilized knowledge.

TRAVEL, RACE, SCIENCE, AND MASCULINITY

Florencio Varela was clearly anxious before his travel to London to negotiate with Lord Aberdeen, as alluded to in the previous section. In September 1843, he wrote that he was "sure that many illusions would be lost" by the end of the diplomatic mission that was sending him to London. He was aware that he would see "men, things, and ideas from many different points of view than those that appeared in books." But he wanted to recognize by himself "all my impressions," registering the facts of his "own experience." He also wished to witness "the progress of truth, intelligence, and work in the field of sciences, literature, and industry," because the memories he was to gather would become a "source of useful applications for my unlucky fatherland" once he returned to Argentina.[74] The insecurity that traveling to Europe presented to Rosas's enemies was that their political project was based on a model they claimed existed in civilized nations through books, but not through personal experience. Finding discrepancies that proved the contrary to what they had read only indicated that their enemy was right and they were mad, imagining things that did not exist.

Reality was itself also a source of anxiety in the case of Alberdi who traveled in this same year. His travel diary, published in 1845 as *Veinte días en Génova* (Twenty Days in Genoa), described his arrival in Europe.[75] He was thirty-three years old at the time but felt "like a 20-year-old," and considered himself "reborn" (63). But he was not sure what his role as narrator should be.

> For a South American, how new the sight of a European capital is! [. . .] Wouldn't it be useful and agreeable for the American reader to find a book containing the fresh and ingenuous expression of the impressions felt by one visiting these cities for the first time? I believe so, though the attempt may cost me a bit of my reputation as a cool observer in the eyes of intelligent and worldly people. I believe that an American would show more prudence in revealing, even at the expense of his pride, his real emotions, than in exhibiting an unfelt nonchalance or wasting breath in vague generalities that say nothing to the listener three thousand leagues away from the things being described. (65)

The reason the book is about Genoa and not Paris, the city that was most attractive to those in the Americas, and one Alberdi visited, was related to Young Italy, which had been inspirational to his generation's own dogma in

1837 and because Genoa was Mazzini's place of birth. Alberdi mentioned meeting with families of Italians who had connections in the River Plate. Also, because the ideas about race that had become important by this time explained that "climate and history" defined the Americas and Europe, there was a split between northern and southern peoples on both continents. According to Alberdi, Spanish America descended from Greco-Roman roots, as did Italy, and thus Celtic and Germanic social organizations would never be useful to them.

Spanish American nations belonged to Mediterranean Europe; it was there where there was and would be the "immediate and natural school." Also, the origins of "our genius" could be found in this region; even when the modern world existed entirely "without traditions and only in the present and future time," Italy could become a good example for nation-building because the mystery of the past was only accessible "to those who inhabited their remains" (122). Alberdi described Spanish American nations as "young and fiery peoples" that would need to wait to acquire material civilization and an intelligent culture to produce good art and culture in the future. They needed organic growth because Spain lived in their "skulls and veins." The "ax of the revolution" had severed the connection between the trunk and the branches of "our genealogical tree," but the resultant "exotic plant" demanded land and "the cultivation of plants like the ones that helped it grow in their original land." The secret of their "current existence" could be revealed in the study of the colonial past, which was connected to Spanish development (124).

Following the image of the tree, Alberdi noted that "science" would make people like him return to "the family that was rejected" during the revolutionary days. Soon, it would not be heard in Spanish that the "Spaniard is barbaric" because every possible bad thing that could be said about "our race" had already been said; by then it was time to do the opposite. The descendants of the conquerors needed to be proud of 1492; they "needed to praise our origin." After expressing admiration for the 1810 revolution, it was now time to praise those who "invented half of the Globe, depopulating it of barbaric races, a kind of human weed, to populate America with the most beautiful race of Europe, which was the noble Spanish race." He pleaded that people should not be "fighting the Spanish race, because we are ourselves part of it," and that it had created "the world inhabited" by those like him; the solution was to "study Spain to know ourselves; and to know Spain well studying in Spain," as it was done before independence (124).

Sensationism and its derivatives had failed, and at this time there was not a system that unified philosophy and science seamlessly in all aspects of life because the domination of politics over all fields made the creation of a universal system impossible. For example, Alberdi explained that the popularity of French philosophy at this time was related to the use of German systems rejected in Italy because of the years of Austrian domination; "abstraction" was hated, and the German doctrines were called "*northern fog*," which explained why only French thinkers were known there. The "transcendental and metaphysical initiative" of Germany was gone; no philosopher or system predominated over others. Hegel's legacy was by then divided into ten schools that "tore each other apart without mercy." Philosophy, art, poetry, and theology had "abdicated" their independence and they were now at the service of politics. Alberdi cited an article about German literature that appeared in the *Revue des Deux Mondes* about the Young Germany movement.[76]

On October 10, 1843, just days before departing to return to the American continent, he wrote: "In four days, I leave Paris for Le Havre, where I will embark for America. How I sigh to be back in those countries! [H]ow beautiful America is! [H]ow soothing! [H]ow sweet! Now I know it; now that I have gotten to know these hellish countries, these selfish and insensitive peoples of gilded vice and titled prostitution. We are worth so much and we do not know it; we value Europe more than it deserves."[77] The trip's cycle had closed in a peculiar way. Alberdi acknowledged that the concept of what was considered civilized had drastically changed; Rosas's criticism of Europe and his praise of the American continent seemed closer to the truth. By the time he arrived in Chile the absence of content or "vision" had become evident. The travelogue prompts Adolfo Prieto to remark, the "Argentina that emerges from this new perception is, curiously, the image of a country suspended in limbo. Its frozen history is the natural environment in which a complete image of the country will be created."[78] In addition, the model of man and the race of the population were also to be determined, which is partly the legacy of this generation.

Domingo F. Sarmiento emerged as an important leader around this time. He also became a traveler around the same time but provided a perspective different from Alberdi's. According to Sarmiento, Hegel's *La filosofía de la historia* (The Philosophy of History) was one of the first readings that influenced him, followed by Paul-Henri Thiry, Baron d'Holbach's *La moral universal* (Universal Morality), and his *Sistema de la naturaleza* (System of Nature),[79] and these books had left him "at least a bit less skeptical," with his soul "rubbed with sweet and bitter herbs, clean as the palm of the hand." He continued, writing that he should pen "a treatise on the cultivation of the soul, to teach the material with which you need to fertilize it to make it produce a hundredfold.[80]

La filosofía de la historia mentioned here may also refer to Herder's writings, or maybe he was just lying about reading Hegel's work, which was not unusual for him to do.[81] This philosopher was mentioned in many of the French journals he read, so he definitely knew about his philosophy.

In 1845, after the publication of *Facundo*, Sarmiento's defense of the conservatives had earned him a set of enemies that put his life in danger. These two situations moved the Chilean senator Manuel Montt to arrange a longed-for trip to Europe, where he would spend two and a half years studying the continent's educational system. In 1846 he was waiting for his ship to sail; he later described his experiences in a letter directed to his exiled friends in Valparaiso while on his way to Montevideo. He noted having discovered himself not walking but "flaneando" (being a flaneur) on the docks with his friend Vélez Sarsfield. As Elizabeth Wilson has shown, the flaneur represented another masculine model that resulted from the culture of modernity. He did not express "the triumph of masculine power, but its attenuation." It was "masculinity as unstable, caught up in the violent dislocations that characterized urbanization."[82] Since many poets and writers were flaneurs, Sarmiento had learned about them from books.

This flaneur character had already been introduced in Buenos Aires by de Angelis, who, from December 19, 1831, to March 3, 1832, had published in Buenos Aires, *Le Flâneur: Ambifu politique et literaire*, in French. Only twelve issues came out, and according to the *British Packet* it was a fun publication, satirical in tone.[83] Its final published words were "Gentlemen, nothing is changed; it's only one Flâneur less."[84] But this character did not originate in the urban life of a South American city, it was in France where the "flâneur" lived "on the boulevards, and made the street and cafes of Paris his drawing room."[85] By the 1840s this masculine model had finally replaced the Romantic bohemian, and unlike the passionate emotion of the latter, he was "a detached observer," often interested in literature, often being a poet, for example. As a literary observer of this figure, Sarmiento understood it as another cosmopolitan fashion that civilized men needed to adopt and experience, in this case to indicate their readiness to fit in in Europe.

The gender ambiguity of this figure was not lost in Sarmiento's narrative, putting him in an uncomfortable position as is evident in the sudden change of topic in the second part of his letter, when he confirmed that he was also a man with very manly desires. He had met with Maria Sánchez de Mendeville, one of the most prominent women in Argentina, supporter of civilized ideas, and possessor of a scandalous personal life that had created controversies back in Buenos Aires. Her second marriage was to Washington Mendeville, the French representative in the River Plate. According to Sarmiento he became fast friends with her, so much so that he recalled how one morning, while they were sitting together on the sofa and she was "talking, lying, with all the grace that she knew how to use" he felt an unexpected urge. "I felt . . . Come on, this could happen to anyone, I surprised myself as the sad victim of an erection, so persistent that I almost interrupted her and, although she is sixty years old, raped her [*violarla*]. Luckily someone came in and saved me from such a crime. I tell you this to convey the degree of our friendship. She overwhelmed me with letters of recommendation."[86] The transition from the Romantic bohemian to the flaneur marked how new forms of representing masculinity replaced older ones in a very short time. As Mary Poovey has demonstrated in her study of mid-nineteenth-century England, representations of gender were in constant flux, as in Argentina among those in exile. Poovey's central conclusions highlight that traditional Victorian masculinity was contested and perpetually under construction. This was part of modernity's cultural framework and its inability to create a universal culture capable of accepting multiple roles for men while establishing the authority to embody them, as envisioned in the late eighteenth century through sensationism.

The unresolved tension between a systemic ideology for the nation and the consequences of the growth of a culture of consumption worsened during this period. An article about the flaneur, published in England in 1841, depicted him as a poor man, perpetually in need of money because "he never has any." He was luxurious and obsessed with consumption, and he used his intelligence to enjoy pleasures without paying for them. He observed everything and was seen everywhere; and though he and others like him were described as "the most indolent," they paradoxically appeared to be "the most industrious of citizens!" As at other critical junctures examined previously, masculine identity became intertwined with consumption and a lack of personal integrity. By 1849, the abundance of travel books, a genre he loved, caused Sarmiento to recognize that this literary genre was stripped of "novelty, [that] civilized life brought the same characters [and] the same means of existence everywhere." People who, like him, were from less civilized climates felt "the inability to observe, lacking the necessary spiritual preparation," a handicap that rendered "the eye bleary and myopic at the vastness of the sights, [and] the multitude of objects [found] therein."[87] Authors like him were less able to represent and classify, which he affirmed, stating: "I wrote, then, what I wrote (for I would not know how else to classify it), obeying the instincts and impulses that come from within, and that sometimes reason itself does not contain" (9). Warning readers about his own limitations, he was open about the uniqueness of his project.

As was the case with *Veinte días en Génova*, this narrative contains many references to nature and trees; America had "deeply rooted, traditional causes from which we must break, if we do not want to let ourselves be dragged into disintegration, into nothingness" that engulfed "the remains of the people and races unable to live, like the primitive and formless creatures that succeed them on Earth when the atmosphere changes," modifications that resulted in "altering the elements" that kept them alive. He had suffered "heartache" when learning about these ideas, feeling that his own were "pulled out one by one to be substituted with others that do little to gratify those instinctive and natural affections of a man's spirit." The new hierarchies that placed populations in fixed categories of superiority and inferiority made him acknowledge that there were "regions too lofty, in whose atmosphere those born of lower lands cannot breathe; and it is mad to stare at the sun, risking certain loss of sight" (12). After warnings and self-doubts, the travel narration began.

A letter written in May 1846 to Carlos Tejedor, who was in Montevideo, includes a narrative of Sarmiento's stay in France. He depicted his arrival as a moment of great anxiety and emotion. He described how, at last, "the coasts of France appeared on the distant horizon." The people arriving were greeted with delight; he greeted them, too, "feeling timid and fearful at the idea of introducing myself into the bosom of European society, lacking their manners and ways, careful not to let my provincial *gaucherie*, the source of so many Parisian jokes, show." While his heart pounded as he approached land, he proceeded to arrange himself to look good, "as when a young lover prepares to meet the ladies" (101). But contrary to expectations, he was received by "an ignoble rabble of tastefully dressed servants" who assaulted and shouted at him, surrounding him "like flies" and "stinking [him up] with their breath." The repugnance that this scene inspired in him prompted a surprising exclamation: "Eh! Europe! [A] sad mixture of grandeur and misery, of wisdom and brutishness at the same time, a sublime and dirty receptacle of all that man raises up or keeps degraded, kings and lackeys, monuments and hospices, opulence, and savagery!" (101). As was the case with Alberdi, the extent of poverty and political instability that existed in France was far from an example of what they were promoting in South America.

The negative feelings did not last long. In September 1846, Sarmiento wrote another letter describing his experience in Paris. Addressed to his close friend from San Juan, Antonio Aberastarain, he described the city of his dreams and how it had left him thoroughly amazed. Its space was completely feminine, with streets as "charming and coquettish as a girlfriend," which excited his attention (116). The city offered itself to him completely, and he quickly became a *flâneur*, strolling and devoting himself to wandering the streets: "*Je flâne*, I walk like a spirit, like an element, like a soulless body in the loneliness of Paris" (117). Only in Paris could "brotherly kindness [be found]. And only in Paris [was] the foreigner the owner, the tyrant of the city" (118). This was the modern space he had imagined—feminized and cosmopolitan.

In his mapping of gender in Parisian society, Sarmiento's careful analysis reveals the multiplicity of masculine identities that existed. The French man was "the boldest warrior, the most ardent poet, the most profound scholar, the most frivolous dandy, the most zealous citizen, the youth most given over to pleasure, the most delicate artist, and the most gentle man in his dealings with others" (141). Regarding women, Sarmiento asserted that "it would take a long time to cover the scale between the prostitute and the married woman," which explained "the culture of French women, the infinite grace of the Parisian, and the way in which all social classes dress alike," which was very positive in his view (144).

At the margins of the geographical space taken in by Sarmiento lies a more complex map of the city, that of the political institutions. On introducing this topic, Sarmiento hastened to communicate the state of mind and distress that Paris inspired in him. The city was "sick with brain fever." "The political world is about to collapse; all the signs point to universal cataclysm; men hurry to record the history of times gone by [. . .] taking one route and abandoning it the next day to start again on another. Nobody is today what he was yesterday" (120). At the center of this end of the world was he who had come hoping to find progress fully formed, functioning, and governing society in every way.

The harshest part of his stay in France was his having to talk to politicians who were not interested at all in his opinion. As previously explained, the River Plate intervention was viewed in France in connection with France's own political problems and failures. When Sarmiento visited M. Dessage, head of the political section of the French Ministry of the Interior and, according to the Argentine, "the eyes and ears of Guizot," the leader of the government, it became clear to him that his side of the conflict was not seen at all. "I tried make to make him understand something, but it was impossible! Everything of which I speak, it's all Greek to him. In summary, for them: Rosas equals Louis Phillipe. The *mazorca* = the moderate party. The gauchos = *la petite propiété*, small landowners. The Unitarians = the opposition. Paz, Varela, etc. = Thiers, Rollín, Odilon-Barrot" (124).

Vice Admiral Mackau was no better: he limited himself to acting as if he were listening, falling asleep in the middle of a fiery speech, Sarmiento's best talk ever in his view. As he tells it, he had better luck with deputy Adolphe Thiers (1797–1877), who at last listened with interest and treated him with respect. As a rule, though, Sarmiento was unable to communicate with anyone politically; civilized men did not consider his opinion and scorned him, thinking that he did not know what he was talking about.

The interweaving of poverty with political and social chaos, ignorance of American historical realities and its men, climaxed when Sarmiento revealed that during the debate between deputy François Guizot (1787–1874) and Adolphe Thiers, both historians, the former had asserted that it was "necessary to stop progress," that there was "too much progress." Sarmiento despaired, seeing that if civilized nations were deciding to halt their efforts to assist in the modernization of uncivilized places, then Argentina was lost: "Poor humanity, what will become of it now!" (135). His hope of finding a civilized political network he could join was destroyed after he realized that the exile community was not recognized as an ally to civilization, or as more enlightened than Rosas.

Once his hopes for a European model faded, he returned to the past idea of the creation of a new republic and political culture on the American continent, an idea that was contradicted by the importance that race had acquired in the United States; but, for Sarmiento, "when reality fades away, the imagination takes over."[88] In a long letter dated November 12, 1847, addressed to Valentín Alsina, Sarmiento sums up his experiences after a two-month stay in the United States.

Proclaiming the delivery of an optimistic message for those in exile, he sought to convince himself and the rest of the world that the Republic existed, "strong, invincible," and that "her lamp [was] lit" in the United States. It was in this country that "someday justice, equality, and law" would come to those like him when the South reflected "the light of the North" (334). The future was in the Americas, in the northern portion of this continent where the strong, virile, but rude North American men were maturing.

In the United States, Sarmiento found a culture that defined its identity in terms of male association, dictating that each man believed "what he wants to, choosing his own leaders, saying and writing what he [thinks]," he was judged by a jury, and had the right "to bail for other than capital crimes"

(379). In short, men in the United States were masters of their own destiny in a way that Sarmiento, at the mercy of Rosas, was not. People living there were "the only ones who can be compared with the ancient Romans," with "the same virile superiority, the same persistence, the same strategy, the same preoccupation with a future of power and of grandeur" (384). This society was superior because it was dominated by a masculine principle, while in France the feminization of life had made men decadent.

American society was so masculine that its effects did not escape the female sex. Following the ideas and model of his idol Alexis de Tocqueville, Sarmiento marvels at this, noting that a woman was a "man of feminine sex," being "free as a butterfly until the moment she [sealed] herself in the domestic cocoon in order to fulfill through marriage her social functions" (348). Unlike the Parisian women whom he presented as oscillating between prostitution and marriage, the virtue of these US women was complete.

Out of need, there is only a brief reference to slavery in the book, and Sarmiento explained this as the "Gordian knot that the sword cannot cut, and it fills the otherwise clear and radiant future of the American Union with gloomy shadows" (355). But among the prospective solutions he did not provide an easy answer. He recognized the seriousness of the problem and the prospects for a civil war, a "race war within a century, a war of extermination, or a backward and vile black nation alongside another white one, the most powerful and cultured on earth" (378). After being at the center of Louis Agassiz's empire, and the institutions that had created ethnology, his racial views became much more racist, affirming the supremacy of the Anglo-Saxon race as reflected in its biological progress, something that is naturalized to be considered inevitable. In this way, the supremacy of one race over another was not the result of human choices but of a natural law that mysteriously chose a favorite race to express itself.

SOCIALIST DOGMA AND THE NEW MAN

As the cases of Sarmiento and Alberdi indicate, they were both attempting to find a way to renew their political plan at a time of great concern about the future of republicanism and democracy. In 1842 Alberdi explained this problem. Knowledge was to be gained by reading the work of "Fichte, Hegel, Stuart [Mill], Kant, Cousin, Jouffroy, Leroux, etc.," because what mattered were "philosophers, but not philosophy."[89] In his view it was no longer necessary "to get oneself killed for any system: in philosophy, tolerance is the rule of our time" (606). He took three important schools into account—the sensualist (sensationism), the mystical (socialist), and the eclectic—and went on to explain that since the American continent practiced what Europe was teaching, its role was "completely positive and applied" (613). The United States had proved "it was not true that it was indispensable to have first a philosophical movement in order to achieve political and social change," which explained why pure abstraction and metaphysics "would not grow roots in America" (613). Philosophy had to be "essentially political and social in its object, ardent and prophetic in its instincts, synthetic and organic in its method, positive and realist in the way it proceed[ed], republican in its spirit and destiny" (615). American philosophy, for Alberdi, had to solve American problems.

Socialism offered a new plan for transforming society, but Alberdi was not comfortable with the gender ideas proposed by socialists, which he mentioned in his writings. In *Tobías*, written in 1844 and published in 1851, Alberdi presented an ironic view about the new radical gender ideas circulating in France, of which he disapproved.[90] George Sand was a woman, but she used a male name to identify herself, the same as the socialists proposing the erasure of boundaries between what it was to be a man or a woman. "In the present century, the sexes have tended to become confused. The anatomical research of certain socialists has revealed that there is not, after all, any real organic difference between woman and man. This doctrine will doubtless lead the women of Paris, at this unlikely date, to renew the famous Tennis Court Oath in protest against their age-old duty to propagate the species. And if the men fail to assume this burden, God only knows how, and by whom, the human race will be renewed.[91]"

Together with the already mentioned Mazzini, Pierre Leroux (1797–1871), a Saint-Simon disciple, became important in the 1840s. Leroux strongly attacked Victor Cousin's conservatism and disagreed with his interpretation of psychology. Published in 1840, *De l'humanité, de son principe et de son avenir* was based on articles previously published in the *Globe*, the *Revue encyclopédique*, and the *Encyclopédie nouvelle*, and offered a synthetic system that again linked philosophy and science with a new conception of humanity.[92] Here man was defined "as a progressive animal" living "in society and in society only." Society could be improved, "and he improved in this improved society."[93] The essence of this "homme-humanité" was virtual, which meant that "in real life the soul continually passes from one manifestation to another, fully feeling its state of virtuality within itself," covering with "ever new manifestations this power of being which causes being" and thus always showing itself visible "while basically its essence is to be virtual." In contrast, the dead had only "the essence of life without having life," they lacked "the power to manifest themselves." Life without manifestation was "living in pure virtuality," an idea inspired by ancient literature.[94]

The transmigration of the soul was partly inspired by Gotthold Ephraim Lessing (1729–81), and in his interest in this poet, Leroux returned to the universalism of the eighteenth century to fix some of the problems present at this time. Those who followed him were called humanitarian socialists with George Sand among the better known members. The most radical revolutionary ideologies in this decade came through derivations of Saint-Simonism, Fourierism, and Communism; the latter's manifesto was published in 1848.[95]

Among the various ideologies focused on social egalitarianism, the primary shared trait was their advocacy for societal regeneration and the application of scientific principles or laws to analyze society. Pierre Leroux's socialist philosophy offered a regenerative vision, portraying humanity as defined by association and mutual acknowledgment, a perspective shared by most socialists. Esteban Echeverría, the Argentine poet, concentrated on revitalizing his country's social fabric, viewing society as an organism that evolved according to natural laws that could not be disregarded. His revolutionary politics aimed to foster realism grounded in an understanding of local conditions, intertwining Romantic history and nature in his analysis.

By the 1840s, it was evident that in the ever-changing realm of ideas and politics, an organic framework was essential to explain transformation on a grand scale. This accounts for the influence of Geoffroy Saint-Hilaire (1772–1844) on the Generation of '37. As Adrian Desmond notes, this scientist and his followers "had come to accept a unitary composition for all animals" by the 1830s. Consequently, "animal life could be arranged into a continuous, interconnected series—rather than divided into distinct 'divisions,'" as Cuvier insisted, and "it was this that enhanced the prospect of evolution." The resulting series "could be used to show the 'history' of each organ rising in complexity from snail to man," and this history "could be transformed into a genuine ancestral lineage." Debates ensued, with differing positions on evolution; Saint-Hilaire championed the transmutationist perspective.[96]

Transmutationist principles were applied to political philosophy, as John Tresch demonstrates in his exploration of Leroux's work, which is very helpful for understanding Echeverría's work in turn. Like the Argentine, the French writer noticed the importance of the work of naturalists for understanding and classifying change and the laws that produced it. These laws could then be extended to the notion of perfectibility and social change under the authority of a philosophical system. In the case of Leroux, the writings of the naturalist Geoffroy Saint-Hilaire provided the foundation of his socialist philosophy. This scientist was one of the most vocal interpreters of Lamarckian transformism and was an advocate "of the unity of composition."[97] This concept revived "aspects of Lamarck's neglected 'zoological philosophy,'" accepting that changes in the milieu were seen "as causes for the appearance of morphological differences." The novelty was that he also emphasized "the mediating role played by habits" because surrounding conditions "could directly accelerate or slow, interrupt or prolong, a universal process of development."[98] This meant that species were modified by material causes in their environment, which was an interpretation that prevailed early in Spanish America as we saw in the case of Francisco Caldas.

Following Leibniz, Saint-Hilaire took two important ideas: nature was "unity in variety," and his universal plan was a set of "virtual conditions" whose "actualizations depended on the circumstances in which the animal was found."[99] Nature and history bonded in this interpretation, leading to a history of natural philosophy that saw society as an organism following laws that regulated and organized transformation. Like Sarmiento in *Facundo*, Echeverría also believed in the importance of the environment and habits to explain the ignorance of Rosas's followers, and changing this environment needed to be a priority for politicians. In 1844 Echeverría gave a speech to celebrate a national holiday in Montevideo. In it he addressed the importance of popular education to transform the social milieu of Argentine society. He repeated that the starting point was the ideology of May 1810, which had created the country with the intention of forming "an emancipated society under a regulating principle that differed from the one that existed in colonial society."[100] In truth, the 1810 revolution lacked clear objectives initially. Republicanism, for instance, was not favored by all; many preferred a monarchy, with figures like Manuel Belgrano advocating a monarchy led by an Indigenous king, while other factions, the most radical being a minority, held differing views. Nonetheless, Esteban Echeverría reinterpreted the past to position his generation's project within a teleological framework that demanded realization.

The ensuing civil wars, in his perspective, arose from a conflict between those championing a "progressive and democratic" principle and caudillos (strongmen) who were "counterrevolutionary," upholding colonial values. These opposing forces needed to coexist, as they were tied to the nation's origins, and that generation of revolutionary "virile men" emerged as either martyrs or victors in the struggle (415). The solution to this persistent conflict was ideological renewal through "the synthetic May thought, the political and social ideal it represented." Independence celebrations displayed mere symbols, lacking "a live expression of a social creed" (419)

In a genuine democracy, "where people [were] the beginning and end of everything," national festivities should serve as "a grandiose temple" to renew and strengthen the creed and "to inspire that serious and virile enthusiasm that fosters and justifies patriotism" (419). In Argentina, however, democracy had faltered because the populace, steeped in ignorance, failed to grasp the true nature of the transformation enveloping them. Notably, in this return to the 1810 creed, the term "virile" reemerged to exalt early republican virtue and patriotism.

At this time, Echeverría was developing an elementary education plan to foster societal regeneration, aiming for "the indispensable preparation of intellectual and moral culture," tailored to Argentina's needs. This meant that implementing the "triad" of liberty, equality, and fraternity, as articulated by Pierre Leroux, would require time (457). The United States, where Leroux's philosophy also garnered interest, served as a model for this approach.

In 1842 the *Boston Quarterly Review* published a book review of *De l'humanité de son principe*, which noted that its author expressed the doctrines and aspirations of "*la jeune France*" (the Young France), and was "a true lover of his race, a firm friend of liberty and equality, and a bold champion of social and religious progress."[101] Leroux's philosophy did not look for "the regeneration and sanctification of individuals" but for "the regeneration and sanctification of the species, of the race, by means of the new life with God" communicated through Jesus (266).

Leroux's understanding of man as made by "*sensation-sentiment-cognition invisibly united*" needed to be translated to the United States environment, reflecting local reality, it should be that man "acts, as well as feels and knows," which explains why this triad changed to "action-sentiment-cognition invisibly united" (274). The difference was explained by Leroux's philosophy's definition of man "phenomenally, rather than ontologically"; the ontological was always "revealed in the phenomenal," and human knowledge of being, "as *the subject of the phenomenon*," was as direct and positive as the knowledge of the phenomenon itself. For this reason, another change then suggested, it should be "activity-sensibility-intelligence indivisibly united" instead of the union previously mentioned; this alteration made it clear that "man is a being who *acts, knows, and feels*, and all these at once, in each and all his phenomena (275).

The reviewer understood that Leroux defined man philosophically, giving "a definition of man not as an abstraction, but as a real being, living and developing himself in the bosom of the race," which meant that man was "defined not from the individual, but the species." While for the ancients, man was meant to be a "'social and political animal'"; the modern definition was "MAN IS PROGRESSIVE, SOCIETY IS PROGRESSIVE, THE HUMAN RACE ITSELF IS PROGRESSIVE" (277). It followed then that masculinity, as a manifestation of man, was also progressive and man should not expect to act and behave the same over time, ending the search for a sense of permanence that constrained man's behavior in a frozen expression of the past time. This doctrine defended primarily two principles: first, "the collective life of humanity; and second, that humanity, as well as individuals, is progressive" (282). Similarly, Echeverría and his generation moved toward this kind of understanding of man by the 1840s.

In this climate it is not unusual that Echeverría decided in 1846 to reprint *Dogma socialista*, a statement that ten years later fit well with the new revolutionary culture dominating progressive politics in Europe. In the introduction that he wrote to contextualize this piece, he recalled that at the time of its creation Argentine society was divided into two factions that could not be reconciled; first, the federal faction that won, "supported by the popular masses that were the genuine expression of their semi-barbarous instincts"; second, the Unitarian faction, a defeated minority that had "good tendencies, but without a local foundation of socialist criteria that was somewhat antipathic because of its proud outbursts of exclusivity and supremacy."[102] Between these two "a *new generation*" was interested in participating in public service due to its age, education, and social standing, but its members were rejected by both sides; the federalists distrusted them because of the books they read and the clothes they wore; the Unitarians because they suspected that the youth were federalist and only concerned with frivolities. But this generation did not belong to anybody, its "virginal heart" anticipated from the crib that it was the legitimate heir of the "religion of the fatherland," a revelation that made them anxious about its realization (iii).

Rosas knew that their sensibilities made the youth closer to the Unitarians and for this reason he tried to "humiliate it staining it with his blood stigma" (iii). Before associating, this "virile youth" was "isolated, unknown in the country, weak, without a connection that united its members to give them strength, consumed by impotent votes," and without a purpose to improve the fatherland (iv). Echeverría explained that, inspired by Giuseppe Mazzini's instructions to the members of Young Italy, he and his friends developed their plan of action. Reading the Italian revolutionary's words, it is clear how much they took from his recommendations. They also believed in the law of Progress and Duty and wanted to turn their country into one nation unified by association.[103]

Echeverría's *Dogma socialista* was a work of "regeneration" aimed at revolution, but its work would be "essentially educational, both before and after the day of revolution. Like Young Italy, the society that these men wanted to create was "*Republican* and *Unitarian*."[104] Because of this connection to Mazzini, the writings of Leroux were perceived correctly by Echeverría as coming from the same root. Mazzini, the founder of Young Italy, was a reader of *Le Globe* and *Revue encyclopédique*, both publications established partly by Leroux, and became known to French socialists; it is within this socialist network that Echeverría's work was also situated by this time.

The Argentine writer recognized that at this moment Argentina was not ready for a "material revolution." Instead, he proposed "a slow and continuous propaganda of a fraternal creed to reignite in the hearts the sentiment of the fatherland." Fraternal unity and the creation of a national party were the main goals of *Dogma socialista* for Argentina.[105] This climate of regeneration needed to be contextualized within the forces present in the capital of Uruguay, Montevideo, in the 1840s, where an international Romantic and socialist revolutionary community gathered in exile. Members of both Young Italy and Young Argentina joined to oppose Rosas with the Uruguayans that supported Rivera led by Giuseppe Garibaldi (1807–82), who arrived at Montevideo in 1841, to help with the defense of the city.

The internationalization of the conflict meant that the capital of Uruguay and the forty thousand people who populated it—four thousand of them Italians and thousands more from France and Spain—were under siege from 1841 to 1850 until a peace treaty with all those involved was signed, ending more than a decade of foreign conflict.

The conflict excited the imagination of many revolutionaries in Europe who used the resistance of Montevideo to exalt the popular imagination; the importance that Argentina's triumph over the European intervention had in France, together with the resistance of Montevideo, cannot be ignored. Both sides of the conflict had important roles in the revolutionary culture of Europe. For example, the famous popular author Alexandre Dumas described Montevideo as a new Troy, publishing a novel about it in 1849 to the satisfaction of those in exile to whom the book was dedicated.[106] The poet Alphonse de Lamartine was another literary figure involved in the conflict. As a Romantic and a liberal, he opposed the role of France in the fight against Rosas, describing the exile community as an entitled minority that did not have the support the governor of Buenos Aires had because of his desire to preserve his country's authenticity. This opinion was shared by socialists; in fact, each French political party saw its interests reflected in the conflict; conservatives were happy with Rosas's emphasis on law and order, and socialists liked his power over the poorest masses who adored him, and the racial and cultural mixing of his supporters.

Echeverría defined his socialism as centered in "*May* [the revolution], *Progress, and Democracy*," and if there was a flag flying higher and more legitimate than theirs, they would "salute it and acclaim it, as the regenerative flag of the fatherland."[107] His generation was "a cursed race, destined by an unfair law to suffer punishment for the crimes and mistakes" committed by the generation that gave them birth. They "desire[d] but could not satisfy ambition," and were by then in their mid-thirties, which made them a very frustrated political group.[108] While they had the strength to wear the "virile Roman gown," the "dominant stupidity" did not let them; they were sterile and unable to reproduce their political ideas in their country. Despotism and the crowning of Rosas as a god kept them looking impotently at their country from outside.[109] The feeling of impotence was so powerful that when this generation finally had access to power, they saw the defeat of Rosas as the moment in which the nation itself had become adult. In 1852, Alberdi wrote about the need to promulgate a constitution to avoid ridicule since the nation had announced "itself its emancipation and virility to the family of nations."[110] The constitution would be written by him and promulgated in 1853.

Man's vital energy was an important idea implicit in Echeverría's *Dogma*; the masculinity of those who had fought in 1810 had decreased in the country because political passions caused men to lose over time "that primitive virility of their powers, that virginity of their hearts, that fire and energy of a robust adolescence. . . . Because disappointment has come to dampen hopes."[111] But, fortunately, "peoples were not subjected to the fatal law that extinguish[ed] little by little the lives and hopes of man" (71). The individual "might disappear, but his works" remained. Each generation brought "new blood" that spread "life to the social body" because while the "flesh of man" was of the earth, his spirit belonged "to humanity." Each generation "inherited the vital spirit of the generation devoured by the grave" but each also sprouted the tree of hope for humanity's progressive future (71).

If the problem was in the culture of men, what was demanded was the spreading of "healthy principles, uniformity in the creed, dissipating the anarchy of the spirits and circulating progressive doctrines to calm so much anguish and so many upheavals, and to satisfy society's more vital needs" (73). This section revealed the philosophy of nature cultivated in the text, which continued emphasizing the role that "perpetual communication" from man to man and generation to generation played for progress; this was the "continuing incarnation of a generation's spirit into another generation, which was what constituted the life and essence of societies." The latter were not a mere accumulation of men; they were a "homogeneous body" animated by "a peculiar life," resulting in the mutual relationship of men among themselves and of one generation with another (72).

As noted in the the *Boston Quarterly Review*, men and masculinity were integral to a nation's advancement.[112] In *Dogma*, Echeverría made the same point: prosperous nations were characterized as those that counted "among their sons many great men," with national glory described as "the most loved patrimony of nations," insofar as it encapsulated the entirety of their enlightenment and progress—"all its intellectual and material wealth, all its civilization and power."[113] The new masculine culture required that men work as "architects to produce or realize social ideas," because men did not have "real value" just for being men.

Only those who were the expression of the "most *perfect virtue* [*acrisolada virtud*] and the highest intelligence" should lead the country.[114] Not all of them were qualified to bring glory to the nation. Despite the importance of democracy in this text, it states clearly that the "radical vice of the Unitarian system" had been the "universal suffrage" for men who were twenty years old or already emancipated, which was approved by law in 1821 in the province of Buenos Aires.[115] This legislation abolished electors and established direct elections, removing prior requirements of education, wealth, or profession for voting eligibility. Thus, while Esteban Echeverría championed socialist democratic principles, he also scrutinized the egalitarian system that was advancing in Europe. The essence of *Dogma socialista* was the pursuit of the "historical law of humanitarian progress," conceived by Leibniz and articulated by Vico in the seventeenth century, substantiated by Herder, Turgot, and Condorcet in the eighteenth century, and recently revived and elucidated by Pierre Leroux.[116]

Instead, Juan Manuel de Rosas had fostered "the deepest ignorance, widespread poverty, stagnation in customs, and the unleashing of brutal passions." The choice was between embracing a constitution to shape a new model of societal bonds or a "Barbaric Dictatorship" incapable of forging a "national sociability" (50). *Dogma socialista* also urged the development of social and political sciences to eradicate the outdated culture of "imitation and plagiarism" that fueled the intellectual chaos among youth (54).

This piece gained immediate attention in Buenos Aires, receiving a critical review by Pedro de Angelis in *Archivo Americano*. As expected, its author, Echeverría, was denounced as an imitator of incomprehensible ideas. Echeverría retorted sharply, labeling de Angelis a "truant or a mazorca bully" (*compadrito mazorquero*) while presenting himself as a "well-educated man" intent on expressing truths in his own manner (5).

Echeverría accused de Angelis of upholding Buenos Aires's reactionary culture, which punished those embracing Romanticism for their perceived "exaggeration and extravagance in everything—clothing, writings, and manners." The term "Romantic" was co-opted by Rosists to mock certain students and women, signifying anything that clashed with reactionary customs. These critics overlooked the term's origin in disputes among German writers over content and form, and the Romantics' rejection of Greek and Latin literature, seeking "emancipation from classicism," a movement de Angelis paradoxically supported, according to Echeverría (13–14). This Italian propagandist of the regime, part of the "hierarchy of the mazorca"—Rosas's vilest police force against civilized foes—had been revived as an editor by the governor to counter attacks from Montevideo in the press, but he was never committed to "any progressive doctrine," only to "infamous speculations" extolling the machinery of Rosas's social control (15, 56).

In this exchange, Echeverría acknowledged that the conflict between the "lomos negros" (black backs), ideological federalists, and the "lomos colorados" (red backs), federalists loyal solely to Rosas, culminated in the governor's victory through a reactionary, conservative despotism. This despotism was not merely ideological or political but also societal, aimed at preserving a social order reliant on Rosas's authority. In his historical analysis, Manuel Dorrego had effectively served as a "tribune of the masses." The word "federation" on his lips carried weight—an echo of a popular instinct for reaction that foreshadowed an uprising (28).

In contrast, Juan Manuel de Rosas eradicated "the spirit of locality" that fostered parochial governance and any trace "of social life in the provinces," a system unrelated to the political concept of federation (26). Rosas's new sociability rested on uniting diverse social classes, bound solely by their submission to the despot to safeguard their collective interests.

Pedro de Angelis accused the exiles of seeking to subordinate the "Republic, established as a modern entity," to the "delirious notions" of French philosophers Charles Fourier and Victor Prosper Considerant, but this was false. These names were absent from Esteban Echeverría's *Dogma socialista*, and its author rejected the most radical socialist doctrines (29). His aim was to forge a more egalitarian and liberated society, a vision shared with socialists, while advocating Christianity to enhance the moral education of youth without endorsing a state religion and championing religious freedom. He acknowledged contradictions in his writings but believed the best approach was "the harmony of two rival principles" to achieve "a legitimate peace" in his nation.[117]

The goal was to "create and centralize Social Power," a task spanning multiple generations and the natural outcome of prior efforts in association, or a convergence of all interests over time through a "slow and difficult" process.[118] In his response to Pedro de Angelis, Esteban Echeverría clarified that the masculinity championed by Young Argentina rested on an aspirational spiritual force, enabling men to rise above instinct, in contrast to the materialism promoted by Juan Manuel de Rosas's regime.

In 1848 numerous revolutions swept across Europe; only Russia, Spain, Portugal, and the Scandinavian nations remained entirely free from political upheaval. Although England lacked a revolutionary movement, Great Britain faced the Irish Revolution, debated by British intellectuals. In France, a profound political shift established the Second Republic with universal male suffrage for those of a certain age. The movement driving this revolution soon split into factions, one embracing a more radical socialist agenda. These developments lent greater strength and legitimacy to Echeverría's *Dogma socialista* regeneration plan. Echeverría himself observed that Republican France had adopted "the democratic triad" of liberty, equality, and fraternity in its flag, ideals his generation had proclaimed eleven years earlier, giving him "reasons to believe they were not so lost."[119]

The French events of 1848 signaled for Esteban Echeverría the culmination of a process begun in 1789, suggesting that the Americas would be influenced by the ideals "that had conceived the Republic in France," reinforcing the renewal of the May Revolution's ideology both locally and internationally (433).

Aligning with this perspective, Echeverría's analysis highlighted that ideas were transmitted across generations until they integrated into the nation's practical and social fabric, fostering humanity's civilization through the continuous "communion and incarnation" of evolving human spirit conceptions, leading to boundless perfectibility (434). However, national leaders must acknowledge the "law of time and space in social progress," as advancement should be structured by "a law of linked and sequential development" derived from historical tradition and societal education (437). Here, Echeverría referenced the "unity of composition" concept, developed by Geoffroy Saint-Hilaire and adopted by Pierre Leroux, which aligned with ideas circulating among Romanticism-influenced scientists advocating social transformation.

In some societies, family, fatherland, and property had evolved into castes, prompting some to believe they were "privileged and racially destined to dominate others, disregarding and usurping their sacred rights." This interpretation clarified an ambiguity in the original *Dogma socialista*, emphasizing the existence of races subjugating others without respecting universal human rights. Echeverría stressed transforming Argentine society into a democracy guided by natural laws and the embodiment of the era's spirit through obedience fostered by education. Thus, while popular education mandated specific content, its purpose was to awaken a spirit that would inspire action in the material world. In contrast, this view of education was in opposition to how students were taught in Buenos Aires. In 1831 Tomás Anchorena reinstated corporal punishment, asserting that it was a potent incentive for youth, signaling how elders' authority should be enforced on young men.[120]

The text concluded with an optimistic vision: nineteenth-century philosophy had opened "the gates of the paradise of perfectibility to humanity." God had just ushered in "the era of humanity's complete emancipation through the voice of the world's foremost people."[121] In this century, peace would prevail, humans would form "one united family," and the "sacred alliance of nations" would be realized.[122] For Argentina, despotism would cease. Yet, the younger generation erroneously overlooked historical lessons and the numerous shortcomings of philosophical systems over time. This republic, proclaiming a new historical era, collapsed in 1851 with a coup d'état, paving the way for the creation of an empire under Louis Napoleon's control a year later. Perhaps it was fortunate for Echeverría that he did not live to see another political disaster; he succumbed to tuberculosis in Montevideo in January 1851.

EPILOGUE

Men in Transit

I began writing this book during the COVID-19 pandemic after years of research. Coincidentally, at this time the relationships among science, politics, and gender changed in unexpected ways in the United States. The masculinity of scientists was belittled to denounce their policies, which some men perceived as intended to emasculate them, reduce their control over their own bodies, and even inject them with substances to make them more docile to authoritarianism. Soon after, the lockdowns and limited socialization imposed in many places became associated with an attack on male autonomy. The fact that this was happening in a country known for its excellent scientists and experience in communicating scientific knowledge was stunning. It showed me how volatile the linkage between scientists, the policies they craft, and the masculinity they are perceived to attack or protect can be. It also demonstrated how this connection can be politicized and gendered. Witnessing these events helped me to appreciate a dynamic like the one I was addressing, albeit in a very different time, place, and political context.

As I explained in the introduction, this book is not about masculinity per se, but about a relationship that defined the role of men in Argentina's society. It explores the link that reveals men struggling to define what they were amid the ever-changing dynamics of modern science and philosophy. The Enlightenment, as discussed in this book, appears as a clear goal deprived of a coherent plan because it consistently failed at its most basic component: the emancipation of the male body and mind from the entanglements of power that gave them meaning. What began as an attempt to emancipate humans from a fixed form of association and sociability to experience modernity, ended with modern cosmopolitanism becoming the enemy of the nation it had partly created, which ignited the final stage of the conflict analyzed in chapter 6.

In the introduction, I asked what men wanted, and the answer was: to experience political simultaneity, which differed from religious simultaneity. The former ensured that individual men were associated as equals in creating a new universal reality, whereas the latter assured them that they were simultaneously living in a world designed for them by God. However, this new experiential world, grounded in philosophical and scientific ideas that were constantly evolving, proved to be politically unstable. In Argentina, it became defined by a small, enlightened class willing to grant foreigners the rights that most men in power reserved for themselves: existence and reproduction.

The proliferation of knowledge and the insecurity caused by the sociopolitical structures that changed as a result meant that the intellectual elites in charge of defining a national ideology never had certainty about their plans. The idea of having a comprehensive philosophical and scientific system that would organize all aspects of society failed shortly after independence because it was defeated by regional, cultural, economic, and social contexts that experienced universality and locality in different ways.

Modern politics required men to change. Every educated man with power needed to understand himself differently according to the constantly emerging knowledge, and this is what created the different models reviewed in this book. The use of scientific ideas to demonstrate the exactitude and precision of political plans attempted to provide the same certainty used by Isaac Newton to demonstrate gravity. But republicanism was not being saved by reason, quite the contrary, it was sometimes destroyed by it. Still, even after spectacular failures, the enlightened class survived, determined to continue because from the beginning the political was intertwined with a knowledge believed to be the only possible foundation on which to build modern nations and their citizens as well as to perfect them over time.

The Enlightenment goals also persisted because they were part of an international and impressive network that communicated knowledge constantly. Nation-building was a universal project that formed brotherhoods suppressing locality and belonging to a single point on the map. This promoted an ideological associationism that differed from the one organized by religion because it was based on change and transformation produced by ideas that could be learned, transmitted, and perfected; what happened in one place could work everywhere. The success of the United States and Haiti in defeating European colonialism on the American continent, followed by the nations that emerged from the Spanish colonies, was limited by the catastrophic defeat of French republicanism, which was what made France such an important source of science and philosophy. This nation was relevant because of its inability to stabilize republicanism, and the plans to do it were helpful for other nations, too.

As we have seen, the evolution of scientific and political ideas happened simultaneously on the American and European continents. They were not unique to Argentina, and it was this focus on the universal that ended up defining politics in this country. However, what was specific to Argentina's case was how the elite split into two groups that ended up fighting about the gendering of the nation through a conflict between men divided by masculine and feminine models of sociability. While Rosas did not support a republicanism defined by the problem of the country's population, by the 1840s the members of the Generation of '37 had concluded that immigration and public mandatory education were the only tools that would allow Argentine men to become civilized. The priority was the formation of the nation's population that needed to be civilized and European, which meant people who could represent whiteness and civilization.

While similar concerns about population were present in the United States, its leaders believed that the country had a modern population defined as Anglo-Saxon. In the case of Argentina, the men who acquired power after Rosas's defeat were not sure whether people like themselves would be present in the future. The body of the nation was foreign and in the process of being formed according to what could be learned from science and philosophy, knowledge would help the national government to achieve the goal of forming the civilization of the future. The enemy of this project was any freeman, an individual who could be outside of the nation's control because he did not abandon his autonomy. This choice was clearly recognized at the time as directly intersecting with the future of Argentina's men. An article published by Rosas's supporters in Montevideo made this point clear. "What does it mean to yield to European dominion, by whatever the method? Do you want to know? Listen: it means to return to colonial abasement, to lose the energetic, male traits of our national physiognomy, to sell our glorious destinies for a moment's rest . . . in short, it means a cowardly suicide, destroying the principle of independence."[1]

The radicalism of the Generation of '37 resided in building a new model of man that was self-destructive in its desire to remain cosmopolitan because knowledge was everywhere. The citizen of Argentina was conceived in a limbo that would end at some point in the future; meanwhile, continuous transformation would be taking place to give the body of citizens final form through science and medicine, which explains the strength of the hygienic state formed after 1852 to take care of the country's new population. Still, Echeverría described the importance of the female form as a sign of modernity, and how the new literature attempted to repress militaristic and violent impulses to create a cult of the feminine. As it was in Europe, according to Maurice Berger and others, the emergence of a similar literary figure was on this continent "a repudiation of a previous model of masculinity that was not only experienced as oppressive, but more importantly, no longer appropriate to the needs of a new and changing collective imaginary and symbolic order."[2] In the case of Argentina, the literary group was minuscule, mostly composed of members of prominent families, and had great intellectual ambition. They established a symbolic order that imposed the experience of the foreign as the national expression.

It is not surprising that the foundational narrative of Argentina depicts exactly this conflict among men. Between 1838 and 1840, Echeverría wrote a piece that remained unpublished until 1871, years after his death. In it, two forms of being a man are contrasted bluntly in a slaughterhouse, an institution presided over by a judge, "an important figure, the caudillo of the butchers exercising complete authority in that tiny republic by delegation of the Restorer" (Rosas).[3] The conflation of the figure of the judge with that of the governor of Buenos Aires, in that both enjoy complete authority in their barbaric domains, was no accident. The similarities extended to the people supporting the leader of the slaughterhouse: persons of "different complexion and race" (223). In one scene, for example, "two African women were dragging the entrails of an animal; over there a mulatta was leaving with a mass of offal," further back "400 black women [could be seen] squatting and untangling guts on their skirts, pulling off, piece by piece, the bits of fat left there by the butcher's greedy knife" (225). But "the most prominent figure in each group was the butcher, with knife in hand, his arms and chest bare, his hair long and wild, his shirt, leggings, and face smeared with blood" (200). Directly after the description of life in this place, an incident captures the attention of everyone in the slaughterhouse.

Figure E.1. The fatherland is an abused woman. "The Fatherland," *El Grito Arjentino*, February 24, 1839 (Montevideo). Publication of the exile community in Montevideo. Text included: "Yes, I hate you, damn fatherland, and with this knife. . . . Hold on, mate, she still has jewels that we can take from her. After that you can finish her." Uruguay's Ministry of Education and Culture, *Anáforas*, Newspapers of Uruguay, https://anaforas.fic.edu.uy/jspui/handle/123456789/35625.

The appearance of a wild bull incited a debate about whether it was young or old, a bull or a steer. The debate provoked "comic and obscene exclamations" from the people there who were all "excited by the spectacle." "Show me the balls!" exclaimed a butcher. "There they are, between the legs. Don't you see them, buddy, bigger than a horse's head?" replied the other. All agreed that the animal's behavior was "as stubborn and hostile as a Unitarian," a comment that once uttered made everybody shout, "Death to the savage Unitarians!" (*Muerte a los salvajes unitarios*) (228). The bull escaped the slaughterhouse to be chased by the mob through the city, sending an unwitting English landowner to the swamp.

Once the bull was finally taken to the slaughterhouse, the "adventure of the gringo in the swamp" provoked laughter and sarcasm as expressions of the mob's rejection of foreigners (232). The bull was eventually caught and killed, and what remained for the butchers was "to resolve the question about the genitalia of the dead animal, provisionally classified as a bull because of its indomitable ferocity. [. . .] Suddenly, a rough voice exclaimed: "'Here are the balls' [. . .] showing the spectators two huge testicles, the unmistakable sign of a bull." After the surprise a round of explanations began, since a "bull in the Slaughterhouse was an extremely rare thing, even forbidden" (233). According to the rules of the institution, only young animals, up to a few years old, could be slaughtered, because adult meat was not edible.

As the people withdraw from the spectacle, the story continues, "The rough voice of a butcher shouted: 'Here comes a Unitarian!' Once again, the word 'Unitarian' made all "that rabble [stop short], as if struck by a sudden idea" (209). Immediately, and much as they had done in discussing the bull's genitalia, they began to analyze the signs present in the body: "Don't you see his U-shaped sideburns? He's wearing neither the insignia on his coat nor mourning ribbon in his hat." Others added that he was "an arrogant fop," he "rides in a saddle like a gringo," and "all these Unitarian snobs are as artsy as the devil" (234). Each successive detail increases the irritation felt toward the youth by the butchers, and so they begin to discuss what they will do to him: should they choose "the *mazorca* for him," or "the shears," or is "a beating called for?" After challenging each other to catch the youth, Matasiete, a "man of few words and much action" is chosen.

This unfortunate subject was "a youth of some 25 years, of elegant and handsome bearing [. . .] trotting toward Barracas, utterly unafraid of any danger" (234–35). Once the youth is captured, some ask Matasiete to cut his throat with his expert knife, but the judge orders the group to carry the captive to his shack and to "prepare the *mazorca* and the shears. Death to the savage Unitarians—Long live the Restorer of the Laws!" (236). The ensuing dialogue foreshadows the sodomization of the young man who is "furious as a wild bull."

> —The stick will tame him soon enough.
>
> —He needs to be beaten.
>
> —For now, the rod and shears.
>
> —No, a candle.
>
> —The *mazorca* would be better. (237)

The youth is enraged, "beside himself," his whole body seized with "convulsions," because he knows what awaits him at the hands of these men (237).

In no time, and to the hilarity of the onlookers, they cut his beard in the federalist style, shaving the sideburns that, in Unitarian fashion, had "continued onto and under his chin." The Unitarian resists and insults them, prompting one of the federalists to remark, "We'll tame him soon enough" (237). In spite of everything, the young man continues to resist and insults those holding him as the judge's patience wears thin: "Insolent brat! They've gotten you worked up. I will have your tongue cut out if you make a sound. Lower that arrogant fool's trousers, tie him down fast to the table and give him the bull's penis [*verga*] on his bare buttocks while he's over the table" (239). In a flash, and surprising the young man, they secure him to the table, "turning his body face down compressing all his members." The Unitarian continues fighting, the veins in his neck and forehead forming a black relief web "over his white complexion," as if he were "overflowing with blood" (240). Tie him down first!" ordered the judge (240).

> In a moment they tied his bent legs to two legs of the table. To do the same with his hands, they untied the ropes binding them against his back. Feeling his hands free, the youth, with a sharp movement that seemed to exhaust all of his strength and vitality, pulled himself up on his arms and then on his knees and then collapsed, muttering:
>
> —Behead me before you strip me, despicable rabble!
>
> His forces had been spent. He was immediately tied spread-eagled and the work of stripping him began. Then a stream of blood burst forth, spurting from his mouth and nose . . .
>
> —The savage Unitarian just burst with rage. (241)

The torturers were surprised by the reaction of the young man and ended up laughing at the fact that he had taken his honor so seriously "when they only wanted to have a little fun with him." Rosas was, in this narration, the chief of a sodomite "brotherhood" and, according to his principles, this was his way of treating any man who was not "an assassin, butcher, savage, or thief," "any decent man with his heart in the right place," an "educated patriot, friend of the Enlightenment and freedom." The account concludes by noting that it was "clear that the core of the Federation was in the Slaughterhouse" (217). The explosion of blood that resulted from the civilized man's rage, which leads him to self-destruction, marked the birth of the nation. As Jorge Salessi has correctly noted, the violation of the Unitarian youth in *The Slaughterhouse* is the inverse of the case presented by Octavio Paz in his analysis of the history of Mexico in *The Labyrinth of Solitude*.[4] Here it is not an Indigenous woman, Malinche, "*chingada*" (screwed) by Europe, but men born on the American continent fighting each other about the creation of a nation alternately envisioned in female and male forms.

This story also presents a variation on what Terry Eagleton has noted about the feminization of civilization represented in the figure of a violated woman.[5] In *The Slaughterhouse* we find that the honor that had to be defended was that of the men who embraced the feminine symbol, civilization, and were violated by the hypermasculinity of brutal men who embodied barbarism. The foreigner was the one expected to defend the honor of the victim because they shared the same values as brothers fighting for a universal civilization. As a result, everything associated with feminization was banned by Rosas and his followers, including the philosophy and science they thought diminished traditional authority and the power of men in society.

The transition from the hypermasculinity of Rosas's times to a modern culture that took hold after his demise were perfectly exemplified in Lucio V. Mansilla's comparison of the male culture of the governor's parents, his own maternal grandparents, and civilized liberal men like his father. Mansilla recounts an exchange between Don León and his wife, in which the husband invited his wife to visit the vegetable garden; once there, they sat, and had a very friendly conversation:

> "Is it true, little Agustina, that I love you very much?" Doña Agustina, who, like all our grandparents, made love [. . .] at set times, [. . .] sat apart and answered: "Rozas, why do you disrespect me so?" "That isn't it, no!" [he said,] and taking some ropes from his pocket said to her: "Do you see this? Well, this is to prove to you that a man is a man, that if I allow you to govern, it is not because I am weak but because of the immense love I have for you, because I think you are faithful." [. . .] He applied quite a number of lashes of the whip, more simulated than forceful, to a certain portion of her anatomy.[6]

Despite her strong personality, so strong that even her son Juan Manuel did not dare contradict her, "Doña Agustina did not resist, she did not speak; Don León left her in that place, walked out of the garden in triumph and they never again talked about the incident, and nothing was changed in the dealings of house and business." Years later, when their daughter Agustina married Mansilla's father, Don León said to his son-in-law: "Look, friend, even though you are a widower and must have experience, I will tell you something because I love you: I think Agustinita is very good, but it could be that sometimes she needs . . . ," and he recounted his encounter with his wife in the vegetable garden. Mansilla proudly explained that his father was a civilized man of his time and that "Agustinita *never had that* need."[7] This marriage was not based on authority, submission, and the need to exercise power and control over the feminine because knowledge had changed their sociability and vice versa.

Liberalism continued with its gender ambiguities and self-destructive acts after Rosas's defeat. One of the more important components in the 1853 constitution written by Alberdi was that it welcomed immigrants. In fact, its preamble was an invitation to "*all the men of the world who wanted to come to inhabit the Argentine soil.*" [8] While the welcoming of foreigners was a good act, Alberdi made it clear that generosity was not the motivation. The purpose of this call was to make "the European foreigner [Argentina's] settler par excellence, the favored worker of its civilization," and it was the government's duty to "promote the European immigration, avoiding any measure" that limited it.[9] The privileges given to immigrants were such that a citizenship law of 1857 gave their children born in the country the opportunity to keep the parents' nationality if they did not want to be Argentines. The reason for this was the profound dislike these men had for the Indigenous peoples, and those who were not white, which, according to race science, included themselves.

It is not surprising that prominent intellectuals who belonged to the generation born around 1850 were particularly vocal in expressing the problems they had dealing with the different demands on them caused by being a man . The death of Lucio V. López (1848–94) partly expressed this anguish. He was one of the intellectual leaders of his generation, an example of civilized man who was engaged in changing his country politically through his knowledge and culture; still, his virility, possibly his sexuality according to the context, was questioned by a rival. The criticism hurt him so deeply that he decided to accept a duel even though he had no experience in this type of challenge. He was killed as result in 1894. López's close friend, the writer, Miguel Cané (1851–1905), read a speech the day of the funeral, giving eloquent voice to the dilemma felt by the men of his generation about "being a man." In his view, López had died "paying tribute to the remnant of barbarism that predominates in our social organism." While all condemned it, they were still in thrall to it, even when their "natural ferocity yields a little bit in the face of reason." His friend, "more than anybody else," was aware of this problem, and "his sweet and clear spirit revolted against these bloody and absurd extremes that do not solve anything and cannot solve anything." Still, he felt that he needed to prove his manhood after a public attack by a political rival.

López had "a patrician name," and "nature had awoken in his brain and in all his soul the feeling, the exquisite taste of the artist." He was refined, devoted to study, and his intelligence and enlightenment were rare among those of his generation. Then, "What did he lack?" asked Cané directly to the audience, replying immediately himself: "Courage, said those whom his public activity had hurt." Some attacked him simply for failing to possess the qualities extolled in the past, as if there were no courage in "elevated and decent activity" and the "high moral concept" that his friend pursued "with firmness through obstacles and perils." There were many who still believed that the possession of "many enviable gifts" was not "compatible with this vulgar and momentary energy that consists in controlling the nerves when faced with material danger."[10] The man of spiritual aspirations and intellectual elevation continued dying even after the aggressive policies to civilize the country.

The model of society built by the Argentine leaders analyzed in this book was effective in framing the meaning of the nation, attracting the men who needed to populate it, and making the country an outlier in the region because of its European identity. Even today, the origin of Argentina is not placed in a definitive spot on the map but is represented as if in transit. In 2021, President Alberto Fernández, elected as a candidate of the Peronist party with a center-left coalition, said publicly that "the Mexicans came from the Indians, the Brazilians came from the jungle, but we Argentines came from the ships," shocking his voters and prompting demands for an apology.[11] The fact that this politically costly statement was made by a progressive politician only indicates the power that this national narrative still holds in the national imaginary. As the science and philosophy that triggered the scientific revolution, the political revolution in Argentina built a nation that represented universality, knowledge, and men's radical self-transformation, all elements that were ultimately unhelpful to building a stable modern nation, though they had originated it.

NOTES

Introduction: What Do Modern Men Want?

1. Some sections of this book have previously been published; see Novoa, "Science, Sensibility and Gender in Argentina"; and Novoa, "Reception of Evolutionary Ideas in Argentina."

2. Abercrombie, *Passing to América*, 8.

3. Pettegrew, *Brutes in Suits*, 161.

4. Geertz, *Interpretation of Cultures*, 14.

5. Connell and Messerschmidt, "Hegemonic Masculinity."

6. Schiebinger, *Nature's Body*, 11.

7. Schiebinger, *Nature's Body*, 12.

8. Schiebinger, "Feminine Icons," 661. The following parenthetical page numbers refer to this source.

9. Conger, "Introduction," 13.

10. Benjamin, "Elbow Room," 27.

11. Ellis, *Masculinity and Science*, 1.

12. Ellis, *Masculinity and Science*, 2.

13. Golinski, *Experimental Self.*

14. Ellis, *Masculinity and Science*, 39.

15. Ellis, *Masculinity and Science*, 7.

16. Salessi, *Medicos maleantes* y *maricas.*

17. Shapin, "Image of the Man of Science," 164.

18. Shapin, "Image of the Man of Science," 168.

19. Shapin, "Image of the Man of Science," 168.

20. G. S. Rousseau, *Enlightenment Crossings*, 226.

21. G. S. Rousseau, *Enlightenment Crossings*, 226

22. O'Neal, *Authority of Experience*, 249.

23. Manning, *Navigating World History*, 182.

24. Mazlish, *New Global History*, 2.

25. Iggers et al., *Global History*, 390.

26. Poskett, *Materials of the Mind*, 3.

27. Secord, "Knowledge in Transit."

28. Secord, "Global Darwin," 37.

29. Secord, "Global Darwin," 46.

30. Raj, "Beyond Postcolonialism," 341.

31. Raj, "Beyond Postcolonialism," 341.

32. Lightman, "Introduction," 3.

Chapter 1. The Citizen of the Republic of Letters and Sensibility

1. Cañizares-Esguerra and Cueto, "Latin American Science."

2. Nieto Olarte, *Remedios para el imperio*, 272.

3. Cañizares-Esguerra, "Iberian Colonial Science," 70.
4. Van Miert, "What Was the Republic of Letters?" 270.
5. Jones Corredera, "Rediscovery," 954.
6. Jones Corredera, "Rediscovery," 954.
7. *Planta y methodo.*
8. Navia-Osorio y Vigil, *Reflexiones militares*, 13.
9. Luzán, *Real Academia*, 80.
10. Luzán, *Real Academia*, 85–86. All translations included in this book are mine unless otherwise indicated.
11. Rodríguez Mohedano and Rodríguez Mohedano, *Historia literaria de España*, 4–5.
12. Giuli, "Arcadia (Accademia)."
13. Fumaroli, *Republic of Letters*, 170.
14. Giuli, "Arcadia (Accademia)," 65.
15. Feingold, *Jesuit Science*, vii.
16. Feingold, *Jesuit Science*, viii.
17. Caesar and Caesar, *Modern Italian Literature*, 24.
18. Lodovico Antonio Muratori, *Riflessioni sopra il buon gusto intorno le scienze et le arti* (Venice: Per Luigi Pavini, 1708), 6–7.
19. Blanchard, "Are Jansenists Among Us?"
20. Strasser, *Missionary Men*, 33. The following parenthetical page numbers refer to this source.
21. Ariew, "Descartes and the Jesuits."
22. Malta Romeiras, *Jesuits and the Book of Nature*, 2–3.
23. Manuel Antonio Valdés, "Mexico," *Gazeta de Mexico* 8, no. 22 (November 11, 1796), 174.
24. Manuel Antonio Valdés, "Encargos," *Gazeta de Mexico* 8, no. 21 (October 28, 1796), 172.
25. Burson, "Between Power and Enlightenment," 41–42.The following parenthetical page numbers refer to this source.
26. Pluche, *Espectáculo de la naturaleza*, 289.
27. Pluche, *Espectáculo de la naturaleza*, 290.
28. Pluche, *Espectáculo de la naturaleza*, 292.
29. Cortesão, *Alexandre de Gusmão*, 292.
30. Malta Romeiras, *Jesuits and the Book of Nature*, 2–3.
31. Malta Romeiras, *Jesuits and the Book of Nature*, 31.
32. Verney, *Verdadero metodo*, 71.
33. Verney, *Verdadero metodo*, 70.
34. Juan Andrés, *Cartas familiares del abate D. Juan Andrés a su hermano* . . . (Madrid: Don Antonio de Sancha, 1793).
35. Aullón de Haro and Mombelli, *Introduction to the Spanish Universalist School*, 4.
36. Vico, *Principi di una scienza nuova.*
37. Hervás y Panduro, *Viage estático*, ix.
38. Hervás y Panduro, *Viage estático*, xi.
39. Clavijero, *Historia antigua*, 48.
40. Aullón de Haro and Mombelli, *Introduction to the Spanish Universalist School*, 152–55.
41. *Coleccion general de las providencias*, 66.
42. *Coleccion general de las providencias*, 66–67.
43. Ricketts, *Who Should Rule?* 30.

44. Safier, *Measuring the New World*, 4.

45. Mayans y Siscar, "Aviso Primero," 104.

46. *Cartel del certamen*, 7.

47. *Cartel del certamen*, 36.

48. Carlos de Sigüenza y Góngora, *Libra astronomica y philosophica* (Mexico: Herederos de la viuda de Bernardo Calderon, 1690), 99.

49. Sigüenza y Góngora, *Libra astronomica*, 94.

50. Laura E. Bland, "Unfriendly Skies: Science, Superstition, and the Great Comet of 1680" (PhD diss., University of Notre Dame, 2016), 4.

51. Zamudio Varela, "Los pintores," 34.

52. Bleichmar, *Visible Empire*, 187, 32.

53. Joseph Antonio de Alzate y Ramírez, *Diario literario de México*, no. 1 (March 12, 1768), n.p.

54. Joseph Antonio de Alzate y Ramírez, *Diario literario de México*, no. 2 (March 18, 1768), n.p.

55. José Ignacio Bertolache, *Mercurio volante: con noticias importantes i curiosas sobre varios asuntos de Fisica i Medicina*, no. 1 (October 17, 1772), n.p.

56. T. A. Smith, *Emerging Female Citizen*, 7.

57. Bertolache, *Mercurio volante*, n.p.

58. de Asúa, *Science in the Vanished Arcadia*, 260.

59. "Informe ál Gobernador del Rio do la Plata," 360

60. Labardén, "Discurso del Doctor Labardén," 75.

61. "Informe al Gobernador del Río de la Plata," 381.

62. Rabin, "Jesuit Science Before 1773."

63. Fleck, "Paraguay Natural Ilustrado."

64. de Asúa, *Science in the Vanished Arcadia*, 261.

65. de Asúa, *Science in the Vanished Arcadia*, 261.

66. Cañeque, *King's Living Image*, 280n18.

67. Portocarrero y Guzmán, *Theatro monarchico de España*, 413. The following parenthetical page numbers refer to this source.

68. Herr, *Eighteenth-Century Revolution*, 84.

69. Herr, *Eighteenth-Century Revolution*, 4.

70. Condillac, *La lógica o los primeros elementos*, trans. Bernardo María de Calzada (Madrid: Joaquín Ibarra, 1784); Condillac, *La lógica*, trans. Valentín de Foronda (Madrid: Imprenta de González 1794); Condillac, *La lengua de los cálculos*, trans. Alvarado y Lezo, Marquesa de Espeja (Madrid: Imprenta de Ruiz, 1805); Étienne Bonnot de Condillac, *Curso de estudios para la instrucción del Príncipe de Parma*, trans. Basilio Antonio Carsi, Basilio Roldán y Godínez, and José Gerosarri (Cádiz: Imprenta de Carreño, 1813); Étienne Bonnot de Condillac, *La lógica, ó, Los primeros elementos del arte de pensar*, Trans. Bernardo María de Calzada (Barcelona: Tomas Gorchs, 1817); Condillac, *La lógica ó los primeros elementos*, trans. Bernardo María de Calzada (Barcelona: Imprenta de V. Sierra y Marti, 1823); Étienne Bonnot de Condillac, *Gramática general escrita en francés por el abad de Condillac, precedida de las lecciones preliminares del mismo autor* (Guanajuato: Imprenta del Supremo Gobierno a cargo de C. José María, 1828).

71. John Torrance, *Karl Marx's Theory of Ideas* (Cambridge University Press, 1995), 40.

72. John Locke, *An Essay Concerning Human Understanding, in Four Books* (London: Printed by Elizabeth Holt, 1690).

73. Condillac, *La lógica o los primeros elementos.*

74. Condillac, *Oeuvres de Condillac.*

75. Herr, *Eighteenth-Century Revolution*, 84.

76. Lloyd, "Reason and Rationality," 156.

77. Haidt, *Embodying Enlightenment.*

78. "Idea de la obra," *Espíritu de los mejores diarios literarios que se publican en Europa*, no. 1 (July 1797) (Madrid: Imprenta Antonio Espinosa, 1787), n.p. For all quotes in this book, emphasis is in the original text unless otherwise noted.

79. François-Martin Poultier d'Elmotte, "Carta á la Señorita Serard, sobre el origen, y formación de las ideas," *Espíritu de los mejores diarios literarios que se publican en Europa*, no. 63 (November 24, 1787), 599.

80. d'Elmotte, "Carta á la Señorita Serard," 600.

81. Tosh, "What Should Historians Do?"

82. J-J. Rousseau, *Emile*, 185.

83. Locke, *Some Thoughts Concerning Education*, 84–85.

84. Fletcher, *Gender, Sex and Subordination*, 95.

85. Barker-Benfield, *Culture of Sensibility*, 139.

86. Vickers, "Introduction," xiii.

87. Williamson, *British Masculinity*, 9.

88. Condillac, *Essay on the Origin*, 28. The following parenthetical page numbers refer to this source.

89. Zárate and Antonio, *De la instrucción pública.*

90. "La langue des Mathematiques, l'Algebre, est la plus simple de toutes les langues," in Rosell, *Instituciones matemáticas*, iii.

91. Rosell, *Instituciones matemáticas.*

92. Lloyd, "Reason and Rationality," 160.

93. Lloyd, "Reason and Rationality," 161.

94. Orain, "Moral Theory of Condillac."

95. de Viegas, *La logica*, 34–35.

96. de Viegas, *La logica*, 37.

97. de Viegas, *La logica*, 38–39.

98. Condillac, *La lógica ò los primeros elementos*, 45.

99. Condillac, *La lógica ò los primeros elementos*, 48.

100. Riskin, *Science in the Age of Sensibility*, 4.

101. Amann, *Dandyism*, 140.

102. Amann, *Dandyism*, 140.

103. Connelly and Higgins, *Diccionario nuevo*, 479.

104. Bolufer, "Gallantry and Sociability."

105. Zaccaria, *Saggio critico*, 116.

106. Haro de San Clemente, *El chichisveo impugnado*, 3–4.

107. T. A. Smith, *Emerging Female Citizen*, 28.

108. Haro de San Clemente, *El chichisveo impugnado*, 12.

109. Feijóo, *Defensa de las mujeres.*

110. Feijóo, *Defensa de las mujeres.*

111. Feijóo, *Defensa de las mujeres.*

112. Royal Spanish Academy, *Diccionario de la lengua castellana*, 460.

113. Herrero y Rubira, *Diccionario universal*, 312.

114. Isidro Bosarte, "Reflexión sobre la majeza," *Diario curioso, erudito, económico y commer-*

cial, no. 360 (June 25, 1787) (Madrid: Imprenta de Manuel Gonzalez, 1787), 719.

115. Isidro Bosarte, "Conclusión de la reflexión sobre la majeza," *Diario curioso, erudito, económico y commercial*, no 362 (June 27, 1787) (Madrid: Imprenta de Manuel Gonzalez, 1787), 726–27.

116. Johan Turesson Oxenstierna, *El philosopho sueco y Lutherano desengañado*, trans. Monsieur Boona (Madrid: Librería de Mons Symond, 1746), 89–90.

117. "Carta de Monsieur Warton, á su amigo Pamberton," *Diario curioso, erudito, económico y commercial*, no. 7 (January 7, 1792) (Madrid: Imprenta de Manuel Gonzalez, 1787), 25–26.

118. Amann, *Dandyism*, 137.

119. Amann, *Dandyism*, 139.

120. Fernández de Rojas, *Libro de moda*. The following parenthetical page numbers refer to this source.

121. Florencio, *Crotalogía*, 1.

122. Florencio, *Crotalogía*.

123. Rodríguez Calderón, *Don Líquido*.

124. Haidt, *Embodying Enlightenment*, 128.

125. Penrose, *Masculinity and Queer Desire*, 130.

126. Berco, "Producing Patriarchy," 357.

127. *Newspaper of the Main News*.

128. Calero y Moreira, *Prospecto del papel periódico*.

129. Calero y Moreira, *Prospecto del papel periódico*.

130. M. García and Pascual, "Baltasar de los Reyes Marrero."

131. Jacsick, *Andrés Bello* (2010), 39.

132. Bello, *Gramática de la lengua castellana*, v–vi.

133. E. O. Wilson and Gómez Durán, *Kingdom of Ants*.

134. Montújar, "Biaje de Quito a la Lima."

135. de Caldas, *Cartas de Caldas* , 164.

136. de Caldas, *Cartas de Caldas*, 153.

137. de Caldas, *Cartas de Caldas*, 168.

138. Riskin, *Science in the Age of Sensibility*, 9.

139. Casaus y Torres, "Parecer."

140. Gómez Marín, *El currutaco por alambique*, 9–10.

141. Gómez Marín, *El currutaco por alambique*, 14.

142. de Mariscal, "'El currutaco por alambique,'" 246.

143. Fasani and Mónica, "La hygiene en el Buenos Aires," 9.

144. Cabello, "Análisis," 5. The following parenthetical page numbers refer to this source.

145. Cabello, "Obgetos principales."

146. Cabello, "Obgetos principales," 14.

147. Cabello, "Continúa la idea general," 25.

148. Cabello, "Continúa la idea general."

149. Cabello, "Continúa la idea general," 26.

150. Cabello, "Continúa la idea general," 28.

151. "Noticias."

152. "Definición del currutaco."

153. Canton, "Satirilla festiva."

154. Canton, "Satirilla festiva." 55.

155. Fernández, "Sr. Editor del Telégrafo."

156. Best et al., "Making the Right Decision."

157. "La virtud del pueblo." The following parenthetical page numbers refer to this source.

158. Humboldt et al., *Political Essay*, 222.

159. Price, *Observations*, 2.

160. Meranze, *Laboratories of Virtue*, 159.

161. Auslander, *Cultural Revolutions*, 101.

162. *Sans Souci*, 19.

163. *Sans Souci*, 5.

Chapter 2. The Patriot

1. Michael Sonenscher, *Before the Deluge: Public Debt, Inequality, and the Intellectual Origins of the French Revolution* (Princeton University Press, 2007), 31.

2. Campbell, "Politics of Patriotism," 551.

3. Campbell, "Politics of Patriotism," 552.

4. Jovellanos, *Elogio de Carlos Tercero*, 1. The following parenthetical page numbers refer to this source.

5. de Caracciolli, *El goce*, ix–x.

6. de Caracciolli, *El goce*, xii.

7. de Caracciolli, *El goce*, xiii.

8. Feijóo y Monenengro, *Cartas eruditas*, 185.

9. Feijóo y Monenengro, *Cartas eruditas*, 187.

10. Feijóo y Monenengro, *Cartas eruditas*, 194.

11. Portugués, *Coleccion general*, 411.

12. Torrubia, *Centinela*, 2.

13. Torrubia, *Centinela*, 34–35.

14. G. Smith, *Use and Abuse*. The following parenthetical page numbers refer to this source.

15. Bleichmar, *Visible Empire*, 33.

16. Anderson, *Constitutions of the Free-Masons*, 8.

17. Entick, *Free Mason's Pocket Companion*, 180.

18. Bullock, *Revolutionary Brotherhood*, 3.

19. Jeffers, *Freemasons*, 68.

20. de Réal de Curban, *La Ciencia del gobierno*, 14–15.

21. Thomson, "Cuando sólo reinasen los indios," 46.

22. Michelena, "Reformas y rebeliones," 42.

23. Michelena, "Reformas y rebeliones," 44.

24. Quoted in Michelena, "Reformas y rebeliones," 48.

25. Quoted in Michelena, "Reformas y rebeliones," 48.

26. Picornell y Gomila, "A las reales sociedades económicas."

27. Mathon de la Cour, *Discurso*, 6.

28. Mathon de la Cour, *Discurso*, 6.

29. Mathon de la Cour, *Discurso*, 7.

30. Tosh, "What Should Historians Do?" 194.

31. Quoted in Homs, "Juan Picornell," 590.

32. Michelena, "Reformas y rebeliones," 141.

33. Miranda, "Carta de Francisco de Miranda."

34. Aizpurúa, "Revolution and Politics," 106.
35. Aizpurúa, "Revolution and Politics," 106.
36. Gual, "Discurso preliminar."
37. Potofsky, "Political Economy," 514.
38. Tavárez and José, "La invención de un imperio," 71.
39. Scipio, "News."
40. "Progresos del papel periódico," 164.
41. "Progresos del papel periódico," 164.
42. Humboldt, *Political Essay*, 1:205. Emphasis added. This translation comes from the first French edition: *Essai politique*, vol. 1.
43. Humboldt, *Political Essay*, 1:211–12.
44. Lyman R. Caswell, "Andrés del Río, Alexander von Humboldt, and the Twice Discovered Element," *Bulletin of the History of Chemistry* 18, no. 1 (2023): 35–41; and Andrea Valencia-Martínez et al., "Emerging of the Mineralogy Discourse in Mexico (1795–1849): Discurso emergente de la mineralogía en México (1795–1849): A Taxonomy of Objects, Procedures, and Instruments," *Boletín de la Sociedad Geológica Mexicana* 74, no. 1 (2022): 1–16.
45. F. Azara, *Diario de un reconocimiento*, 3.
46. F. Azara, *Diario de un reconocimiento*, 37.
47. Dillon, *Travels Through Spain*, 75.
48. Garriga, *Descripción del esqueleto.*
49. Belgrano, *Autobiografía*, 176.
50. Belgrano, "Solicita permiso."
51. Belgrano, *Autobiografía*, 176.
52. D'Israeli, *Domestic Anecdotes*, 14.
53. Meek, *Economics of Physiocracy*, 19.
54. Beer, *Inquiry into Physiocracy*, 108.
55. d'Abadal, "Physiocrats," 147.
56. Astigarraga, "Ramón de Salas."
57. Shovlin, *Political Economy of Virtue*, 107.
58. Shovlin, *Political Economy of Virtue*, 68.
59. M. F. López and Orellana, "Manuel Belgrano," 116; and F. Quesnay, "Maximes générales du government."
60. d'Abadal, "Physiocrats," 147.
61. Lluch and d'Abadal, "Physiocracy in Spain."
62. Belgrano, *Compendio de los principios*, 1:82–83.
63. T. A. Smith, *Emerging Female*, 77.
64. Belgrano, "Memoria," 112. The following parenthetical page numbers refer to this source.
65. D'Israeli, *Domestic Anecdotes*, 4.
66. Zeballos, *Cancionero popular*, 59.
67. "Historia del Dr. Buñuelos," 218.
68. "Nuebo renombre de Apolo," 264.
69. "Nuebo renombre de Apolo," 264.
70. Cabello y Mesa, "Política," 22. The following parenthetical page numbers refer to this source.
71. Vieytes, "Prospecto." The following parenthetical page numbers refer to this source.
72. Velarde, "Cartas," 42. The following parenthetical page numbers refer to this source.

73. "Agricultura," 1.
74. "Agricultura," 2.
75. "Agricultura," 8.
76. "Real cedula."
77. Franklin, "Letter to Peter Collinson."
78. "Commerce," 19.
79. "Commerce," 19.
80. "Commerce," 26.
81. Velarde, "Cartas," 42.
82. Orden, "Carta," 202.
83. Orden, "Vetoño en orden," 211.
84. Orden, "Vetoño en orden," 213.
85. "Concluye la materia" (1802) , 27.
86. W. Guthrie, *Nouvelle géographie universelle*, 405; the comments on Creoles and the Portuguese are on pages 283 and 415, respectively.
87. G. Guthrie, *Geografía universal descriptiva.*
88. W. Guthrie, *New Geographical*, 704.
89. W. Guthrie, *New Geographical*, 718.
90. Pauw, *Recherches philosophiques.*
91. Pauw, *Recherches philosophiques*, 20.
92. Valverde, *La America vindicada.*
93. Hughes, *Forging Napoleon's Grande Armée*, 12.
94. Hughes, *Forging Napoleon's Grande Armée*, 13.
95. Hughes, *Forging Napoleon's Grande Armée*, 13.
96. Duckenfield, *Battles Over Free Trade*, 70.
97. Miranda, "Exhaustivo y documentado alegato."
98. Miranda, "Exhaustivo y documentado alegato," 367.
99. Cabildo de Buenos Aires al de Tucumán.
100. "Foreign Official Paper."
101. "Foreign Official Paper."
102. Belgrano, "Proclama del real consulado," 50.
103. Gillespie, *Gleanings and Remarks*, 62.
104. Alzaga, "El muy ilustre Ayuntamiento,"82.
105. Nafría, "La América española," 324.
106. Quoted in Nicolás Aguirreche, *Obras del Dr. D. Francisco Majesté*, vol. 1 (Barcelona: P. Riera, 1867), 71.
107. "Prospecto del periódico."
108. "Concluye la materia" (1810), 18–19.
109. "Concluye la materia," (1810), 20.
110. "Rara temporum felicitate, ubi sentire quæ velis, et quæ sentias dicere licet," *Correo de Comercio* 1, no. 1 (June 7, 1810), 5. Hume, *Treatise of Human Nature.*
111. "Buenos Ayres 7 de Junio," 12.
112. "Buenos Ayres 7 de Junio," 13.
113. "Sobre la libertad," 31.
114. "El Consejo de Regencia."
115. "El Consejo de Regencia."

116. Belgrano, "Concluye la materia," 137. The following parenthetical page numbers refer to this source.

117. Belgrano, "Señores editores," 111.

118. Belgrano, "Comercio."

119. Moreno, "Prologo," 376–77.

120. Vélez, *Preservativo contra la irreligión*, 3.

121. Vélez, *Preservativo contra la irreligion*, 4.

122. Raynal, *Historia política*. This edition was translated by the Duke of Almodóvar, who used the pseudonym Malo de Luque.

123. Pelliza, *Monteagudo*, 31.

124. Yates, *Rosicrucian Enlightenment*, xi.

125. Yates, *Rosicrucian Enlightenment*, xiv.

126. Posadas, "Logias masónicas."

127. Mancini, *Bolívar*, 181.

128. Zúñiga, *La Logia*, 34.

129. Mancini, *Bolívar*, 131.

130. Zúñiga, *La Logia*, 37.

131. Quoted in Mitre, *Historia de San Martín*, 163.

132. J. Miller, *Memoirs of General Miller*, vii.

133. de Monteagudo, "Censura política," April 13, 1812, 20.

134. de Monteagudo, "Censura política," May 11, 1812, 53.

135. de Monteagudo, "Oración inaugural" 248.

136. de Monteagudo, "Oración inaugural," 257.

137. de Monteagudo, "Oración inaugural," 258.

138. de Monteagudo, "El siglo XIX," 182.

139. de Monteagudo, "El siglo XIX," 183.

Chapter 3. Man of Ideology

1. Colwill, "Women's Empire," 265.

2. Condillac, *Treatise on the Sensations*.

3. Gross, "Lessened Locus of Feelings," 233.

4. Gross, "Lessened Locus of Feelings," 235.

5. Gross, "Lessened Locus of Feelings," 265.

6. Colwill, "Women's Empire," 272.

7. Colwill, "Women's Empire," 272.

8. Colwill, "Women's Empire," 277.

9. Lesch, "Systematics," 88.

10. Lesch, "Systematics," 86.

11. Destutt de Tracy, *Projet d'éléments*, 1.

12. Richards, "Ideology," 103.

13. Destutt de Tracy, *Projet d'éléments*, 4.

14. Destutt de Tracy, *Treatise on Political Economy*, 35.

15. Destutt de Tracy, *Treatise on Political Economy*, 162–63.

16. Cabanis, *Coup d'oeil sur les révolutions*. The citations included come from the English translation of 1806: Cabanis, *Sketch of the Revolutions*. The following parenthetical page numbers refer to this source.

17. Peaden, "Condillac," 140–41.

18. Derrida, "Introduction."

19. Head, *Politics and Philosophy.*

20. Chilcoat, "Legacy of Enlightenment," 4.

21. Lajer-Burcharth, *Necklines*, 176.

22. Lajer-Burcharth, *Necklines*, 177.

23. Lajer-Burcharth, *Necklines*, 288.

24. Cabanis, *Oeuvres Philosophiques*, 299; cited in Lajer-Burcharth, *Necklines*, 322.

25. Lajer-Burcharth, *Necklines*, 289–90.

26. Mosse, *Image of Man* (1996), 61. The following parenthetical page numbers refer to this source.

27. Weiner, "Mind and Body," 339.

28. Weiner, "Mind and Body," 342.

29. Pinel, *Nosographie philosophique*, 43.

30. Pinel, *Nosographie philosophique*, 1.

31. Pinel, *Nosographie philosophique*, 13.

32. Sheriff, *Exceptional Woman*, 21.

33. Sheriff, *Exceptional Woman*, 21.

34. Pinel, *Nosographie philosophique*, 40. The following parenthetical page numbers refer to this source.

35. Riskin, *Science in the Age of Sensibility.*

36. "Jean Baptiste Say to Thomas Jefferson, 2 November 1803."

37. *Commentary and Review of Montesquieu's "Spirit of Laws."*

38. Jefferson, "Letter from Monticello, October 25, 1818," i.

39. Jefferson, "Prospectus," viii.

40. Destutt de Tracy, "Destutt de Tracy to Thomas Jefferson, 11 April 1818."

41. Cabanis, "Pierre Jean Georges Cabanis to Thomas Jefferson, 20 October 1802."

42. For the original, see Cabanis, *Du degré de certitude.*

43. Cabanis, *Compendio histórico.*

44. M. M. Gutiérrez, "Introduction," xxi. The following parenthetical page numbers refer to this source.

45. Teulon, "Fiasco théorique."

46. Bello, "Sobre la enseñanza," 75.

47. Levasseur, *Lafayette in America*, 7.

48. Lafayette, "Letter from Lafayette to Jefferson," 247.

49. Gallo, "Jeremy Bentham."

50. Carrasquilla, "Oscar Saldarriaga," 495.

51. di Pasquale, "Entre la experimentación política."

52. "Manifesto of Independence," *Morning Chronicle* [1801], August 5, 1818.

53. "Destutt de Tracy to Thomas Jefferson, 10 March 1819."

54. Jefferson, "Thomas Jefferson to Destutt de Tracy, 26 December 1820."

55. Núñez, *Noticias históricas.* The following parenthetical page numbers refer to this source.

56. Rodríguez, *Diario.*

57. Strangford, "Letter of Strangford to the Government."

58. Strangford, "Exmo. Señor Lord Visconde Strangford."

59. "Bienes de la Iglesia."

60. Avazoa, *Observaciones*, 10–11.

61. Avazoa, *Observaciones*, 10–11.

62. de Asúa, *La ciencia de mayo*, 174.

63. "Felipe Senillosa."

64. Senillosa, *Gramática española*, ii.

65. "Articulo comunicado," 6.

66. "Articulo comunicado," 6–7.

67. "Sobre mujeres."

68. "De la disciplina militar."

69. "Academia de matemáticas."

70. "Señor Editor," 2.

71. "Señor Editor," 4.

72. "Señor Editor," 18.

73. "Soliloquio."

74. "Soliloquio," 37.

75. Gez, *El Dr. Juan Crisóstomo Lafinur*, 193.

76. Movsichoff, *Juan Crisóstomo Lafinur*. He was also directly related to the writer Jorge Luis Borges.

77. Gez, *El Dr. Juan Crisóstomo Lafinur*, 21.

78. He meant Cabanis, *Rapports du physique*; and Gez, *El Dr. Juan Crisóstomo Lafinur*, 49. The following parenthetical page numbers refer to the latter source.

79. Lafinur, "Ideolojia," 78. All emphases are original unless otherwise indicated.

80. Lafinur, "Ideolojia," 79.

81. Lafinur, "Ideolojia," 79.

82. Mably, *Principes de morale*; Mably, Observations sur le; and Mably, *De la législation*.

83. Baker, *Inventing the French Revolution*, 93–94.

84. Lafinur, "Ideolojia," 80.

85. Cabanis, *On the Relations*.

86. Lafinur, "Fragmentos ineditos," 91. The following parenthetical page numbers refer to this source.

87. Gez, *El Dr. Juan Crisóstomo Lafinur*, 56.

88. Gez, *El Dr. Juan Crisóstomo Lafinur*, 185.

89. Gez, *El Dr. Juan Crisóstomo Lafinur*, 61.

90. F. Vidal, *Sciences of the Soul*, 17.

91. Harris, "Bernardino Rivadavia."

92. Harris, "Bernadino Rivadavia," 137

93. James Thomson, *Letters on the Moral and Religious State of South America; Written During a Residence of Nearly Seven Years in Buenos Aires, Chile, Peru, and Colombia* (London: James Nisbet, 1827).

94. Hudson, *Recuerdos históricos*, 206.

95. Hudson, *Recuerdos históricos*, 206.

96. Caldcleugh, *Travels in South America*, 175.

97. Thomson, *Letters on the Moral and Religious State*, 294.

98. Sociedad de Beneficiencia de la Capital, "Origen y fundación, 15.

99. Sociedad de Beneficiencia de la Capital, "Origen y fundación," 16.

100. Sociedad de Beneficiencia de la Capital, "Acta de instalación," 32.

101. Sociedad de Beneficiencia de la Capital, "Acta de instalación de la Sociedad," 33.

102. Rivadavia, *Mensaje del Gobierno*, 5.
103. Rengger and Longchamps, *Reign*, 196.
104. Rengger and Longchamps, *Reign*, 197.
105. Rengger and Longchamps, *Reign*, 158–59.
106. Bentham, *Works of Jeremy Bentham*, vol. 4
107. O'Neal, *Authority of Experience*, 7.
108. Vila, *Enlightenment and Pathology*, 1.
109. Mill, *Elements of Political Economy*, iii.
110. "Creación de la clase de economía política," 438.
111. "Creación de la clase de economía política," 505.
112. "Fundación de las escuelas," 513.
113. "Fundación de las escuelas," 513.
114. "Medicina: origen," 463.
115. Pinel, *Nosografía filosófica*, xxx.
116. "Reglando los estudios preparatorios."
117. *Crónica politica y literaria de Buenos Aires*, no. 56 (June 19, 1827), 8.
118. Isabelle, *Voyage á Buénos Ayres*, 31.
119. di Pasquale, "Prensa, política y medicina," 119–36, 133.
120. Myers, "Identidades porteñas," 39–63.
121. J. M. Gutiérrez, "Don Esteban Luca."
122. Humboldt and Bonpland, *Personal Narrative*, 252.
123. "Origen y estado," 24.
124. "Origen y estado," 25.
125. "Medicina," *La Argentina*, May 15, 1822.
126. "La balanza de los poderes," 227.
127. "Sofismas politicos, 236."
128. "Ciencias: Concluye el discurso, 358."
129. "Escuela de música, 4."
130. Quoted in A. Williams, "Estética músical," 264.
131. "Memoria sobre esta cuestion, 1."
132. "Ciencias: Concluye el discurso."
133. "Articulo comunicado."
134. "Ciencias," September 15, 1822, 246.
135. "Enseñanza pública," July 15, 1822.
136. "Enseñanza pública," July 15, 1822.
137. Gorriti, *Reflecciones*, 234. The following parenthetical page numbers refer to this source.

Chapter 4. Eclectic and Romantic Man

1. McGrath, *Making Spirit Matter*, 5.
2. Lacordaire, "Un voyageur."
3. Pacheco, *Una revolucion*, 4. The following parenthetical page numbers refer to this source.
4. Haines, "Inter-Relations," 21.
5. Löwy and Sayre, *Romanticism*, 1.
6. Löwy and Sayre, *Romanticism*, 14.
7. Shepperson, "Intellectual Background," 30.
8. Tanghe and Kestemont, "Edinburgh," 585.

9. B. Jenkins, *Evolution Before Darwin*, 191.
10. Bell, *Life in Shadow*.
11. Richards, *Romantic Conception*.
12. Richards, *Romantic Conception*, 539.
13. McGrath, *Making Spirit Matter*, 7.
14. McGrath, *Making Spirit Matter*, 7.
15. Mignet, "Notice historique," 460–61.
16. Davie, "Victor Cousin,"193–94.
17. Hamilton, "On the Philosophy of the Unconditioned," 198. The following parenthetical page numbers refer to this source.
18. Hamilton, "Cours de Philosophie," 9.
19. Buck-Morss, "Hegel and Haiti," 865.
20. Janet, *Victor Cousin*, 157.
21. Janet, *Victor Cousin*, 68.
22. Urbas, "In Praise."
23. Hamilton, "On the Philosophy of the Unconditioned," 194.
24. Scott, "Against Eclecticism," 118.
25. McGrath, *Making Spirit Matter*, 5.
26. Echeverría, "Pensamientos: Afectos íntimos," in Echeverría, *Obras completas* 5:441–54, 442.
27. Echeverría, "Pensamientos," 449.
28. McGrath, *Making Spirit Matter*, 3.
29. Vilain, "'Excess of Savage Force?'" 313; and Goethe, *Faust*.
30. Echeverría, "Á Mr. F. Stapher," *Obras completes*, 5:415.
31. Echeverría, "Á Mr. F. Stapher," *Obras completas*, 5:417.
32. Echeverría, "Argument," 415.
33. Maine de Biran and Cousin, *Leçons de philosophie*. This was a new edition published because of the triumph of spiritualism over the ideologues.
34. Echeverría, "Argument," 419.
35. Echeverría, "Argument," 421. I am grateful to Roger Ariew for his assistance with the translation, and the explanation of Descartes's context.
36. Ampère, *Théorie des phénomènes*.
37. Maine de Biran, *Influence de l'habitude*.
38. Barkhofft, "Romantic Science," 209–10.
39. B. G. Smith, "Rise and Fall," 377–78.
40. Cousin, "Kant et sa philosophie," 382.
41. Cousin, "Kant et sa philosophie," 382.
42. Echeverría, "Clasicismo y romanticismo," 97.
43. Echeverría, "Clasicismo y romanticismo."
44. Echeverría, "Melancolía."
45. Echeverría, "Clasicismo y romanticismo," 97.
46. Piazza, "Maine de Biran," 5.
47. Echeverría, "Sistemas," 372.
48. Echeverría, "Sistemas," 373.
49. Lefebvre, *Sociology of Marx*, 22.
50. Goldstein, "'Official Philosophies,'" 261.
51. Cousin, *Cours de philosophie*, vi.

52. Morell, "Socialistic Mystics," 495.

53. Morell, "Socialistic Mystics," 494.

54. Strube, "Socialism and Esotericism," 198.

55. Taylor, "Saint-Simon," 59.

56. Taylor, "Saint-Simon," 59–60.

57. Strube, "Religion, Politics and Utopia," 199.

58. Taylor, "Saint-Simon," 59.

59. Taylor, "Saint-Simon," 61.

60. Taylor, "Saint-Simon," 63.

61. Beecher, *Charles Fourier*, 413.

62. Poldervaart, "Theories About Sex," 57–58.

63. Poldervaart, "Theories About Sex," 45.

64. Poldervaart, "Theories About Sex," 49.

65. Fourier, *Passions of the Human Soul*, 109.

66. Fourier, *Passions of the Human Soul*, 110.

67. Doherty, "Introduction," xviii.

68. Strube, "Socialist Religion."

69. Breckman, "Politics in a Symbolic Key," 63.

70. Olson, *Science and Scientism*, 80.

71. Comte, *General View of Positivism*, 234. This is a translation of the 1948 French edition.

72. Herder, *Outlines of a Philosophy* (1803), 1:257. The following parenthetical page numbers refer to this source.

73. Camper, "Petrus Camper on the Origin and Color of Blacks," 6.

74. Camper, "Petrus Camper on the Origin and Color of Blacks," 6.

75. Caldas, "Del influjo del clima."

76. Mensch, "Kant and the Skull Collectors," 205.

77. Grindle, "Our Own Imperfect Knowledge," 147.

78. Potts, *Flesh and the Ideal*, 161.

79. Siebers, "Kant and the Politics of Beauty," 31.

80. Herder, *Outlines of a Philosophy* (1803), 1:277

81. Herder, *Outlines of a Philosophy* (1803), 1:289–91.

82. Letter from Henning Gotfried Linberg, Brooklyn, New York, to Theophilus Parsons, Junior, July 31, 1823. Brooklyn, New York. Boston Public Library, "Personal Correspondence," https://bpl.bibliocommons.com/v2/record/S75C7988844.

83. Linberg, "Preface," iii–vi.

84. Cousin, État de *l'instruction*.

85. Cousin and Taylor, *Digest*.

86. Cousin, *Report on the State*.

87. "Victor Cousin and the United States," Paris, December 17, 1894, *The Nation* 60, no. 1541 (January 10, 1895), 28.

88. Cousin, *Curso de filosofía*, trans. N. R. de Losada.

89. Echeverría, "Pensamientos," 447.

90. C. Darwin, *Journals and Remarks*, 140. The following parenthetical page numbers refer to this source.

91. [Letter to Charles Darwin] "From Edward Lumb, 8 May 1834." Darwin Correspondence Project, "Letter no. 245," https://www.darwinproject.ac.uk/letter/DCP-LETT-245.xml.

92. C. Darwin, *Journals and Remarks*, 86.
93. Lynch, *Argentine Dictator*, 11.
94. Lynch, *Argentine Dictator*, 154
95. Dorrego, "Guerra civil," 345.
96. Dorrego, "Guerra civil," 343.
97. Dorrego, "Guerra civil," 346.
98. Dorrego, "Discurso sobre reformas," 38.
99. Dorrego, "Guerra civil," 347.
100. Dorrego, "Guerra civil," 348.
101. Dorrego, "Guerra civil," 352.
102. "Los editores," 233.
103. Lynch, *Argentine Dictator*, 155
104. Myers, *Orden y virtud*, 47.
105. Myers, *Orden y virtud*, 31.
106. Halperín-Donghi, *Politics, Economics and Society*, 57.
107. de Angelis, *¡Viva La Federacion!* 31.
108. Halperín-Donghi, "Una nación," xiv.
109. Halperín-Donghi, "Una nación," xiv.
110. Alberdi, "Mi vida privada," 63.
111. J. M. Gutiérrez, "Noticias biograficas," xxix.
112. Wright, *Religion of Humanity*, 9.
113. J. M. Gutiérrez, "Noticias biográficas," xc.
114. "Etudes d' histoire," 267–68.
115. Schell, *Sociable Sciences*, 90.
116. Andrés Bello, "Observaciones sobre el terremoto de 20 de febrero," *El Araucano*, no. 447 (1839). Original text: P. P. King, P. Stokes, and R. Fitz-Roy, "Sketch of the Surveying Voyages of His Majesty's Ships *Adventure* and *Beagle*, 1825–1836," *Journal of the Royal Geographical Society of London* 6 (1836): 311–43; "Del *Edinburgh Review*, narrativas de los viajes de los buques de guerra S. N. B. Adventure y Beagle, por los capitanes King y Fitz-Roy, de la marina naval británica, y por Charles Darwin, Escudero, naturalista de la Beagle, 3 tomos, Londres, 1839," *El Araucano*, no. 494 (1840). Original text: "Narrative of the Voyages of H. M. S. *Adventure* and *Beagle*. . . . " London (1839). *Edinburgh Review*, no. 140 (July1839): 467–93.
117. Bello, *Philosophy of Understanding*; and Bello, "Prólogo," viii.
118. Jaksic, *Andrés Bello: la pasión por el orden*, 101.
119. Bello, *Philosophy of Understanding*, 110.
120. El Entreacto, "Análisis del curso."
121. El Entreacto, "Prospecto," *El Entreacto, Periódico Semanal* 1, no. 11, May 11, 1845, 3.
122. Echeverría, *Obras completas*, 4:246.
123. J. M. Gutiérrez, "Noticias biograficas," 5:xvii. The following parenthetical page numbers refer to this source.
124. Berman, *Experience of the Foreign*, 14.
125. Berman, *Experience of the Foreign*, 188.
126. Jervis, *Transgressing the Modern*, 128.
127. Davies et al., *South American Independence*, 83.
128. Davies et al., *South American Independence*, 83.
129. Echeverría, *Obras completas*, 5:91–92.

130. Rotger, *Captive Women.*

131. Echeverría, "La Cautiva," in *Obras completas,* 1:33. The following parenthetical page numbers refer to this source.

132. Echeverría, *Obras completas,* 5:105.

133. Mosse, *Image of Man,* 75.

134. Echeverría, "Manual de ensenanza moral," *Obras completes,* 4:329.

135. V. F. López, *Evocaciones históricas*; cited in H. E. Davis, *Latin American Thought,* 72.

136. Echeverría, "Discurso de introducción," *Obras completas,* 5:313–14.

137. Katra, *Argentine Generation,* 48.

138. Katra, *Argentine Generation,* 56.

139. Bergman-Carton, *Woman of Ideas,* 30.

140. Bergman-Carton, *Woman of Ideas,* 30.

141. Sommer, *Foundational Fictions,* 88.

142. Quesada, "El movimiento intelectual."

143. Myers, *Orden y virtud.*

144. Alberdi, "Novedad inteligente."

145. Juan B. Alberdi, "Teoremas fundamentales del arte moderno," *La Moda, Gacetín Semanal,* no. 2 (Buenos Aires, November 25, 1837), 3.

146. Corvalán, Untitled, 5.

147. S. S. El Regañon, "Diálogo sobre alguna cosa importante."

148. "Modas de señoras," *La Moda, Gacetín Semanal,* no. 3 (Buenos Aires, December 2, 1837), 3–4.

149. Echeverría, *Dogma socialista,* v.

150. Mazzini, *Joseph Mazzini,* 63.

151. Mazzini, *Joseph Mazzini,* 65.

152. Frost, *Secret Societies,* 236–37.

Chapter 5. Racial Man

1. Sarmiento, "Argirópolis."
2. Sarmiento, "Colonización inglesa," 23:332–33.
3. Sarmiento, "Argirópolis," 23:71.
4. Sarmiento, "Colonización inglesa," 23:334.
5. Sarmiento, "Colonización inglesa," 23:336.
6. Sarmiento, "Colonización inglesa," 23:336–37.
7. Sarmiento, "Decreto del gobernador," 23:228–29.
8. Sarmiento, "Las filípicas de los Andes," 23:203.
9. Sarmiento, "Argirópolis," 23:86.
10. Sarmiento, "Argirópolis," 23:94.
11. Sarmiento, "Argirópolis," 23:86–87.
12. Ballantyne, *Orientalism and Race.*
13. Lamarck, *Philosophie zoologique.*
14. Biddiss, "Gobineau," 258.
15. Blumenbach, *Elements of Physiology,* 564.
16. Cantor, "Edinburgh Phrenology Debate," 195.
17. J. Gordon, "Doctrines of Gall," 263.
18. Spurzheim, *Phrenology,* 252.

19. Spurzheim, *Phrenology*, 552.
20. Spurzheim, *Phrenology*, 518.
21. Combe, *Essays on Phrenology* (1819), 361.
22. Combe, *Constitution of Man*, 46.
23. Combe, *Constitution of Man*, 69.
24. Combe, *Constitution of Man*, 70.
25. Kidd, "Race, Empire, and the Limits."
26. C. Darwin, Letter to W. D. Fox, January 3, 1830.
27. Tiedemann, "On the brain of the Negro," 398.
28. Turner, *History of the Anglo-Saxons* 1:1.
29. Turner, *History of the Anglo-Saxons*, 1:2.
30. Turner, *Sacred History of the World*, x.
31. A. Thierry, *Histoire de la conquête.*
32. Gossman, "Augustin Thierry," 3.
33. Gossman, "Augustin Thierry," 4.
34. Pomian, "Franks and Gauls," 57.
35. Roger, *American Enemy*, 169.
36. Melman, "Claiming the Nation's Past," 581.
37. Horsman, *Race and Manifest Destiny*, 4.
38. Michelet, *Histoire de France*, vol. 1, 29. The following parenthetical page numbers refer to this source.
39. Michelet, "Letter to Charles Darwin, 15 November 1872."
40. J. Davis, "Anthropology and Ethnology," 395. See also W. F. Edwards, *De l'influence des agens physiques*, an important book on physiology.
41. Société ethnologique, "Liste des membres," xvi–xxi.
42. Société ethnologique, "Liste des membres," xvi–xxi.
43. W. F. Edwards, *Des caractères physiologiques.*
44. Stewart, "William Frédéric Edwards," 276.
45. W. F. Edwards, *Des caractéres physiologiques*, 62. The following parenthetical page numbers refer to this source.
46. "Physical Evidences." The following parenthetical page numbers refer to this source.
47. Sera-Shriar, *Making of British Anthropology.*
48. C. Darwin, *Journal of Charles Darwin*, 525.
49. C. Darwin, *Journal of Charles Darwin*, 534.
50. C. Darwin, *Journal of Charles Darwin*, 520.
51. Combe, "Varieties of the Human Species," 4.
52. Strang, *Frontiers of Science*, 209.
53. Strang, *Frontiers of Science*, 170.
54. "Phrenology," 1.
55. "Edwards on the Physiological Characters."
56. Capen, *Reminiscences of Dr. Spurzheim*, 119. The following parenthetical page numbers refer to this source.
57. Branson, "Phrenology," 175.
58. "Mr. Clay's Speech," 2.
59. "Mr. Clay's Speech," 2.
60. W. F. Edwards, *De l'influence des agens physiques.*

61. Combe, "Physiological Characters of Races," 364.

62. Combe, "Physiological Characters of Races," 367.

63. Combe, "Physiological Characters of Races," 378.

64. Combe, "Crania Americana," 341.

65. Combe, "Crania Americana," 349.

66. Morton, "Introductory Essay," 11.

67. Malte-Brun, *Universal Geography*, 484. When Malté-Brun died, others completed the final two volumes.

68. Malte-Brun, *Universal Geography*, 484.

69. "Extract from a Recent Work," 72.

70. Channing, *Letter on the Annexation*, 18.

71. Channing, *Letter on the Annexation*, 15.

72. Channing, *Letter on the Annexation*, 20.

73. "Mr. Fenimore Cooper."

74. "Morning Chronicle."

75. "Ethnography."

76. "Races of Men."

77. "Ethnological Society," 396.

78. R. King, "Address to the Ethnological Society," 14.

79. R. King, "Address to the Ethnological Society" 14.

80. Lamennais, *Words of a Believer.*

81. Lamennais, *Palabras de un creyente.*

82. Lamennais, *Words of a Believer*, 94.

83. Lamennais, *Words of a Believer*, 3.

84. Humboldt, *Cosmos essai d'une*; and Humboldt, *Cosmos ó ensayo de una descripción.*

85. C. Darwin, Letter to J. D. Hooker, September 18, 1845.

86. Humboldt, *Kosmos*, viii.

87. C. Darwin, Letter to J. D. Hooker, October 28, 1845.

88. Dana, "Cosmos."

89. Morton, *Types of Mankind.*

90. Gliddon, "Preface," ix.

91. Patterson, "Notice of the Life," xxx.

92. Nott, "Introduction to *Types of Mankind*," 49. The following parenthetical page numbers refer to this source.

93. "Express from Paris," *Morning Chronicle* [1801], January 21, 1845. *British Library Newspapers*, link.gale.com/apps/doc/BC3207315864/BNCN?u=tamp44898&sid=bookmark-BNCN&xid=bd917510.

94. Lacroix, *Patagonie*, 144. In addition, a complete book was translated to Spanish; see Frédéric Lacroix, *Historia de la Patagonia, Tierra de Fuego, é Isla Malvinas* (Barcelona: Imprenta del Liberal Barcelones, 1841).

95. Lacroix, *Géographie politique*144.

96. Whitington, *Falkland Islands*, 11.

97. Whitington, *Falkland Islands*, 11.

98. Whitington, *Falkland Islands*, 12.

99. Jackson, "Message."

100. *La Gaceta Mercantil*, 2403, February 9, 1832.

101. Balcarce and García, "Latest from Buenos Ayres."
102. Jackson, "Message of the President."
103. C. Darwin, *Journal of Charles Darwin*, 245.
104. Tornel y Mendívil, *Tejas y los Estados Unidos*, 3.
105. Mirón, "Un orizabeño distinguido," 234.
106. Tornel y Mendívil, *Tejas y los Estados Unidos*, 44–45.
107. Calvo y O'Farrill, "Examen de los distintos caracteres," 11.
108. José Antonio Saco, *Ideas sobre la incorporacion de Cuba en los Estados-Unidos* (Paris: Imprenta de Panckoucke, 1848), 3.
109. *Recopilación de las leyes* (1836), 1345.
110. Rosas, *Manifiesto de las razones.*
111. Rosas, *Manifiesto de las razones*, n.p.
112. Santa Cruz, *Contra-Manifesto.*
113. "Latest from Buenos Ayres."
114. "Commerce and Industry."
115. De Angelis, *De la conducta de los agentes*, 16. The following parenthetical page numbers refer to this source.
116. Myers, *Orden y virtud*, 60.
117. Bonvini, "Los exiliados."
118. Alberdi, "Figarillo en Montevideo," 53.
119. Viola, "La abolición," 567.
120. Mackau–Arana Treaty, article 3, 155.
121. Sarmiento, *Facundo*, 7:219.
122. Rock, "British Communities," 156.
123. Letter cited in Manuel Bilbao, *Vindicación y memorias*, 300–301.
124. Sarmiento, "La cuestión del Plata," 6:70.
125. Lamas, *Escritos politicos*, 65.
126. Lamas, *Escritos politicos*, 68.
127. Lamas, *Escritos politicos*, 68.
128. Sarmiento, "La cuestión del Plata," 2:69.
129. Sarmiento, "Avíos y monturas," 1:9.
130. Sarmiento, "El romanticism,"1:287–92.
131. Sarmiento, "Continúa el exámen," 1:305.
132. Sarmiento, "Concluye el análisis," 1:313.
133. Sarmiento, "El comunicado," 1:227.
134. Lamennais, *Le livre du people*, 2:647.
135. Sarmiento, "Fisiolojia del Paquete," 2:9–10. The following parenthetical page numbers refer to this source.
136. Sarmiento, "Cuadros de Monvoisin," 2:213.
137. Sarmiento, "Las procesiones," 2:144.
138. Sarmiento, "Investigaciones," 2:213. The following parenthetical page numbers refer to this source.
139. Michelet, *Principes de la philosophie*, 156.
140. Sarmiento, *Facundo.*
141. Sarmiento, *Facundo*, 7:6.
142. T. Lacordaire, "La bataille de la Tablada," 275.

143. Albert Gallatin Mackey, Charles McClenachan, and Charles Thompson, *An Encyclopædia of Freemasonry and Its Kindred Sciences. . . .* (Philadelphia: L. H. Everts, 1889), 1031.

144. Sarmiento, *Facundo*, 209.

145. Sarmiento, *Facundo*, 27.

146. Cooper, *Last of the Mohicans*.

147. Sarmiento, *Facundo*, 35.

148. Sarmiento, *Facundo*, 25.

149. Sarmiento, *Facundo*, 9, 141.

150. Sarmiento, "Al señor H. Southern," 6:272.

151. MacGregor, *Progress of America*, 1:v.

152. MacGregor, *Progress of America*, 958.

153. MacGregor, *Progress of America*, 954.

154. "Progress of America," 591.

Chapter 6. Sodomized Man

1. Myers, *Orden y virtud*.

2. "Carta de José de San Martín a Juan Manuel de Rosas."

3. Domínguez, "Carta del doctor," 451.

4. Domínguez, "Carta del doctor," 458.

5. Bonvini, "'Good Christians.'"

6. McGrath, *Making Spirit Matter*, 7.

7. Bonvini, "Los exiliados."

8. Quoted in *Anales de la liga de estudiantes americanos* 2, no. 3 (April 1915) (El Siglo Ilustrado Montevideo, 1915), 309.

9. Bell, *Life in Shadow*, 141.

10. Bell, *Life in Shadow*, 143. The life of the French botanist in Corrientes became very difficult once he returned to his estate after his visit to Montevideo, but he stayed there until he died in 1858 at age eighty-five. The news was communicated in Europe by Humboldt, and Bonpland's life and extraordinary works were celebrated in France. His archive was sent by his family to France, where it arrived in 1859, but another part remained in his state. In 1905 Amado Bonpland, his son, helped to donate his collection of books, notes, and papers to the Medical School of the University of Buenos Aires. Juan A. Domínguez and Eugenio Autran, "Archivos inéditos de Aimé Bonpland," in *Actas del XVII Congreso Internacional de Americanistas* (Buenos Aires: Imprenta de Coni, 1912), 599–602.

11. de Pena, "Por vía de Introducción."

12. "Memoire," *Le Patriote Français* 7, no. 2448 (August 26, 1843), 3.

13. Quoted in "Larrañaga," 2.

14. Domingo F. Sarmiento, *Vida y escritos del Coronel D. Francisco J. Muñiz* (Buenos Aires: F. Lajouane, 1885), 33.

15. Rivera Indarte, *La intervención*, 50.

16. Macintyre, "Corinne in the Andes," 187.

17. *Recopilación de las leyes* (1858), 133–34.

18. *Recopilación de las leyes* (1858), 134.

19. José Antonio Wilde, *Buenos Aires desde setenta años atras* (Buenos Aires: C. Casavalle, 1881), 161.

20. Frías, *La gloria del tirano Rosas*, 35.

21. Lamas, *Apuntes históricos*, lv.
22. Halperín-Donghi, *Politics, Economics and Society*, 50.
23. Szuchman, *Order, Family, and Community*, 38–39.
24. Halperín-Donghi, *Revolución y guerra*; and Halperín-Donghi, *De la revolución*.
25. Andrews, *Afro-Argentines*, 64–77.
26. Andrews, *Afro-Argentines*, 98.
27. Andrews, "Afro-Latin America."
28. Halperín-Donghi, *Politics, Economics and Society*, 165.
29. Interview of Vicente and Ernesto Quesada with Rosas, 1873, in Sampay, *Las ideas políticas*, 215.
30. Mansilla, *Rozas* (1899), 104. The following parenthetical page numbers refer to this source.
31. Godbeer, *Overflowing of Friendship*, 184.
32. Salessi, *Médicos maleantes y maricas*.
33. Rodríguez Molas, *Historia de la tortura*, 56.
34. Rodríguez Molas, *Historia de la tortura*, 35. Possibly written by Rivera Indarte.
35. "Cielito, cielo, cielito, / Cielo de los maricones, / un decreto debe darse, / para que usen calzones. / En un momento hace un sastre/un unitario decente, / pues ellos se juzgan serlo / con tener levita y lente." Quoted in Rodríguez Molas, *Historia de la tortura*, 33.
36. Blomberg, *Los poetas*, 42–43.
37. Anonymous, "Decima," in Blomberg, *Los poetas*, 38.
38. Quoted in J. M. Paz, *Memorias postumas*, 3:69–70.
39. Varela, *Observations*, 8.
40. Mitre, "Estudio Biográfico."
41. Scotto, *Las diabluras*, 141.
42. Rivera Indarte, *El voto de America*.
43. Rivera Indarte, "Un opúsculo."
44. Rivera Indarte wrote for *El imparcial*, and *La lanza federal* in 1834. In 1835 he collaborated on *Diario de anuncios y publicaciones oficiales*.
45. Rivera Indarte, *La intervención*; and Rivera Indarte, *Poesías*.
46. Brambilla, *José Mármol*; and Mitre, "Estudio Biográfico."
47. Rivera Indarte, *Rosas y sus opositores* (2nd ed.). The following parenthetical page numbers refer to this source.
48. Rivera Indarte, *Las tablas de la sangre*, 121.
49. Pufendorf, *Of the Law of Nature*, 192.
50. Rivera Indarte, *Las tablas de la sangre*, 122.
51. "Affairs of the River Plate."
52. "Saturday, July 5, 1845."
53. "British Aggression on the La Plata."
54. "Siege of Montevideo."
55. Quoted in Lynch, *Argentine Caudillo*, 94.
56. de Angelis, "Infamous Libel," 59.
57. de Angelis, "Infamous Libel," 56.
58. "La más horrenda," 89.
59. "La más horrenda," 103.
60. Avila, *Ordenes privadas*, 94. The following parenthetical page numbers refer to this source.

61. Muñoz, "La Sociedad de la Mazorca, 285.

62. Quoted in J. M. Gutiérrez, "Noticias biograficas," lxv.

63. C. *Darwin Correspondence Project*, "Letter no. 886."

64. C. Darwin, *Journal of Researches*, 73.

65. Agrelo, *Causa criminal*, 24.

66. Agrelo, *Causa criminal*, 117

67. Aquiles B. Oribe, *Brigadier General D. Manuel Oribe*, vol. 1, 2nd ed. (A. Barreiro y Ramos, 1912), 331.

68. Garibaldi, *Garibaldi*,154.

69. Lamas, *Apuntes históricos*, 27–28.

70. "Mister Southern," 4.

71. Pedro de Angelis, "Al editor del Nacional," 20–21.

72. de Angelis, "Al editor del *Nacional*," 21.

73. Delacour, "Le Rio de la Plata," 509.

74. Varela, *Auto-biografía*, 16.

75. Alberdi, *Veinte días*, 63. The following parenthetical page numbers refer to this source.

76. Taillandier, "La littérature politique," January 15, 1845; and Taillandier, "La littérature politique," March 1, 1845.

77. Alberdi, *Impresiones y recuerdos*, 3:228.

78. Prieto, *Los viajeros ingleses*, 124.

79. Hegel, *Philosophy of History*; Thiry, *La moral universal*; and Thiry, *System of Nature.*

80. Sarmiento, *Viajes por Europa*, 89.

81. Herder, *Outlines of a Philosophy.*

82. Elizabeth Wilson, "Invisible Flâneur."

83. Zinny, "Bibliografia periodistica."

84. "Messieurs, rien n'est changé; ce n'est qu' un Flaneur de moins." Zinny and Funes, *Efemeridografia*, 113.

85. Elizabeth Wilson, *Sphinx in the City*, 54.

86. Sarmiento, *El gran Sarmiento*, 72–73.

87. Sarmiento, *Viajes en Europa*, 8. The following parenthetical page numbers refer to this source.

88. Sarmiento, *Travels*, 211. The following parenthetical page numbers refer to this source.

89. Alberdi, "Ideas para presidir," 605. The following parenthetical page numbers refer to this source.

90. Alberdi, *Tobías*, 2:343–87.

91. Alberdi, *Tobías*, 2:347.

92. Leroux, *De l'humanité*, 1:xi.

93. Leroux, *De l'humanité*, 1:142.

94. Leroux, *De l'humanité*, 1:307.

95. Engels and Marx, *Manifest.*

96. Desmond, *Politics of Evolution*, 8.

97. Tresch, *Romantic Machine*, 226.

98. Tresch, *Romantic Machine*, 162.

99. Tresch, *Romantic Machine*, 226–27.

100. Echeverría, "MAYO," 413. The following parenthetical page numbers refer to this source.

101. "De l'humanité de son principe," 258. The following parenthetical page numbers refer to this source.

102. Echeverría, "Introduction." The following parenthetical page numbers refer to this source.

103. Mazzini, "General Instructions," 98–99.

104. Mazzini, "General Instructions," 100.

105. Echeverría, "Introduction," vi.

106. Dumas, *Montevideo*.

107. Echeverría, *Dogma socialista*, 5.

108. Echeverría, *Dogma socialista*, 7–8.

109. Echeverría, *Dogma socialista*, 8.

110. Alberdi, *Derecho público*, 172.

111. Echeverría, *Dogma socialista*, 71. The following parenthetical page numbers refer to this source.

112. "De l'humanité de son principe."

113. Echeverría, *Dogma socialista*, 42.

114. Echeverría, *Dogma socialista*, lxxxvii.

115. Echeverría, *Dogma socialista*, xxvi.

116. Echeverría, *Cartas á D. Pedro de Angelis*, 47. The following parenthetical page numbers refer to this source.

117. Echeverría, *Dogma socialista*, 84.

118. Echeverría, *Cartas á D. Pedro de Angelis*, 37.

119. Echeverría, "Revolución de Febrero," 450. The following parenthetical page numbers refer to this source.

120. Groussac, "Noticia biografica," xiv.

121. Echeverría, "Revolución de Febrero," 453.

122. Echeverría, "Revolución de Febrero," 453.

Epilogue: Men in Transit

1. The Rosist article was published by *El defensor de la independencia americana*, 272 (El Miguelete, Uruguay, January 17, 1848), in Herrera y Obes and Berro, *El caudillismo*, 107–20.

2. Berger et al., *Constructing Masculinity*, 73.

3. Echeverría, *El matadero*, 222. The following parenthetical page numbers refer to this source.

4. O. Paz, *Labyrinth of Solitude*; and Salessi, *Médicos maleantes y maricas*.

5. Eagleton, *Rape of Clarissa*.

6. Mansilla, *Rozas*, 12-14.

7. Mansilla, *Rozas*, 13.

8. Alberdi, "La diplomacia de Buenos Aires," 6:249.

9. Alberdi, "La diplomacia de Buenos Aires," 6:249.

10. Cané, "La cultura Argentina," 161–62.

11. "'Los mexicanos salieron de los indios, los brasileros salieron de la selva': la polémica frase del presidente de Argentina por la cual tuvo que disculparse," *BBC News Mundo*, June 9, 2021, https://www.bbc.com/mundo/noticias-57422159.

SELECTED BIBLIOGRAPHY

Aarsleff, Hans. "Pufendorf and Condillac on Law and Language." *Journal of the Philosophy of History* 5, no. 3: 308–21.

Abercrombie, Thomas Alan. *Passing to América: Antonio (née María) Yta's Transgressive, Transatlantic Life in the Twilight of the Spanish Empire*. Pennsylvania State University Press, 2019.

"Academia de matemáticas del estado." *Los Amigos de la Patria y de la Juventud*, no. 5 (April 1816).

"Advertisement to the American Edition." In E. Darwin, *Botanic Garden*, n.p.

"Affairs of the River Plate." *Times* (London), October 3, 1845: 6.

Agrelo, Emilio. *Causa criminal seguida contra el ex-gobernador Juan Manuel de Rosas ante los tribunales ordinarios de Buenos Aires*. Editor J. Palumbo, 1908.

"Agricultura." *Semanario de Agricultura, Industria y Comercio* 1, no. 1 (September 1, 1802): 1–8.

Aizpurua, Ramón. "Revolution and Politics in Venezuela and Curaçao, 1795–1800." In *Curaçao in the Age of Revolutions, 1795–1800*, edited by Wim Klooster and Gert Oostindie, 97–123. Brill, 2014.

"Al patriotism de las más sensibles." *Gaceta del Gobierno del Perú Independiente* 2, no. 4 (January 12, 1822).

Alberdi, Juan B. *Derecho público provincial argentino*. Edited by Joaquín V. González. Vol. 11, *Obras selectas: Nueva edicion*. La Facultad, 1920.

Alberdi, Juan B. "Figarillo en Montevideo." *El Iniciador* 3, no. 2 (November 15, 1838): 52–55.

Alberdi, Juan B. "Ideas para presidir a la confección del curso de filosofía contemporánea." In *Escritos póstumos de J. B. Alberdi*. Vol. 15. Imprenta Europea, 1900.

Alberdi, Juan B. *Impresiones y recuerdos*. In *Obras selectas*. Librería la Facultad, 1920.

Alberdi, Juan B. "Mi vida privada." In *Autobiografía*. Colección Grandes Escritores Argentinos II. Buenos Aires: W. M. Jackson, n.d.

Alberdi, Juan B. *Tobías*. In *Obras completas de Juan B. Alberdi*. Vol. 2. Buenos Aires: La Tribuna Nacional, 1886.

Alberdi, Juan B. *Veinte días en Génova*. Valparaíso: Imprenta del Mercurio, 1845. https://biblioteca.org.ar/libros/158438.pdf.

Alberdi, Juan Bautista. "La diplomacia de Buenos Aires y los intereses americanos y europeos en el Plata." In *Obras completas de J. B. Alberdi*. Vol. 6, 219–67. Buenos Aires: Imprenta de la Tribuna Nacional, 1886.

Alberdi, Juan Bautista. "La generación presente á la faz de la generación pasada." In *Obras completas de Juan B. Alberdi*. Vol. 1. Buenos Aires: Imprenta de la Tribuna Nacional, 1886.

Alberdi, Juan Bautista. "Mi nombre y mi plan." In *Obras completas de J. B. Alberdi*. Vol. 1, 288–91. Buenos Aires: Imprenta de la Tribuna Nacional, 1886.

Alberdi, Juan Bautista. "Modas de señoras." In *Obras completas de Juan B. Alberdi*. Vol. 1, 273–74. Buenos Aires: Imprenta de la Tribuna Nacional, 1886.

Alberdi, Juan Bautista. "Novedad inteligente." *La Moda, Gacetín Semanal*, no. 2 (Buenos Aires), November 25, 1837: 3–4.

Alberdi, Juan Bautista. "Reglas de urbanidad para una visita." In *Obras completas de Juan B. Alberdi*. Vol. 1. Buenos Aires: Imprenta de la Tribuna Nacional, 1886.

Alberdi, Juan Bautista. *Tobías o la cárcel a la vela*. Argentina: Biblioteca Cervantes Virtual, n.d. In *Obras completas de Juan B. Alberdi*. Vol. 2, 343–87. Buenos Aires: Imprenta de la Tribuna Nacional, 1886.

Alvarez y Thomas, Don Ignacio. "Recuerdos historicos sobre la provincia de Cuyo, capitulo 4, de 1822 a 1825." *La Revista de Buenos Aires* 6, no. 68 (December 1868): 481–96. http://www.saij.gob.ar/docs-f/biblioteca_digital/libros/revista-buenos-aires_t17_1868/revista-buenos-aires_t17_1868.pdf.

Alzaga, Martín de. "El muy ilustre Ayuntamiento de Buenos Ayres al Rey Nuestro Señor." Buenos Aires, August 10, 1806. *Invasiones inglesas al Rio de la Plata: Documentos inéditos para servir a la historia del Rio de la Plata durante las invasiones de los generales ingleses Beresford y Whitelocke en los años de 1806 y 1807*, 88–94. Buenos Aires: Imprenta Republicana, 1870.

Amann, Elizabeth. *Dandyism in the Age of Revolution: The Art of the Cut*. University of Chicago Press, 2015.

Amar y Borbón, Josefa. *Discurso sobre la educacion fisica y moral de las mugeres*. Madrid: Benito Cano, 1790.

Ampère, André-Marie, *Théorie des phénomènes électro-dynamiques, uniquement déduite de l'expérience*. Paris: Méquignon-Marvis, 1826.

Anderson, James. *The Constitutions of the Free-Masons: Containing the History, Charges, Regulations, &c. of that Most Ancient and Right Worshipful Fraternity*. For the Use of the Lodges. London: William Hunter, 1734.

Andrade, Carlos Valderrama. "Relación polémica de Miguel Antonio Caro con el benthamismo." *Ideas y Valores* 38, no. 80 (1989): 121–44.

Andrews, George Reid. "Afro-Latin America: Five Questions." *Latin American and Caribbean Ethnic Studies* 4, no. 2 (2009): 191–210.

Andrews, George Reid. *The Afro-Argentines in Buenos Aires, 1800–1900*. University of Wisconsin Press, 1980.

Anglés, Matías, and Matías Gortari. *Coleccion general de documentos, que contiene los sucesos tocantes á la segunda época de las conmociones de los regulares de la compañía en el Paraguay, [. . .]*. 3 vols. Madrid: Imprenta Real de la Gaceta, 1769.

"Apology." In Erasmus Darwin, *The Botanic Garden* [American edition]. New York: Published in the Faculty of Physics of Columbia College, 1798, n.p.

Ardao, Arturo. *Andrés Bello, filósofo*. Biblioteca de la Academia Nacional de la Historia, 1986.

Arellano, Ignacio, and Antonio Lorente Medina, eds. *Poesía satírica y burlesca en la Hispanoamérica colonial*. Vol. 18. Universidad Iberoamericana, Iberoamericana Editorial, 2009.

Ariew, Roger. "Descartes and the Jesuits: Doubt, Novelty, and the Eucharist." In Feingold, *Jesuit Science*, 157–94.

Arráiz, Francisco Jiménez. *Panegírico del generalísimo Francisco de Miranda: Precursor de la independencia latino-americana*. Lit. y Tip. del Comercio, 1916.

"Articulo Ccomunicado." *Los Amigos de la Patria y de la Juventud*, no. 2 (December 1815): 6.

Astigarraga, Jesús. "Ramón de Salas y la difusión de la fisiocracia en España." *Historia Agraria*, 52 (December 2010): 75–102.

Aullón de Haro, Pedro, and Davide Mombelli. *Introduction to the Spanish Universalist School*. Brill, 2020.

Auslander, Leora. *Cultural Revolutions: Everyday Life and Politics in Britain North America, and France*. University of California Press, 2009.

Avazoa, José Manuel. *Observaciones contra el dictamen que el Dr. D. José Eusebio Agüero dió á la cúria provisoral de Buenos-Ayres sobre la facultad de dispensar en el impedimento para el matrimonio por la diversidad de religion*. Buenos Aires: n.p., 1827.

Avila, Pedro C. *Ordenes privadas del general D. Juan Manuel Rosas en la revolución de 1840 y abril de 1842*. Lima: Imprenta y Litografía de J. Montoya, 1847.

Azara, Félix de. *Diario de un reconocimiento de las guardias y fortines, que guarnecen la linea de frontera de Buenos-Aires*. Buenos Aires: Imprenta del Estado, 1837.

Baeza, Rafael Sagrado. *Ciencia-Mundo: Origen republicano, arte y nación en América*. Editorial Universitaria, 2010.

Baines, Thomas. *Observations on the Present State of Affairs of the River Plate*. Liverpool: Liverpool Times Office, 1845.

Baker, Keith Michael. *Inventing the French Revolution: Essays on French Political Culture in the Eighteenth Century*. Cambridge University Press, 1990.

Bakunin, Jack S. "Pierre Leroux on Democracy, Socialism, and the Enlightenment." *Journal of the History of Ideas* 37, no. 3 (1976): 455–74.

Balbín, José Celedonio. "Señor coronel don Bartolomé Mitre." In Museo Mitre, *Documentos del Archivo de Belgrano*. Vol. 1, 241–44. Imprenta de Coni hermanos, 1913.

Balcarce, Juan Román, and Manuel Jose García. "Latest from Buenos Ayres." *Providence Patriot, Columbian Phenix*, April 28, 1832, 2.

Ballantyne, Tony. *Orientalism and Race: Aryanism in the British Empire*. Palgrave, 2016.

"Bando de Pedro de la Cruz Condori a los vecínos de Cotagaita." In *Colección documental de la independencia del Perú*. Vol. 2, *La rebelión de Túpac Amaru*, edited by Carlos Daniel Valcárcel, 585–86. Talleres Gráficos Cecil S.A., 1971. [Edición digital Narra la Independencia: Fuentes Documentales, https://archivocolectivo.org/narra/cdip-02283.]

Barker-Benfield, G. J. *Abigail and John Adams: The Americanization of Sensibility*. University of Chicago Press, 2010.

Barker-Benfield, G. J. *The Culture of Sensibility: Sex and Society in Eighteenth-Century Britain*. University of Chicago Press, 1992.

Barkhofft, Jürgen. "Romantic Science and Psychology." In *The Cambridge Companion to German Romanticism*, edited by Nicholas Saul, 209–26. Cambridge University Press, 2009.

Batticuore, Graciela, and Klaus Gallo. *Resonancias románticas: Ensayos sobre historia de la cultura argentina (1820–1890)*. Eudeba, 2005.

Bayly, C. A., and E. F. Biagini, eds. *Giuseppe Mazzini and the Globalisation of Democratic Nationalism, 1830–1920*. Oxford University Press, 2008.

Beecher, Jonathan. *Charles Fourier: The Visionary and His World*. University of California Press, 1990.

Beer, Max. *An Inquiry into Physiocracy*. Routledge, (1939) 2014.

Belgrano, Manuel. *Autobiografía del General Manuel Belgrano*. In Museo Mitre, *Documentos del Archivo de Belgrano*, 173–94. Imprenta de Coni Hermanos, 1913.

Belgrano, Manuel. "Comercio." *Correo de Comercio* 1, no. 27 (September 1, 1810): 204. In Museo Mitre, *Documentos del Archivo de Belgrano*. Vol. 2, 316. Imprenta de Coni Hermanos, 1913.

Belgrano, Manuel. *Compendio de los principios de economía política*, 1796. In Museo Mitre, *Documentos del archivo de Belgrano*. Vol. 1. Imprenta de Coni hermanos, 1913.

Belgrano, Manuel. "Concluye la materia del numero anterior." *Correo de Comercio* 1, no. 18 (June 30, 1810): 137. In Museo Mitre, *Documentos del Archivo de Belgrano*. Vol. 2, 139–216. Imprenta de Coni Hermanos, 1913.

Belgrano, Manuel. "Memoria escrita por el licenciado Manuel Belgrano, abogado de los Reales Consejos y Secretario por su Majestad Real Consulado del Virreynato de Buenos Aires en 1797. In Museo Mitre, *Documentos del Archivo de Belgrano*, 81–98. Imprenta de Coni Hermanos, 1913.

Belgrano, Manuel. "Memoria que leyó el licenciado don Manuel Belgrano . . . en la session que celebró su Junta de gobierno a 14 de Junio de 1798." In Museo Mitre, *Documentos del Archivo de Belgrano*. Vol. 1, 99–115. Imprenta de Coni Hermanos, 1913.

Belgrano, Manuel. "Proclama del real consulado." In Museo Mitre, *Documentos del Archivo de Belgrano*. Vol. 1, 50–53. Imprenta de Coni Hermanos, 1913.

Belgrano, Manuel. "Señores editores del *Correo de Comercio*." *Correo de Comercio* 1, no. 14 (June 2, 1810): 111. In Museo Mitre, *Documentos del Archivo de Belgrano*. Vol. 2, 163–65. Imprenta de Coni Hermanos, 1913.

Belgrano, Manuel. "Solicita permiso al sumo Pontífice para leer libros prohibidos"; "Consentimiento del papa." In Museo Mitre, *Documentos del Archivo de Belgrano*. Vol. 1, 18–19. Imprenta de Coni Hermanos, 1913.

Bell, Stephen. *A Life in Shadow: Aimé Bonpland in Southern South America, 1817–1858*. Stanford University Press, 2010.

Bello, Andrés. *Gramática de la lengua castellana: Destinada al uso de los americanos*. Santiago: Imprenta del Progreso, 1847.

Bello, Andrés. "Narrativa de los viajes de los buques de guerra de su majestad británica 'Advantage' i 'Beagle,' por los capitanes King i Fitz Roi, de la Marina Real Británica, i por Carlos Darwin, escudero, naturalista de la 'Beagle.'" Translated by Andrés Bello. In Andrés Bello, *Obras completes*. Vol. 15, 217–45. Santiago: Imprenta Cervantes, 1893.

Bello, Andrés. *Philosophy of the Understanding*. Translated by O. Carlos Stoetzer. General Secretariat, Organization of American States, (1881) 1984.

Bello, Andrés. "Prólogo." In *Obras completas de Don Andrés Bello*. Vol. 1, i–xvii. Santiago: Pedro G. Ramírez, 1881.

Bello, Andrés. "Sobre la enseñanza del canto." In *Biblioteca Americana*. Vol. 1, 72–77. London: G. Marchant, 1823.

Bello, Andrés. "Sobre la jeografía de la extremidad sur de la América, la Tierra del Fuego I el Estrecho de Magallanes hechos en la visita de estas costas por los buques de S. M. B. 'Adventure' i 'Beagle' en 1826 i 1830, por el capitan Philip Parker, comandante de la espedicion. Leídas a la Real Sociedad Jeográfica de Lóndres el 15 i 31 de mayo de 1831." Translated by Andrés Bello. In *Obras completas*. Vol. 15, 135–50. Santiago: Imprenta Cervantes, 1893.

Beltrán, Oscar. *Historia del periodismo argentino*. Sopena, 1943.

Benjamin, Marina. "Elbow Room: Women Writers on Science, 1740–1840." In *Science and Sensibility: Gender and Scientific Inquiry, 1780–1945*, edited by Marina Benjamin, 27–59. Basil Blackwell, 1991.

Bentham, George. *Outline of a New System of Logic: With a Critical Examination of Dr. Whately's Elements of Logic*. London: Hunt and Clarke, 1827.

Bentham, Jeremy. *The Collected Works of Jeremy Bentham*. 33 vols. Edited by J. H. Burns, J. R. Dinwiddy, F. Rosen, and others. Clarendon Press, 1968–.

Bentham, Jeremy. *The Works of Jeremy Bentham*. Vol. 4. London: W. Trait, 1843.

Berco, Cristian. "Producing Patriarchy: Male Sodomy and Gender in Early Modern Spain." *Journal of the History of Sexuality* 17, no. 3 (September 2008): 351–76.

Berger, Maurice, Brian Wallis, Simon Watson, and Carrie Mae Weems. *Constructing Masculinity*. Routledge, 1995.

Bergman-Carton, Janis. *The Woman of Ideas in French Art, 1830–1848*. Yale University Press, 1995.

Berman, Antoine. *The Experience of the Foreign: Culture and Translation in Romantic Germany*. State University of New York Press, 1992.

Bernal, Martin. *Black Athena Writes Back: Martin Bernal Responds to His Critics*. Duke University Press, 2001.

Best, Mark, Achilles Katamba, and Duncan Neuhauser. "Making the Right Decision: Benjamin Franklin's Son Dies of Smallpox in 1736." *BMJ Quality and Safety* 16, no. 6 (December 2007): 478–80.

Biddiss, Michael D. "Gobineau and the Origins of European Racism." *Race* 7, no. 3 (January 1966): 255–70.

"Bienes de la Iglesia, quedan á disposicion del gobierno, quien se obliga á costear los gastos del Departamento eclesiástico," January 17, 1823. In Aurelio Prado y Rojas, *Leyes y decretos promulgados en la provincia de Buenos Aires desde 1810 á 1876: 1810 a 1823*, 374–75. Buenos Aires: Imprenta del Mercurio, 1877.

Bilbao, Manuel. *Vindicación y memorias de Don Antonino Reyes*. Vol. 1. Buenos Aires: Imprenta del Porvenir, 1883.

Bindman, David. *Ape to Apollo: Aesthetics and the Idea of Race in the 18th Century*. Cornell University Press, 2002.

Blanchard, Shaun. "Are Jansenists Among Us?" *Church Life Journal*, October 4, 2019. https://churchlifejournal.nd.edu/articles/are-jansenists-among-us/.

Bleichmar, Daniela. *Visible Empire Botanical Expeditions and Visual Culture in the Hispanic Enlightenment*. University of Chicago Press, 2012.

Blomberg, Héctor Pedro, ed. *Los poetas de la tiranía: Cancionero federal*. Ediciones Anaconda, 1934.

Blumenbach, Johann Friedrich. *The Elements of Physiology*. 4th ed. Translated by John Elliotson. London: Longman, Rees, Orme, Brown, and Green, 1828.

Bolufer, Mónica. "Gallantry and Sociability in the South of Europe: Shifting Gender Relationships and Representations." In *European Modernity and the Passionate South*, edited by Xavier Andreu-Miralles and Mónica Bolufer-Peruga, 14–35. Brill, 2022.

Bonvini, Alessandro. "'Good Christians, Good Citizens, Good Patriots': Young Italy and the Atlantic Struggle for the Nation, 1835–1848." *Diasporas* 34 (2019). http://journals.openedition.org/diasporas/4259.

Bonvini, Alessandro. "Los exiliados del Risorgimento: El mazzinianesismo en el Cono Sur." *Memoria y Sociedad* (Bogotá) 22, no. 44 (June 2018): 42–65. .

Brambilla, Alberto Blasi. *José Mármol y la sombra de rosas*. Pleamar, 1970.

Branson, Susan. "Phrenology and the Science of Race in Antebellum America." *Early American Studies* 15, no. 1 (2017): 164–93.

Braun, Carlos Rodríguez. "Early Smithian Economics in the Spanish Empire: J. H. Vieytes and Colonial Policy." *Journal of the History of Economic Thought* 4, no. 3 (1997): 444–54.

Braun, Harald E. *Juan de Mariana and Early Modern Spanish Political Thought*. Taylor and Francis, 2016.

Breckman, Warren. "Politics in a Symbolic Key: Pierre Leroux, Romantic Socialism, and the Schelling Affair." *Modern Intellectual History* 2, no. 1 (2005): 61–86.

Brissot, Jacques Pierre. *New Travels in the United States of America: Performed in 1788*. Edited and translated by Durand Echeverria. Harvard University Press, (1792) 1964.

"British Aggression on the La Plata." *Daily Ohio Statesman*, November 26, 1845. Nineteenth Century U.S. Newspapers.

Brown, Jonathan C. "The Bondage of Old Habits in Nineteenth-Century Argentina." *Latin American Research Review* 21, no. 2 (1986): 3–26.

Buci-Glucksman, Christine. *Baroque Reason: The Aesthetics of Modernity*. Translated by Patrick Camiller. Sage, 1994.

Buck-Morss, Susan. "Hegel and Haiti." *Critical Inquiry* 26, no. 4 (2000): 821–65.

Buckingham, Joseph Tinker. *Specimens of Newspaper Literature: With Personal Memoirs, Anecdotes, and Reminiscences*. Vol. 2. Boston: Little and J. Brown, 1850.

"Buenos Ayres 7 de Junio." *Gazeta de Buenos-Ayres*, no. 1 (June 7, 1810): 12–13.

Buffon, Leclerc de. *Histoire naturelle des oiseaux, par Leclerc de Buffon; augmentée de notes, d'additions considèrables, et mise au courant des connoissances actuelles . . .* Vol. 9. Paris: F. Dufart, 1803.

Bullock, Steven C. *Revolutionary Brotherhood: Freemasonry and the Transformation of the American Social Order, 1730–1840*. University of North Carolina Press, 1996.

Cabanis, Pierre Jean Georges. *Compendio histórico de las revoluciones y reforma de la medicina*. Madrid: Imprenta de Repullés, 1820.

Cabanis, Pierre Jean Georges. *Coup d'oeil sur les révolutions et sur la réforme de la médecine*. Paris: De l'Imprimerie de Crapelet, 1804.

Cabanis, Pierre Jean Georges. *Du degré de certitude dans la medicine*. Paris: Firmin Didot, 1798.

Cabanis, Pierre Jean Georges. *Essay on the Certainty of Medicine*. Translated by René La Roche. Philadelphia: R. Desilver, 1823.

Cabanis, Pierre Jean Georges. *Oeuvres philosophiques de Cabanis premiere partie*. Presses Universitaire de France, 1956.

Cabanis, Pierre Jean George. *On the Relations Between the Physical and Moral Aspects of Man*. Translated by Margaret Duggan Saidi. Johns Hopkins University Press, 1981 (translation of *Rapports du physique et du moral de l'homme*, 1805).

Cabanis, Pierre Jean Georges. "Pierre Jean Georges Cabanis to Thomas Jefferson, 20 October 1802." *Founders Online*, National Archives. https://founders.archives.gov/documents/Jefferson/01-38-02-0490. [Original source: *The Papers of Thomas Jefferson*. Vol. 38, July 1–November 12, 1802, edited by Barbara B. Oberg, 524–25. Princeton University Press, 2011.]

Cabanis, Pierre Jean Georges. *Rapports du physique et du moral de l'homme*. Paris: Crapelet, 1805.

Cabanis, Pierre Jean Georges. *Sketch of the Revolutions of Medical Science, and Views Relating to Its Reform*. Translated with notes by A. Henderson. London: J. Johnson, 1806.

Cabello, Francisco Antonio. "Continúa la idea general del comercio." *Telégrafo Mercantil, Rural, Político-Económico, e Historiógrafo del Río de la Plata* 1, no. 4 (April 11, 1801): 25–28.

Cabello, Francisco Antonio. "Política: Circunstancias en que se halla la provincial de Buenos Aires, é Islas Malvinas, y modo de repararse." *Telégrafo Mercantil, Rural, Político-Económico, e Historiógrafo del Río de la Plata* 2, no. 5 (October 8, 1802): 21–27.

Cabello y Mesa, Francisco Antonio. "Analysis del papel periódico intitulado *Telégrafo Mercantil*." *Telégrafo Mercantil, Rural, Político-Económico, e Historiógrafo del Río de la Plata (1801–1802)*. Vol. 4, reprint, 3–18. Compañía Americana de Billetes de Banco, 1914.

Cabello y Mesa, Francisco Antonio. "Obgetos principales de esta obra." In *Analysis del papel periódico intitulado Telégrafo Mercantil*. Buenos Aires: Real Imprenta de Niños Expósitos, 1800: 12–14.

Cabildo de Buenos Aires al de Tucumán, marzo 26 de 1807, LADAGT, 77–78. Cited in Gabriela Lupiañez, "Usos de la noción de 'pueblo' en Tucumán en tiempos de las invasiones inglesas." *Humanidades: Revista de la Universidad de Montevideo*, no. 3 (June 2018): 101–23, 109. https://doi.org/10.25185/3.4.

Caesar, Ann Hallamore, and Michael Caesar. *Modern Italian Literature*. Polity Press, 2007.

Caldas, Francisco José de. *Cartas de Caldas*. Imprenta Nacional, 1917.

Caldas, Francisco José de. "Del influjo del clima sobre los seres organizados." In *Semanario de la Nueva Granada: Miscelanea de ciencias literatura, artes é industria publicada por una sociedad de patriotas granadinos*, 110–54. Paris: Librería Castellana, 1849.

Caldcleugh, Alexander. *Travels in South America, During the Years 1819–20–21: Containing an Account of the Present State of Brazil, Buenos Ayres, and Chile*. Vol. 1. London: J. Murray, 1825.

Calero y Moreira, Jacinto. *Prospecto del papel periódico intitulado Mercurio peruano de historia, literatura, y noticas públicas*. Lima: Imprenta Real de los Niños Expositos, 1790. *Mercurio Peruano: Edición facsimilar*. Biblioteca Nacional del Peru, 1964.

Calvo y O'Farrill, José María. "Examen de los distintos caracteres distintivos de la raza aborígen de la América, por el Dr. Samuel Jorge Morton." Real Sociedad Económica de Amigos del País (Cuba). *Memorias de la Real Sociedad Económica de la Habana*. Vol. 2. Havana: Oficina del Gobierno y de la Real Sociedad Patriótica, 1846.

Campbell, Peter R. "The Politics of Patriotism in France 1770–1788." *French History* 24, no. 4 (2010): 550–75.

Camper, Petrus. "Petrus Camper on the Origin and Color of Blacks." Translated by Miriam Claude Meijer. *History of Anthropology Newsletter* 24, no. 2 (1997): 3–9.

Campomanes, Pedro Ramón Rodríguez. "To Benjamin Franklin from Campomanes, 26 July 1784." *Founders Online*, National Archives. https://founders.archives.gov/documents/Franklin/01-42-02-0281. [Original source: *The Papers of Benjamin Franklin*. Vol. 42, *March 1 Through August 15, 1784*, edited by Ellen R. Cohn, 448–51. Yale University Press, 2017.]

Cané, Miguel. "La cultura Argentina." *Discursos y conferencias* (Casa Vaccaro, 1919).

Cañeque, Alejandro. *The King's Living Image: The Culture and Politics of Viceregal Power in Colonial Mexico*. Taylor and Francis, 2013.

Cañizares-Esguerra, Jorge. "Iberian Colonial Science." *Isis* 96, no. 1 (2005): 64–70.

Cañizares-Esguerra, Jorge, and Marcos Cueto. "Latin America." In *An Introduction to the History of Science in the Non-Western Tradition*, edited by Douglas Allchin and Robert DeKosky, 49–62. History of Science Society, 1999.

Cañizares-Esguerra, Jorge, and Marcos Cueto. "Latin American Science: The Long View." *NACLA Report on the Americas* 35, no. 5 (March 2002): 18–22.

Canter, Juan. *Monteagudo, Pazos Silva y el censor de 1812*. J. Peuser, 1924.

Canton, D. Narciso Fellobio [Francisco Cabello]. "Satirilla festiva." *Telégrafo Mercantil, Rural, Político-Económico, e Historiógrafo del Río de la Plata* 3, no. 4 (January 24, 1802): 55.

Cantor, Geoffrey N. "The Edinburgh Phrenology Debate: 1803–1828." *Annals of Science* 32, no. 3 (1975): 195–218.

Capen, Nahum. *Annals of Phrenology*. Boston: Marsh, Capen and Lyon, 1835.

Capen, Nahum. *Reminiscences of Dr. Spurzheim and George Combe: And a Review of the Science of Phrenology, from the Period of Its Discovery by Dr. Gall, to the Time of the Visit of George Combe to the United States, 1838, 1840*. New York: Fowler & Wells, 1881.

Caracciolli, Louis Antoine de. *El goce ó posesion de sí mismo*. Madrid: Miguel Escribano, 1777. Translation of *La jouissance de soi-même*. Amsterdam: E. van Harrevelt Berghen, 1759.

Carrasquilla, Rafael María. "Oscar Saldarriaga." In *Pensamiento colombiano del siglo XX*. Vol. 1, edited by Santiago Castro Gómez, 479–525. Pontificia Universidad Javeriana, 2007.

"Carta de José de San Martín a Juan Manuel de Rosas Desde el exilio le agradece y felicita por los esfuerzos realizados a favor de la Patria. Boulogne Sur Mer. 6 de mayo de 1850." Archivo General de la Nación. Documentos escritos. Sala VII. Legajo 191.

Carte, Philip. "Tears and the Man." In Knott and Taylor, *Women, Gender and Enlightenment*, 156–73.

Cartel del certamen "el nuevo heroe de la fama": En el solemne triumphal recibimiento del Exmo. Sor. don Manuel de Amat y Junient. . . . Perú: Real Universidad de San Marcos, 1762.

Casaus y Torres, Fray Ramón. "Parecer del R. P. Dr. Fr. Ramón Casaus y Torre." In Gómez Marín, *El currutaco por alambique*.

Channing, William Ellery. *A Letter on the Annexation of Texas to the United States*. London: John Green, 1837.

Chazin, Maurice, "Quinet an Early Discoverer of Emerson." *PMLA/Publications of the Modern Language Association of America* 48, no. 1 (1933): 147–63. https://doi.org/10.2307/457977.

Chilcoat, Michelle. "The Legacy of Enlightenment Brain Sex." *Eighteenth Century* 41, no. 1 (2000): 3–20.

"Ciencias." *La Abeja Argentina* 1, no. 6 (September 15, 1822): 246–47.

"Ciencias: Concluye el discurso suspendido en el numero sexto." *La Abeja Argentina* 1, no. 9 (December 15, 1822): 358.

Clak, Bertoldo. "Carta de Bertoldo Clak." *Telégrafo Mercantil*, no. 26 (June 27, 1801): 201–7.

Clavigero, Francisco. *Storia antica del Messico cavata da migliori storici spagnuoli*. Cesena: G. Biasini, 1780.

Clavigero, Francisco. *Storia della California*. Venice: Modesto Fenzo, 1789.

Clavijero, Francisco Javier. *Historia antigua de México y de su conquista*. Mexico: Imprenta de Lara, 1844.

Coleccion general de las providencias hasta aqui tomadas por el gobierno sobre el estrañamiento y ocupacion de temporalidades de los regulares de la Compañia, que existian en los dominios de S. M. de España, Indias, e islas Filipinas á consequencia del real decreto de 27 de febrero, y pragmáticasancion de 2 de abril de este año. Vol. 1 Madrid: Imprenta Real de la Gazeta, 1767.

Colwill, Elizabeth. "Women's Empire and the Sovereignty of Man in la Décade Philosophique, 1794–1807." *Eighteenth-Century Studies* 29, no. 3 (1996): 265–89. https://muse.jhu.edu/article/10675.

Combe, George. *The Constitution of Man Considered in Relation to External Objects* . . . 4th ed., revised, corrected, and enlarged. Edinburgh: William & Robert Chambers, 1835.

Combe, George. "Crania Americana." *American Journal of Science* 38 (April 1840): 341–75.

Combe, George. *Essays on Phrenology: Or an Inquiry into the Principles and Utility of the System of Drs. Gall and Spurzheim, and into the Objections Made Against It*. Edinburgh Longman, Hurst, Rees, Orme & Brown, 1819.

Combe, George. *Essays on Phrenology: Or an Inquiry into the Principles and Utility of the System of Drs. Gall and Spurzheim, and into the Objections Made Against It*. Philadelphia: H. C. Curry and I. Lea, 1822.

Combe, George. "The Physiological Characters of Races of Mankind Considered in Their Relations to History; Being a Letter to M Amédée Thierry Author of the History of the Gauls." *Annals of Phrenology* 2: 364–68. Boston: Marsh, Capen and Lyon, 1835.

Combe, George. "Varieties of the Human Species." In Samuel George Morton, *Crania*

Americana; Or, A Comparative View of the Skulls of Various Aboriginal Nations of North and South America: To Which Is Prefixed an Essay on the Varieties of the Human Species, 1–95. London: J. Dobson, 1839. https://archive.org/details/Craniaamericana00Mort.

A Commentary and Review of Montesquieu's "Spirit of Laws": To which are annexed, Observations on the Thirty First Book by the late M. Condorcet; and Two Letters of Helvetius, on the Merits of the same Work. Translated by Thomas Jefferson. Philadelphia: William Duane, 1811.

"Commerce." *Semanario de Agricultura, Industria y Comercio* 3, no. 1 (October 6, 1802): 1–24.

"Commerce and Industry." *Era*, December 22, 1839. British Library Newspapers.

Companys, Isabel, and Núria Montardit, "El mestre pintor tres-centista Joan de Tarragona." *Universitas Tarraconensis: Revista de Geografia, Història i Filosofia* 4 (2018): 23–36.

Comte, Auguste. *A General View of Positivism*. Translated by J. H. Bridges. George Routledge, 1908.

"Concluye la materia del numero anterior." *Correo de Comercio* 1, no. 3 (March 17, 1810): 17–20.

"Concluye la materia del numero anterior." *Semanario de Agricultura, Industria y Comercio* 4, no. 1 (October 13, 1802): 25–27.

Condillac, Etienne Bonnot de. *Essay on the Origin of Human Knowledge*. Translated by Hans Aarsleff. Cambridge: Cambridge University Press, (1746) 2001.

Condillac, Etienne Bonnot de. *La lógica o los primeros elementos del arte de pensar*. Translated by Bernardo María de Calzada, Madrid: D. Joachin Ibarra, 1784.

Condillac, Etienne Bonnot de. *Logica de Condillac*. Translated by Valentin de Foronda. Madrid: Imprenta de Gonzalez, 1794.

Condillac, Etienne Bonnot de. *The Theory of Agreeable Sensations . . . Including Likewise a Dissertation upon Harmony of Stile*. London: W. Owen, 1774.

Condillac, Etienne Bonnot de. *Treatise des animaux, oeuvres complètes de Condillac, revues, corrigées par l'auteur et imprimées sur ses manuscrits autographes*. Vol. 5. Paris: Chez Dufart, Imprimeur-Libraire, 1803.

Condillac, Étienne Bonnot de. *Treatise on the Sensations*. Translated by Geraldine Carr. London: Favil Press, (1754) 1930.

Condillac, Etienne Bonnot de, and Francisco Guerra. *La lengua de los cálculos*. Madrid: Imprenta de Ruiz, 1805.

Condillac, Etienne Bonnot de, Francisco Guerra, Tomás Gorchs, and Bernardo María de Calzada. *La lógica ó los primeros elementos del arte de pensar*. Barcelona: Imprenta de Gorchs, 1827.

Condillac, Etienne Bonnot de, and Ch. Houel. *Oeuvres de Condillac: Cours d'études pour l'instruction du Prince de Parme, histoire modern*. Vol. 4. Paris: de l'Imprimerie de Ch. Houel, 1798.

Conger, Syndy McMillen, "Introduction." In *Sensibility in Transformation: Creative Resistance to Sentiment from the Augustans to the Romantics: Essays in Honor of Jean H. Hagstrum*, edited by Syndy McMillen Conger, 13–25. Associated University Presses, 1990.

Connell, R. W., and James W. Messerschmidt, "Hegemonic Masculinity: Rethinking the Concept." *Gender & Society* 19, no. 6 (2005): 829–59.

Connelly, Thomas, and Thomas Higgins. *Diccionario nuevo de las dos lenguas española é inglesa*. Vol. 2. Madrid: P. J. Pereyra, 1798.

Cooper, James Fenimore. *The Last of the Mohicans*. Philadelphia: H. C. Carey & I. Lea, 1826.

Cooter, Roger. *The Cultural Meaning of Popular Science: Phrenology and the Organisation of Consent in Nineteenth-Century Britain*. Cambridge University Press, 2005.

Cortesão, Jaime. *Alexandre de Gusmão e o tratado de Madrid*. Ministério das Relações Exteriores, Instituto Rio-Branco, 1950.

Corvalán, Rafael. Untitled. *La Moda, Gacetín Semanal*, no. 9 (Buenos Aires), January 13, 1838: 4.

Coss, Stephen. *The Fever of 1721: The Epidemic That Revolutionized Medicine and American Politics*. Simon and Schuster, 2016.

Cousin, Victor. *Cours de philosophie: Introduction à l'histoire de la philosophie*. Paris: Pichon et Didier, 1828.

Cousin, Victor. *Curso de filosofía sobre el fundamento de las ideas absolutas de lo verdadero, lo bello y lo bueno*. Translated by N. R. de Losada. Madrid: Imprenta de Repullés, 1847.

Cousin, Victor. *Curso de filosofía sobre el fundamento de las ideas absolutas de lo verdadero, lo bello y lo bueno*. Translated by José María Repullés Víctor and Ojea y Compañía. Lima: Calleja Llama, Íñigo Sánchez and Co., 1847.

Cousin, Victor. *État de l'instruction primaire dans le royaume de Prusse à la fin de l'année 1831*. Paris: F. G. Levrault, 1833.

Cousin, Victor. *Introduction to the History of Philosophy*. Translated by Henning Gotfried Linberg. Boston: Hilliard, Gray, Little, and Wilkins, 1832.

Cousin, Victor. "Kant et sa philosophie." *Revue des Deux Mondes* 21, no. 3 (1840): 382–414.

Cousin, Victor. *Report on the State of Public Instruction in Prussia*. Translated by Sarah Austin. London: Effingham Wilson, 1834.

Cousin, Victor, and J. Orville Taylor. *A Digest of M. Victor Cousin's Report on the State of Public Instruction in Prussia*. Albany, NY: Office of the Common School Assistant, 1836.

Cowles, William. "Los afro-descendientes de Buenos Aires: Mitos y realidades." *ISP Collection*, Paper 185 (2007). http://digitalcollections.sit.edu/isp_collection/185.

"Creación de la clase de economía política en la universidad." In *Noticias históricas sobre el orígen y desarrollo de la Enseñanza Pública Superior en Buenos Aires: Desde la época de la extinción de la compañía de Jesús en el año 1767, hasta poco después de fundada la universidad en 1821*, edited by Juan María Gutiérrez: 504–505. Buenos Aires: Imprenta del Siglo, 1868.

D'Abadal, Lluis Argemí. "The Physiocrats." In *Handbook of the History of Economic Thought*, edited by Jürgen G. Backhaus, 131–61. Springer, 2011.

Dana, Charles A. "Cosmos: Or a Survey of the Physical History of the Universe." *Harbinger* 1, no. 17 (October 4, 1845): 263.

Darton, Robert. *The Great Cat Massacre: And Other Episodes in French Cultural History*. Basic Books, 2009.

Darwin, Charles. *The Autobiography of Charles Darwin*. In *The Works of Charles Darwin*. Vol. 29, edited by Paul H. Barrett, 1–153. Routledge, 2016.

Darwin, Charles. *Journals and Remarks, 1832–1836: In Narrative of the Surveying Voyages of His Majesty's Ships Adventure and Beagle, Between the Years 1826 and 1836*. Vol. 3, edited by Robert Fitzroy. London: Henry Colburn, 1839.

Darwin, Charles. *Journal of Researches into the Natural History and Geology of the Countries Visited During the Voyage of H. M. S. Beagle Round the World*. London: John Murray, 1845.

Darwin, Charles. Letter to J. M. Herbert, June 2, 1833. *Darwin Correspondence Project*, "Letter no. 209." https://www.darwinproject.ac.uk/letter/?docId=letters/DCPLETT-209.xml.

Darwin, Charles. Letter to J. D. Hooker, October 28, 1845. *Darwin Correspondence Project*, "Letter no. 922." https://www.darwinproject.ac.uk/letter/?docId=letters/DCPLETT-922.xml.

Darwin, Charles. Letter to J. D. Hooker, September 18, 1845. *Darwin Correspondence Project*, "Letter no. 917." https://www.darwinproject.ac.uk/letter/?docId=letters/DCPLETT-917.xml.

Darwin, Charles. Letter to W. D. Fox, January 3, 1830. *Darwin Correspondence Project*, "Letter no. 75." https://www.darwinproject.ac.uk/letter/?docId=letters/DCPLETT-75.xml.

Darwin, Erasmus. *The Botanic Garden: A Poem, in Two Parts: Part I. Containing the Economy of Vegetation: Part II. The Loves of the Plants: with Philosophical Notes.* London: J. Johnson, 1789.

Davie, George Elder. "Victor Cousin and the Scottish Philosophers." *Journal of Scottish Philosophy* 7, no. 2 (2009): 193–214.

Davies, Catherine, Claire Brewster, and Hilary Owen. *South American Independence.* Liverpool University Press, 2006.

Davis, Harold Eugene. *Latin American Thought.* Louisiana State University Press, 1972.

Davis, Joseph B. "Anthropology and Ethnology." *Anthropological Review* 6, no. 23 (1868): 394–99.

de Angelis, Pedro. "Al editor del Nacional." In *Archivo americano y espiritu de la Prensa del mundo* 1, no. 12, 1844, 20–21. https://archive.org/details/archivoamerican00unkngoog.

de Angelis, Pedro. *Coleccion de obras y documentos relativos á la historia antigua y moderna de las provincias del Rio de la Plata: Ilustrados con notas y disertaciones por P. de Angelis*, Part 3, "José Gabriel Tupac Amaru, en las provincias del Peru." Vol. 5. Buenos Aires: Imprenta del Estado, 1836.

de Angelis, Pedro. *De la conducta de los agentes de la Francia durante el bloqueo del Rio de la Plata: Por un observador imparcial.* Buenos Aires: Imprenta del Estado, 1839.

de Angelis, Pedro. "Documentos Oficiales." *Archivo americano y espiritu de la prensa del mundo* 1, no. 12 (May 31, 1844): 295–305. First reprint of the Spanish text according to the original edition, vol. 1. Buenos Aires: Editorial Americana, 1946–47.

de Angelis, Pedro. "El General Rosas y los Salvajes Unitarios." In *Archivo americano y espiritu de la prensa del mundo* 1, no. 12 (May 31, 1844): 337. First reprint of the Spanish text according to the original edition, 1843–1851, vol. 1. Editorial Americana, 1946–47.

de Angelis, Pedro. "Infamous Libel Published in the 'Dublin Review' Against the Supreme Chief of the Argentine Confederation, Against That Republic, and Against the Past and Actual State of the Other South-American Republics." *Archivo americano y espíritu de la prensa del mundo*, no. 19 (1850): 59.

de Angelis, Pedro. *¡Viva la federacion! Rasgos de la vida publica de S. E. El Sr. Brigadier General D. Juan Manuel de Rosas, ilustre restaurador de las leyes, heroe del desierto, defensor heroico de la independencia americana, gobernador y capitan general de la provincia de Buenos Aires.* Buenos Aires: Imprenta del Estado, 1842.

de Arteaga, Esteban. *Investigaciones filosóficas sobre la belleza ideal.* Madrid: A. de Sancha, 1789.

de Asúa, Miguel. *La ciencia de mayo, 1800–1820.* Fondo de Cultura Económica, 2010.

de Asúa, Miguel. *Science in the Vanished Arcadia: Knowledge of Nature in the Jesuit Missions of Paraguay and Rio de la Plata.* Brill, 2014.

"Déclaration des droits de l'homme et du citoyen de 1789." https://www.conseil-constitutionnel.fr/le-bloc-de-constitutionnalite/declaration-des-droits-de-l-homme-et-du-citoyen-de-1789.

"Definición del currutaco." *Telégrafo Mercantil, Rural, Politico-económico, e Historiográfico del Río de la Plata* 2, no. 29 (November 15, 1801): 223.

Delacour, Adolphe. "Le Rio de la Plata." In *La revue indépendante.* Vol. 20. Paris: La Revue indépendante, 1845, 497–520.

"De la disciplina militar." *Los Amigos de la Patria y de la Juventud*, no. 5 (April 1816): 34–35.

"De l'humanité de son principe et de son avenir où se trouve exposée la vraie définition de la religion et où l on se explique le sens la suite et l'enchaînment du Mosaïsme et du Christianisme par Pierre Leroux." *Boston Quarterly Review* 5: 257–322. Boston: Benjamin H. Greene, 1842.

de Mariscal, Blanca López. "'El currutaco por alambique' de Manuel Gómez marín: Un texto satírico del siglo XVIII." In *Poesía satírica y burlesca en la Hispanoamérica colonial*, edited by Ignacio Arellano and Antonio Llorente Medina, 239–53. Iberoamericana, 2009.

de Monteagudo, Bernardo. "A las Americanas del sur." In Monteagudo and Rojas, *Obras políticas de Bernardo Monteagudo*, 98–99.

de Monteagudo, Bernardo. "A Pueyrredón." In Monteagudo and Rojas, *Obras políticas de Bernardo Monteagudo*, 278–79.

de Monteagudo, Bernando. "Censura política." *Mártir, o Libre*, no. 3, April 13, 1812.

de Monteagudo, Bernando. "Censura política." *Mártir, o Libre*, no. 3, May 11, 1812.

de Monteagudo, Bernardo. "'El siglo XIX y la revolución': *El censor de la revolución*, April 30, 1820." In *Monteagudo, su vida y sus escritos*. Vol. 1, edited by Mariano A. Pelliza, 181–87. Buenos Aires: Imprenta y Librería de Mayo, 1880.

de Monteagudo, Bernardo. "Oración inaugural de la 'Sociedad Patriótica.'" In Monteagudo and Rojas, *Obras políticas de Bernardo Monteagudo*, 245–60.

de Monteagudo, Bernardo, and Ricardo Rojas, eds. *Obras políticas de Bernardo Monteagudo*. J. Roldán, 1916.

de Pena, Carlos M. "Por vía de Introducción." In *Anales del Museo Nacional de Montevideo*. Vol. 1, v–xlvii. Montevideo: Imprenta Artística, 1894.

de Viegas, Simón. *La logica o el órden natural del raciocinio*. Madrid: Lastra y Compañia, 1799.

de Ovalle, Alonso. *An Historical Relation of the Kingdom of Chile by Alonso de Ovalle*. Translated from Spanish, in *A Collection of Voyages and Travels, some now first printed from original manuscripts*. Vol. 3, 1–153. London: Awnsham and John Churchill, 1704.

de Réal de Curban, Gaspar. *La ciencia del gobierno: Obra de moral, de derecho, y de politica*. Translated by Mariano Joseph Sala. Barcelona: Carlos, Gibert y Tutó, 1775.

Derrida, Jacques. "Introduction." In *The Archeology of the Frivolous: Reading Condillac*. Duquesne University Press, 1973.

Desmond, Adrian. *The Politics of Evolution: Morphology, Medicine, and Reform in Radical London*. University of Chicago Press, 1992.

Destutt de Tracy, Antoine. *Comentario sobre el espíritu de las leyes de Montesquieu*. Translated by Ramón Salas. Madrid: Imprenta de Fermín Villalpando, 1821.

Destutt de Tracy, Antoine. *Commentary and Review of Montesquieu's Spirit of the Laws*. Translated by Thomas Jefferson. Philadelphia: John Duane, 1811.

Destutt de Tracy, Antoine. *Élémens d'idéologie: troisième partie, Logique*. Paris: Courcier libraire-éditeur, 1805.

Destutt de Tracy, Antoine. *Elementos de verdadera lógica: compendio ó sea estracto de los elementos de ideología del senador Destutt-Tracy*. Translated by Juan Justo García. Madrid: Imprenta de don Mateo Repullés, 1821.

Destutt de Tracy, Antoine. *Éléments d'idéologie: Première partie- Idéologie proprement dite*. Paris: Chez Courcier, 1803.

Destutt de Tracy, Antoine. "Letter from Destutt de Tracy, April 1818." *The Papers of Thomas Jefferson: Retirement Series*. Vol. 12, 623–25. English translation by the editor, J. Jefferson Looney. September 1, 1817, to April 21, 1818, Retirement series, edited by J. Jefferson Looney et al. Princeton University Press, 2004–17.

"Destutt de Tracy to Thomas Jefferson, 10 March 1819." *Founders Online*, National Archives. https://founders.archives.gov/documents/Jefferson/03-14-02-0105. [Original source: *The Papers of Thomas Jefferson: Retirement Series*. Vol. 14, February 1 to August 31, 1819, edited by J. Jefferson Looney, 114–17. Princeton University Press, 2017.]

Destutt de Tracy, Antoine. *Principios de economia politica: Considerados por las relaciones que tienen con la voluntad humana*. Translated by Manuel Maria Gutierrez. 2 vols. Madrid: Cano, 1817.

Destutt de Tracy, Antoine. *Projet d'éléments d'idéologie*. Paris: Pierre Didot, 1801.

Destutt de Tracy, Antoine. *A Treatise on Political Economy; To Which Is Prefixed a Supplement to a Preceding Work on the Understanding, or Elements of Ideology; With an Analytical Table, and an Introduction on the Faculty of the Will*. Translated from the unpublished French original, revised and corrected by Thomas Jefferson. Georgetown, DC: Joseph Milligan, 1817.

de Terreros y Pando, Esteban. *Diccionario castellano con las voces de ciencias y artes y sus correspondientes en las tres lenguas francesa, latina é italiana*. Vol. 2. Madrid: Imprenta de la Viuda de Ibarra, 1787.

Diario de noticias sobresalientes en Lima y Noticias de Europa (1700–1711). Edición y estudio de Paul Firbas y José A. Rodríguez Garrido. Vol. 1 (1700–1705). IDEA, 2017. https://dadun.unav.edu/entities/publication/ba5064b7-3bb0-4047-be1c-60201840440b

Díaz de Yraola, Gonzalo. *La vuelta al mundo de la expedición de la vacuna*. Consejo Superior de Investigaciones Científicas, Instituto de Historia, 2003.

Dillon, John-Talbot. *Travels through Spain, with a View to Illustrate the Natural History . . . in a Series of Letters*. London: G. Robinson, 1780.

di Pasquale, Mariano. "Entre la experimentación política y la circulación de saberes: la gestión de Bernardino Rivadavia en Buenos Aires, 1821–1827." *Secuencia* (México), no. 87 (December 2013): 51–65. http://www.scielo.org.mx/scielo.php?script=sci_arttext&pid=S0186-03482013000300003&lng=es&nrm=iso.

di Pasquale, Mariano. "La recepción de la idéologie en la Universidad de Buenos Aires: El caso de Juan Manuel Fernández de Agüero (1821–1827)." *Prismas* 15, no. 1 (2011): 63–86.

di Pasquale, Mariano. "Prensa, política y medicina en Buenos Aires: Un estudio de *La Abeja Argentina, 1822–1823. Estudios de Teoría Literaria* 5, no. 9 (April 2016): 119–36.

Dirk Van Miert, "What Was the Republic of Letters? A Brief Introduction to a Long History." *Groniek: Historisch Tijdschrift* 204, no. 5 (2014).

D'Israeli, Isaac. *Domestic Anecdotes of the French Nation, During the Last Thirty Years, Indicative of the French Revolution*. London: C. and G. Kearsley, 1794.

Doherty, Hugh. "Introduction." In Charles Fourier, *The Passions of the Human Soul*. Translated by Hugh Doherty and John Reynell Morell. London: H. Bailliere, 1851.

Domínguez, Luis L. "Carta del doctor Luis L. Domínguez." In Gregoria F. Rodríguez, *Contribución histórica y documental*, 451–59. Casa Jocobo Peuser, 1922.

Dorrego, Manuel. "Guerra Civil." *El Tribuno*, no. 41–42 (February 28 and March 3, 1827). In *Dorrego: Tribuno y periodista*, edited by Alberto del Solar. Imprenta de Coni Hermanos, 1907.

Dorrego, Manuel. "Reformas fundamentales en la milicia," August 8, 1826. In *Dorrego: Tribuno y periodista*, edited by Alberto del Solar, 25-49. Imprenta de Coni Hermanos, 1907.

Duckenfield, Mark. *Battles Over Free Trade, Volume 1: The Advent of Free Trade, 1776–1846*. Routledge, 2017.

Dumas, Alexandre. *Montevideo, ou une nouvelle Troie*. Paris: Imprimerie centrale de N. Chaix, 1850.

Eagleton, Terry. *The Rape of Clarissa: Writing, Sexuality, and Class Struggle in Samuel Richardson*. University of Minnesota Press, 1982.

Echeverría, Esteban. "Á Mr. F. Stapher." In *Obras completes*, 5:413–19.

Echeverría, Esteban. "Argument que j'ai pose a un spiritualiste partisan outre des doctrines de Laromiguiere." In *Obras completes*, 5:419–21.

Echeverría, Esteban. *Cartas á D. Pedro de Angelis*. Montevideo: Imprenta del 18 de Julio, 1847.

Echeverría, Esteban. "Clasicismo y romanticism." In *Obras completes*, 5:97.

Echeverría, Esteban. "Discurso de introducción á una serie de lecturas pronunciadas en el 'Salón Literario' en setiembre de 1837." In *Obras completes*, 5:309–36.

Echeverría, Esteban. *Dogma socialista de la Asociación de Mayo, precedido de una ojeada retrospectiva sobre el movimiento intelectual en el Plata desde el año 37.* Montevideo: Imprenta del Nacional, 1846. https://archive.org/details/EstebanEcheverria1846DogmaSocialista/page/n7/mode/2up.

Echeverría, Esteban *El matadero.* Colección Grandes Escritores Argentinos. Vol. 12. Buenos Aires: W. M. Jackson Inc., 1871.

Echeverría, Esteban. "Introduction." In *Dogma socialista de la Asociacion Mayo, precedido de una ojeada retrospectiva sobre el movimiento intelectual en el Plata desde el año 37.* Montevideo: Imprenta del Nacional, 1846, iii. https://autores.uy/obra/13503.

Echeverría, Esteban. "La Cautiva." In *Obras completes, de D. Esteban Echeverría: Escritores argentinos, Poemas varios.* Vol. 1. Carlos Casavalle, 1970.

Echeverría, Esteban. "Manual de ensenanza moral: para las escuelas primarias del estado oriental." In *Obras completas de D. Esteban Echeverria: Escritos en prosa.* Vol. 4. Buenos Aires: Imprenta y Librería de Mayo, 1873.

Echeverría, Esteban. "MAYO y la enseñanza popular." In *Obras completes*, 4:412–30.

Echeverría, Esteban. "Melancolía." In Esteban Echeverría, *Obras completes*, 5:195–99.

Echeverría, Esteban. *Obras completas de D. Esteban Echeverría: Escritores Argentinos.* Buenos Aires: Carlos Casavalle, (1870) 1874.

Echeverría, Esteban. "Pensamientos. Afectos íntimos." In *Obras completes*, 5:441–54.

Echeverría, Esteban. "Revolución de Febrero en Francia." In *Obras completes*, 4:431–61.

Echeverría, Esteban. "Sistemas." In *Obras completes*, 5:370–73.

"Edwards on the Physiological Characters of the Human Race." *North American Medical and Surgical Journal* 8, no.15: 223–24. Philadelphia: Philadelphia Medical Society, 1829.

Edwards, Paul, ed. *The Encyclopedia of Philosophy.* Macmillan 1967, 3:50.

Edwards, William F. *Des caractères physiologiques des races humaines considérés dans leurs rapports avec l'histoire; lettre à Amédéc Thierry.* Paris: Compère, 1829. Reprinted in *Mémoires de la Société ethnologique.* Vol. 1. Paris: Mme. ve. Dondey Dupré, 1841, 61–168. https://babel.hathitrust.org/cgi/pt?id=uc1.b3392615&view=1up&seq=61&skin=2021.

Edwards, William Frédéric. *De l'influence des agens physiques sur la vie.* Paris: Chez Crochard libraire, 1824.

"El Consejo de Regencia de España é Indias á los Americanos Españoles," February 14, 1810. Cadiz: En la Oficina de D. Nicolas Gomez de Requena, Impressor del Govierno, 1810.

El defensor de la independencia americana 272 (El Miguelete, Uruguay: January 17, 1848). In Herrera y Obes and Berro, *El caudillismo y la revolución americana*, 107–20.

El Entreacto. "Análisis del curso de filosofía de D. Andrés Bello." *El Entreacto, Periódico Semanal* 1 (May 11, 1845).

El Entreacto. "Filosofía del entendimiento umano, por A. B." *El Entreacto, Periódico Semanal* 1 (May 11, 1845): 7–11.

"El gobierno ha concedido." *El Argos de Buenos Aires* 1, no. 81 (October 26, 1822): 4.

Ellis, Heather. *Masculinity and Science in Britain, 1831–1918.* London: Palgrave Macmillan, 2017.

Enfantin, Prosper. *Life Eternal: Past-Present-Future.* Open Court, 1920.

Engels, Friedrich, and Karl Marx. *Manifest der Kommunistischen Partei: Veröffentlicht im February 1848.* London: Bildungs-Gesellschaft für Arbeiter: von J. E. Burghard, 1848.

"Enseñanza pública." *La Abeja Argentina* 1, no. 4 (July 15, 1822): 141.

Entick, John. *The free mason's pocket companion; containing, the origin, progress, and present state of that ancient fraternity; the institution of the Grand Lodge of Scotland; lists of the grand masters and other officers of the Grand Lodge of Scotland* [. . .]. Glasgow: John & Peter Wilson et al.,1792.

"Escuela de música." *El Argos de Buenos Aires* 1, no. 75 (October 8, 1822).

"Ethnography." *National Intelligencer* 31, no. 9575 (October 26, 1843): 1.

"Ethnological Society: A Scientific Society for Investigating the Natural History of Civilized as well as Uncivilized Man." *Provincial Medical Journal and Retrospect of the Medical Sciences* 5, no. 124 (1843): 395–96.

"Etudes d' histoire et de philosophic par E. Lerminier professeur de l' histoire generale des legislations compares au College de France." *Monthly Review* 2: 264–72. London: G. Henderson, 1840.

"Extract from a Recent Work of That Celebrated Philosopher, Professor Charles Caldwell, Entitled 'Phrenology Vindicated.'" *Mississippi Free Trader and Natchez Gazette* 1, no. 18 (December 4, 1835): 72.

Fasani, González, and Ana Mónica. "La higiene en el Buenos Aires del siglo XVIII." *Libros de la Corte.es* 11 (2015): 7–26. http://hdl.handle.net/10486/669332.

Faucci, Riccardo, and Antonella Rancan. "Transforming the Economy: Saint-Simon and His Influence on Mazzini." *History of Economic Ideas* 17, no. 2 (2009): 79–105.

Fay, Bernard. *La francmasonería y la revolución intellectual del siglo XVIII.* Editorial Huemul, 1958. https://archive.org/details/LaFrancmasoneraYLaRevoluciinIntelectualDelSigloXVIIIBernardFay/mode/2up.

Feijoó, Benito Jerónimo. *Defensa de las Mujeres.* In *Teatro crítico universal.* Vol. 1, Discourse 15, 1726. Texto tomado de la edición de Madrid, por D. Joaquín Ibarra, a costa de la Real Compañía de Impresores y Libreros, 1778. http://www.filosofia.org/bjf/bjft116.htm.

Feijóo y Monenengro, Benito Jerónimo. *Cartas eruditas y curiosas em que por la mayor parte, se continua el designio del Theatro Critico Universal* . . . . Vol. 4. Madrid: Imprenta Real de la Gaceta, 1753.

Feingold, Mordechai, ed. *Jesuit Science and the Republic of Letters.* MIT Press, 2003.

"Felipe Senillosa." *Anales de la Universidad de Buenos Aires.* Vol. 2, 697. Buenos Aires: Obras Clásicas, 1877.

Fernández, Pedro Juan. "Sr. Editor del Telégrafo." *Telégrafo Mercantil, Rural, Político-Económico, e Historiógrafo del Río de la Plata* 11 (May 6, 1801): 82.

Fernández de Rojas, Juan Osa. *Libro de moda, ó ensayo de la historia de los currutacos, pirracas, y madamitas del nuevo cuño/escrito por un Filósofo currutaco; y corregido nuevamente por un Señorito pirracas.* 3rd ed. Madrid: Imprenta de Don Blas Román, 1798.

Feros, Antonio. *Speaking of Spain: The Evolution of Race and Nation in the Hispanic World.* Harvard University Press, 2017. http://www.jstor.org/stable/j.cttln2tvln.

Figarillo [J. B. Alberdi]. "Señales del Hombre Fino." *La Moda, Gacetín Semanal,* no. 15 (Buenos Aires), February 24, 1838: 2–3.

Fleck, Eliane. "Paraguay Natural Ilustrado by José Sánchez Labrador SJ: Between the American Experience and Exile." *HoST-Journal of History of Science and Technology* 15 (2021): 121–48. https://doi.org/10.2478/host-2021-0015.

Fletcher, Anthony. *Gender, Sex and Subordination in England, 1500–1800.* Yale University Press, 1995.

Florencio, Francisco Agustín [Juan Fernández de Rojas]. *Crotalogía, ó, Ciencia de la castañuelas:*

Instruccion científica del modo de tocar las castañuelas para baylar el bolero. Madrid: Imprenta Real, 1792.

"Foreign Official Paper." *Cobbett's Weekly Political Register* 12 (September 12, 1807): 411. https://www.google.com/books/edition/Cobbett_s_Weekly_Political_Register/oq0TAAAAYAAJ?hl=en&gbpv=1.

Fourier, Charles. *The Passions of the Human Soul*. Translated by Hugh Doherty and John Reynell Morell. London: H. Bailliere, 1851.

France's National Constituent Assembly. "Declaration des droits de l'homme et du citoyen," 1789. https://www.conseil-constitutionnel.fr/le-bloc-de-constitutionnalite/declaration-des-droits-de-l-homme-et-du-citoyen-de-1789.

Franklin, Benjamin. "Letter to Peter Collinson," May 9, 1753. *Teaching American History*. https://teachingamericanhistory.org/document/letter-to-peter-collinson/.

Fregeiro, Clemente L. *Don Bernardo de Monteagudo: Ensayo biográfico*. Buenos Aires: Igon Hermanos, 1879.

Frías, Félix. *Escritos y discursos de Félix Frías*. Vol. 3. Buenos Aires: Imprenta de Mayo, 1884.

Frías, Félix. *La gloria del tirano Rosas: Y otros escritos polémicos*. W. M. Jackson, 1945.

Frigerio, Alejandro. "'Negros' y 'Blancos' en Buenos Aires: Repensando nuestras categorías raciales." *Temas de Patrimonio Cultural* 16 (2006): 77–98.

Frigerio, Alejandro, and Eva Lamborghini. "Criando um movimento negro em um país 'Branco': Ativismo político e cultural afro na Argentina." *Afro-Asia* 39 (2010): 153–81.

Frost, Thomas. *The Secret Societies of the European Revolution, 1776–1876*. London: Tinsley Bros., 1876.

Fumaroli, Marc. *The Republic of Letters*. Yale University Press, 2018.

"Fundación de las escuelas de Medicina y Cirugía." *Noticias históricas sobre el orígen y desarrollo de la Enseñanza Pública Superior en Buenos Aires: Desde la época de la extinción de la compañía de Jesús en el año 1767, hasta poco después de fundada la universidad en 1821*, edited by Juan María Gutiérrez: 509-519. Buenos Aires: Imprenta del Siglo, 1868.

Funes, Don Gregorio. "Biografía del doctor Don Gregorio Funes." In *Ensayo de la historia civil de Buenos Aires, Tucumán y Paraguay*. Vol. 1, edited by Gregorio Funes, 9. Buenos Aires: Imprenta Bonaerense, 1856.

Galería de escritoras isabelinas: La prensa periódica entre 1833 y 1895. Editorial Cátedra / Universitat de València, 2000.

Galimberti, V. A. "Las leyes de 1821 y sus implicancias en las dinámicas político-electorales de los pueblos rurales bonaerenses." *Trabajos y Comunicaciones, Época* 54, no. 153 (July–December 2021). https://doi.org/10.24215/23468971e153.

Gallo, Klaus. "A la altura de las luces del siglo: El surgimiento de un clima intelectual en la Buenos Aires postrevolucionaria." In *Historia de los intelectuales en América Latina I*, edited by Carlos Altamirano, 184–204. Katz Editores, 2008.

Gallo, Klaus. "El exilio forzado de un ideologue Rioplatense: El pensamiento republicano de Lafinur y sus traumas." *Estudios de Teoría Literaria-Revista Digital: Artes, Letras y Humanidades* 3, no. 5 (2014): 187–200.

Gallo, Klaus. "Jeremy Bentham y la 'Feliz Experiencia.'" *Prismas: Revista de Historia Intelectual* 6 (2002): 79–96.

Gambarota, Paola. *Irresistible Signs: The Genius of Language and Italian National Identity*. University of Toronto Press, 2011.

García, Mora, and José Pascual. "Baltasar de los Reyes Marrero (1752–1809): Primer educador de la enseñanza de la física moderna en la Universidad de últimas décadas del siglo XVIII." *Revista Historia de la Educación Latinoamericana* 13 (2009): 148–65.

García, Servando, Carlos Molina Arrotea, Apolinario Casabal et al. *Diccionario biográfico nacional, que contiene: La vida de todos los hombres de estado, escritores, poétas, militares, etc. fallecidos que han figurado en el país desde el descubrimiento hasta nuestros dias*, 231–32. Buenos Aires: M. Sanchez & cia, 1877.

Garibaldi, Giuseppe. *Garibaldi: An Autobiography*. Edited by Alexandre Dumas and translated by William Robson. London: Routledge, Warne, and Routledge, 1860. [Original publication: Giuseppe Garibaldi, *Mémoires de Joseph Garibaldi*. Publiés par Alexandre Dumas. Lausanne: J. L. Borgeaud, 1860.]

Garriga, José. *Descripción del esqueleto de un quadrúpedo muy corpulento y raro que se conserva en el Real Gabinete de Historia Natural de Madrid*. Madrid: Imprenta de la Viuda de don Joaquín Ibarra, 1796.

Geertz, Clifford. *The Interpretation of Cultures*. Basic Books, 1973.

Gez, Juan Wenceslao. *El Dr. Juan Crisóstomo Lafinur: Estudio biográfico y recopilación de sus poesías*. Cabaut, 1907.

Gianello, Leoncio. *Florencio Varela*. Guillermo Kraft, 1948.

Gillespie, Alexander. *Gleanings and Remarks: Collected During Many Months of Residence at Buenos Ayres, and Within the Upper Country*. Leeds: B. Dewhirst, 1818.

Gilman, Sander L. "The Figure of the Black in German Aesthetic Theory." *Eighteenth-Century Studies* 8, no. 4 (1975): 373–91.

Giuli, Paola. "Arcadia Accademia." In *Encyclopedia of Italian Literary Studies*, edited by Gaetana Marrone, Paolo Puppa, and Luca Somigli. Routledge, 2007.

Gliddon, George Robins. "Preface." In Morton, *Types of Mankind*.

Godbeer, Richard. *The Overflowing of Friendship: Love Between Men and the Creation of the American Republic*. Johns Hopkins University Press, 2009.

Goethe, Johann W. *Versuch die Metamorphosen der Pflanzen*. Gotha: Carl Wilhem Ettinger, 1790.

Goethe, Johann Wolfgang von. *Faust [première partie]* (in French). Translated by Philipp Albert Stapfer, illustrated by Eugène Delacroix. Paris: Chez Ch. Motte, 1816. https://www.gutenberg.org/files/54202/54202-h/54202-h.htm.

Goldstein, Doris S. "'Official Philosophies' in Modern France: The Example of Victor Cousin." *Journal of Social History* 1, no. 3 (1968): 259–79.

Golinski, Jan. *The Experimental Self: Humphry Davy and the Making of a Man of Science*. University of Chicago Press, 2018.

Gómez Marín, Manuel. *El currutaco por alambique*. Mexico City: Mariano Zúñiga y Ontiveros, 1799.

Gonzalez, Johnhenry. *Maroon Nation: A History of Revolutionary Haiti*. Yale University Press, 2019.

González Claverán, Virginia. *La expedición científica de Malaspina en Nueva España 1789–1794*. El Colegio de Mexico, 1993.

Gordon, Daniel. *Citizens Without Sovereignty*. Princeton University Press, 1994.

Gordon, John. "The Doctrines of Gall and Spurzheim." *Edinburgh Review* 25 (June 1815): 227–68.

Gorriti, Juan Ignacio de. *Reflecciones sobre las causas morales de als convulsiones interiores en los Nuevos Estados Americanos y examen de los medios eficaces para reprimirlas*. Valparaiso: Imprenta del Mercurio, 1836.

Gossman, Lionel. "Augustin Thierry and Liberal Historiography." *History and Theory* 15, no. 4, (1976): 3–6.

Grattan-Guinness, Ivor. *The Search for Mathematical Roots, 1870–1940: Logics, Set Theories and the Foundations of Mathematics from Cantor Through Russell to Gödel.* Princeton University Press, 2000.

Grindle, Nicholas. "Our Own Imperfect Knowledge: Petrus Camper and the Search for an 'Ideal Form.'" *RES: Anthropology and Aesthetics* 31, no. 1 (1997): 139–48.

Gross, Michael. "The Lessened Locus of Feelings: A Transformation in French Physiology in the Early Nineteenth Century." *Journal of the History of Biology* 12, no. 2 (1979): 231–71.

Groussac, Paul. "Noticia biografica del Dr. Diego Alcorta." *Anales de la Biblioteca* 2 (1902).

Gual, Manuel. "Discurso preliminar dirigido a los Americanos, 1797." *Constitution Web.* http://constitucionweb.blogspot.com/2010/04/discurso-preliminar-dirigido-los.html.

Guasti, Niccolò. "Campomanes' Civil Economy and the Emergence of the Public Sphere in Spanish *Ilustration.*" In *L'économie politique et la sphère publique dans le débat des Lumières*, edited by Jesús Astigarraga and Javier Usoz, 229–45. Casa de Velázquez, 2013.

Guido, José Tomás. *Biografía de Manuel Dorrego.* Buenos Aires: Imprenta y Librerías de Mayo, 1877.

Guthrie, Guillermo. *Geografía universal descriptiva, histórica, industrial y comercial, de las cuatro partes del mundo.* Madrid: Imprenta de Villalpando, 1804.

Guthrie, William. *A new geographical, historical, and commercial grammar; and present state of the several kingdoms of the world. Illustrated with a new and correct set of maps, engraved by Mr. Kitchin. A new edition, improved and enlarged; the astronomical part by James Ferguson.* 6th ed. London: Charles Dilly, 1779.

Guthrie, William. *A New System of Modern Geography: Or, A Geographical, Historical, and Commercial Grammar; and Present State of the Several Nations of the World.* 1st ed. Vol. 2. Philadelphia: Mathew Carey, 1795.

Guthrie, William. *Nouvelle géographie universelle, descriptive, historique, industrielle et commerciale des quatre parties du monde* Translated by Fr. Noel. Vol. 6, part I. Paris: H. Langlois, 1802.

Gutiérrez, Juan María. "Don Esteban Luca: Noticias sobre su vida y escritos." *Revista del Rio de la Plata: Periódico Mensual de Historia y Literatura de América* 13 (1877): 3–229.

Gutiérrez, Juan María. "Fragmentos de un studio." In Marcos Sastre, Juan B. Alberdi, Juan María Gutiérrez, and Esteban Echeverría, *El Salón Literario*, xlviii. Hachette, 1958.

Gutiérrez, Juan María. "Juan Crisostomo Lafinur." In *Apuntes biograficos de escritores, oradores y hombres de Estado de la Republica Arjentina.* Buenos Aires: Mayo, 1860.

Gutiérrez, Juan María. "Noticias biograficas sobre Don Esteban Echeverría." In *Obras completas de D. Esteban Echeverría: Escritos en prosa.* Vol. 5, edited by Juan María Gutiérrez, i–xcix. Buenos Aires: C. Casavalle, 1874.

Gutiérrez, Manuel María. "Introduction." In Destutt de Tracy, *Principios de economia politica*, vol. 2.

Guy, Donna. "La verdadera historia de la Sociedad de Beneficencia." In *La política social antes de la política social: Caridad, beneficencia y política social en Buenos Aires, Siglos XVIII a XX*, edited by José Luis Moreno, 321–41. Prometeo-Trama, 2000.

Guzman, Rodolfo M. "Welcoming Alexander von Humboldt in Santa Fé de Bogotá, or the Creoles' Self-Celebration in the Colonial City." *Atlantic Studies* 7, no. 2 (2010): 143–62.

Haggard, Robert F. "The Politics of Friendship: Du Pont, Jefferson, Madison, and the Physiocratic Dream for the New World." *Proceedings of the American Philosophical Society* 153, no. 4 (2009): 419–40.

Haidt, Rebecca. *Embodying Enlightenment: Knowing the Body in Eighteenth-Century Spanish Literature and Culture*. St. Martin's Press, 1998.

Haines, Barbara. "The Inter-Relations Between Social, Biological, and Medical Thought, 1750–1850: Saint-Simon and Comte." *British Journal for the History of Science* 11, no. 1, (1978): 19–35.

Hallamore Caesar, Ann, and Michael Caesar. *Modern Italian Literature*. Polity Press, 2007.

Halperín-Donghi, Tulio. *De la Revolución de la Independencia a la Confederación Rosista*. Paidós, 1980.

Halperín-Donghi, Tulio. *Politics, Economics and Society in Argentina in the Revolutionary Period*. Translated by Richard Southern. Cambridge University Press, 1975.

Halperín-Donghi, Tulio. *Revolución y guerra: Formación de una elite dirigente en la Argentina criolla*. Siglo XXI, 1972.

Halperin-Donghi, Tulio. *Revolución y guerra: Formación de una élite dirigente en la Argentina criolla*. Siglo XXI Editores, 2014.

Halperín-Donghi, Tulio. *Una nación para el desierto argentino*. Prometo, 2005.

Hamilton, William. "Cours de Philosophie." *Edinburgh Review* 1, no. 99 (1829): 194–221.

Hamilton, William. "On the Philosophy of the Unconditioned: In Reference to Cousin's Infinito-Absolute." In *Discussions on Philosophy and Literature, Education, and University Reform*. New York: Harper & Bros., 1863.

Hamnett, Brian. *The Enlightenment in Iberia and Ibero-America*. University of Wales Press, 2017.

Haro de San Clemente, José O. *El chichisveo impugnado/por el RPM Fr. Joseph Haro de S. Clemente, Religioso del Sagrado Orden de Nra. Sra. del Carmen . . . ; sacado a luz, i lo costea un Amigo del author*. Seville: Imprenta del Dr. D. Geronymo de Castilla, 1729.

Harris, Jonathan. "Bernardino Rivadavia and Benthamite 'Discipleship.'" *Latin American Research Review* 33, no. 1 (1998): 129–49. https://doi.org/10.1017/S0023879100035780.

Head, Brian W. *Politics and Philosophy in the Thought of Destutt de Tracy*. Routledge, 2019.

Hegel, G. W. F. *The Philosophy of History*. Translated by J. Sibree. Dover Publications, (1840) 2012.

Heibron, J. L. "The Measure of Enlightenment." In *The Quantifying Spirit in the Eighteenth Century*, edited by Tore Frangsmyr, J. L. Heilbron, and Robin E. Rider, 207–44. University of California Press, 1990. http://ark.cdlib.org/ark:/13030/ft6d5nb455/.

Herder, Johann Gottfried. *Outlines of a Philosophy of the History of Man*, vols. 1 and 2. Translated by T. Churchill. London: J. Johnson, 1803.

Herder, Johann Gottfried. *Outlines of a Philosophy of the History of Man*. Bergman, (1784) 1966. https://archive.org/details/outlinesaphilos00churgoog.

Hernández, Pedro. *Origen y descubrimiento de la vaccina*. Madrid: B. García y Cía., 1801.

Herr, Richard. *The Eighteenth-Century Revolution in Spain*. Princeton University Press, 2015.

Herrera y Obes, Manuel, and Bernardo Berro. *El caudillismo y la revolución americana: Polémica*. Biblioteca Artigas, 1966.

Herrero y Rubira, Antonio. *Diccionario universal, francés, y español: Mas copioso que quantos hasta ahora se han visto*. Vol. 1. Madrid: Imprenta del Reyno, 1744.

Herschel, John. *Preliminary Discourse on the Study of Natural Philosophy*. London: Printed For Longman, Brown, Green & Longmans, Paternoster Row, 1830.

Hervás y Panduro, Lorenzo. *Viage estático al mundo planetario*. Madrid: Imprenta de Aznar, 1793.

"Historia de la universidad." *Anales de la Universidad de Buenos Aires* 3 (1888): 5–435.

"Historia del Dr. Buñuelos, escrita en francés por Mr. Boudein y traducida al castellano por

D. Sancho Rabioles." *Telégrafo Mercantil, Rural, Politico-Económico, e Historiográfico del Río de la Plata* 29, no. 2 (November 15, 1801): 218–22.

Hocquellet, Richard. *Resistencia y revolución durante la Guerra de la Independencia: Del levantamiento patriótico a la soberanía nacional.* Universidad de Zaragoza, 2012.

Homs, Román Piña . "Juan Picornell: de maestro reformista a líder revolucionario." In *Masonería política y sociedad.* Centro de Estudios Históricos de la Masonería en España, 1989.

Horsman, Reginald. *Race and Manifest Destiny: The Origins of American Racial Anglo-Saxonism.* Harvard University Press, 2009.

Hudson, Damián. *Recuerdos históricos sobre la Provincia de Cuyo.* Vol. 1. Buenos Aires: Imprenta de J. A. Alsina, 1898.

Hughes, Michael J. *Forging Napoleon's Grande Armée: Motivation, Military Culture, and Masculinity in the French Army, 1800–1808.* Vol. 7. New York University Press, 2012.

Humboldt, Alexander von. *Cosmos: A Survey of the General Physical History of the Universe.* New York: Harper & Brothers, 1845.

Humboldt, Alexander von. *Cosmos essai d'une description physique du monde.* Vol. 1. Paris: Gide et J. Baudry, 1846.

Humboldt, Alexander von. *Cosmos ó ensayo de una descripción fisica del mundo.* Madrid: Est. tip. de Ramón Rodriguez de Rivera, 1851.

Humboldt, Alexander von. *Essai politique sur le royaume de la Nouvelle-Espagne.* 4 vols. 1st ed. Paris: F. Schoell, 1811.

Humboldt, Alexander von. *Kosmos: A General Survey of Physical Phenomena of the Universe.* London: H. Baillière, 1845.

Humboldt, Alexander von. *Political Essay on the Kingdom of New Spain.* Vol. 1. Translated and edited by John Black. 4 vols. London: Longman, Hurst, Rees, Orme, and Brown, 1811.

Humboldt, Alexander von, and Aimé Bonpland. *Personal Narrative of Travels to the Equinoctial Regions of the New Continent, During the Years 1799–1804.* Translated by Helen Maria Williams. Vol. 6, part 1. London: Longman, Rees, Orme, Brown, and Green, 1826.

Humboldt, Alexander von, Vera M. Kutzinski, and Ottmar Ette. *Political Essay on the Kingdom of New Spain: A Critical Edition.* Vol. 1. University of Chicago Press, 2020.

Hume, David. *A Treatise of Human Nature: being an attempt to introduce the experimental method of reasoning into moral subjects.* Vol. 1. London: John Noon, 1739.

Iggers, Georg G., Q. Edward Wang, and Supriva Mukherjee. *A Global History of Modern Historiography.* Taylor and Francis, 2013.

"Industria." *Semanario de Agricultura, Industria y Comercio* 2, no. 1 (September 8, 1802): 1–16.

"Informe al Gobernador del Río de la Plata dado por el Cabildo Eclesiástico de Buenos Aires sobre el destino que debe darse a las fincas de Temporalidades y sobre el establecimiento de un colegio y de una Real Pública Universidad, Diciembre 5 de 1771." In *Noticias históricas sobre el orígen y desarrollo de la Enseñanza Pública Superior en Buenos Aires: Desde la época de la extinción de la compañía de Jesús en el año 1767, hasta poco después de fundada la universidad en 1821,* edited by Juan María Gutiérrez: 350-390. Buenos Aires: Imprenta del Siglo, 1868.

"Informe del cabildo secular sobre la misma materia." In *Noticias históricas sobre el orígen y desarrollo de la Enseñanza Pública Superior en Buenos Aires: Desde la época de la extinción de la compañía de Jesús en el año 1767, hasta poco después de fundada la universidad en 1821,* edited by Juan María Gutiérrez: 368-390. Buenos Aires: Imprenta del Siglo, 1868.

Insúa, Mariela. "La falsa erudición en la Ilustración española y novohispana, Lizardi." *Estudios Filológicos* 48 (2011): 61–79. http://dx.doi.org/10.4067/S0071-17132011000200005.

Isabella, Maurizio. "Mazzini's Internationalism in Context: From the Cosmopolitan Patriotism of the Italian Carbonari to Mazzini's Europe of the Nations." In Bayly and Biagini, *Giuseppe Mazzini*, 37–58.

Isabelle, Arséne. *Voyage á Buénos Ayres*. Havre: J. Morlent, 1837.

Jackson, Andrew. "Message." *Washington Globe*, December 7, 1831: 2.

Jackson, Andrew. "Message of the President of the United States to Both Houses of Congress." *Washington Globe*, December 4, 1833: 2.

Jacsick, Iván. *Andrés Bello: La passion por el order*. Editorial Universitaria de Chile, 2010.

Janet, Paul. *Victor Cousin et son œuvre*. Paris: Alcan, 1893.

"Jean Baptiste Say to Thomas Jefferson, 2 November 1803." *Founders Online*, National Archives. https://founders.archives.gov/documents/Jefferson/01-41-02-0486. [Original source: *The Papers of Thomas Jefferson*, vol. 41, July 11–November 15, 1803, edited by Barbara B. Oberg, 652–53. Princeton University Press, 2014.]

Jeffers, Harry Paul. *Freemasons: A History and Exploration of the World's Oldest Secret Society*. Citadel Press, 2005.

Jefferson, Ann. *Genius in France: An Idea and Its Uses*. Princeton University Press, 2014.

Jefferson, Thomas. "Letter from Monticello, October 25, 1818." In Destutt de Tracy, *Treatise on Political Economy*.

Jefferson, Thomas. "Notes on Virginia." In *The Life and Selected Writings of Thomas Jefferson* 1781. https://thefederalistpapers.org/wp-content/uploads/2012/12/Thomas-Jefferson-Notes-On-The-State-Of-Virginia.pdf.

Jefferson, Thomas. "Preface." In Destutt de Tracy, *A Treatise on Political Economy: To Which Is Prefixed a Supplement*.

Jefferson, Thomas. "Prospectus." In Destutt de Tracy, *Treatise on Political Economy*, viii–xii.

Jefferson, Thomas. "Thomas Jefferson to Destutt de Tracy, 26 December 1820." *Founders Online*, National Archives. https://founders.archives.gov/documents/Jefferson/03-16-02-0396. [Original source: *The Papers of Thomas Jefferson: Retirement Series*, vol. 16, June 1, 1820 to February 28, 1821, edited by J. Jefferson Looney et al., 485–88. Princeton University Press, 2019.]

Jenkins, Bill. *Evolution Before Darwin: Theories of the Transmutation of Species in Edinburgh, 1804–1834*. Edinburgh University Press, 2019.

Jenkins, Bill. "Phrenology, Heredity and Progress in George Combe's *Constitution of Man*." *British Journal for the History of Science* 48, no. 3 (September 2015): 455–73.

Jenkins, S. P. "Buenos Ayres and the Republic of Banda Oriental." *American Review: A Whig Journal, Devoted to Politics and Literature* 3, no. 2 (February 1846): 160–67. New York: Wiley and Putnam, 1846.

Jervis, John. *Transgressing the Modern*. Wiley-Blackwell, 1999.

Jolly, Claude. "Introduction." In Antoine Louis Claude Destutt de Tracy, *De l'Amour*. Vren, 2006.

Jones Corredera, Edward. "The Rediscovery of the Spanish Republic of Letters." *History of European Ideas* 45, no. 7 (2019): 953–71.

Jovellanos, Gaspar Melchor de. *Elogio de Carlos Tercero: Leido á la Real Sociedad de Madrid por el socio D. Gaspar Melchor de Jove Llanos [sic], en la Junta plena de sábado 8 de Noviembre de 1788 . . .* Madrid: Vda. de Ibarra, 1789.

Julius, N. H. "Zustand der Heilkunde im Spanischen America." *Magazin der ausländischen Literatur der gesammten Heilkunde und Arbeiten des Ärztlichen Vereins zu Hamburg* 14 (1827).

Katra, William H. *The Argentine Generation of 1837: Echeverria, Alberdi, Sarmiento, Mitre*. Associated University Press, 1996.

Kelley, Donald R. "Eclecticism and the History of Ideas." *Journal of the History of Ideas* 62, no. 4 (2001): 577–92.

Kerber, Linda K. *Women of the Republic: Intellect and Ideology in Revolutionary America*. University of North Carolina Press, 1980.

Kidd, Colin. "Race, Empire, and the Limits of Nineteenth-Century Scottish Nationhood." *Historical Journal* 46, no. 4 (2003): 873–92.

King, Philip Parker. *Narrative of the Surveying Voyages of His Majesty's Ships Adventure and Beagle: Between the Years 1826 and 1836* . . . Vol. 3. London: H. Colburn, 1839. https://openlibrary.org/books/OL25476840M/Narrative_of_the_surveying_voyages_of_His_Majesty%27s_Ships_Adventure_and_Beagle_between_the_years_182.

King, Richard. "Address to the Ethnological Society of London Delivered at the Anniversary, 25th May 1844." *Journal of the Ethnological Society of London* (1848–1856) (1850): 9–42.

Knell, Rebecca. "Reevaluating Science and Romanticism: The Case of Erasmus Darwin's the Loves of Plants." In *Restoring the Mystery of the Rainbow*, vol. 1, edited by Valeria Tinkler-Villani and C. C. Barfoot, 111–31. Brill, 2011.

Knight, Richard Payne. *The Progress of Civil Society: A Didactic Poem, in Six Books*. London: Printed by W. Bulmer and Co. for G. Nicol, 1796.

Knott, Sarah. "Benjamin Rush's Ferment: Enlightenment Medicine and Female Citizenship." In Knott and Taylor, *Women, Gender and Enlightenment*, 649–66.

Knott, Sarah. "Sensibility and the American War for Independence." *American Historical Review* 109 (2004): 19–40.

Knott, Sarah, and Barbara Taylor, eds. *Women, Gender and* Enlightenment. Palgrave Macmillan, 2005.

L'Ardèche, Laurent de. "Fedéralistés et Unitariennes." *Republique* 7926 (January 9, 1850).

L'Ardèche, Laurent de. "Les Unitaires et the Federalistes." *Archivo Americano Espíritu de la Prensa del Mundo* 20 (September 21, 1850).

"La balanza de los poderes." *La Argentina* 1, no. 6 (September 15, 1822): 227.

Labardén, Manuel José. "Discurso del Doctor Labardén." In *Noticias historicas sobre el oríjen y desarrollo de la enseñanza publica superior en Buenos Aires desde la época de la estincion de la Compañia de Jesus en el año 1767* [. . .], edited by Juan María Gutiérrez, 75–76. Buenos Aires: Imprenta del Siglo, 1868.

Lacordaire, Jean-Théodore. *Mémoires du baron Georges Cuvier*. Paris: H. Fournier, 1833.

Lacordaire, Jean-Théodore. "Un voyageur, 'Une révolution dans la république Argentine.'" *Revue des Deux Mondes* 1, no. 1 (1835): 23–40.

Lacordaire, Théodore. "Chronique de la Quinzaine." *Revue des Deux Mondes*, Deuxième Série 1, no. 3 (1833): 285–96.

Lacordaire, Théodore. "La bataille de la Tablada: Episode des guerres civiles de Buenos-Ayres (Extrait d'un journal de voyages dans l'Amerique du sur)." *Revue des Deux Mondes* 7, no. 3 (1832): 273–96.

Lacordaire, Théodore. "Une estancia." *Revue des Deux Mondes*, Deuxième Série 1, no. 6, (1833): 503–20.

Lacroix, Frédéric. *Géographie politique: Les iles Malouines*. N.p., 1840. https://www.google.com/books/edition/G%C3%A9ographie_politique/Q0NDAAAAcAAJ?hl=en.

Lacroix, Frédéric. *Patagonie, Terre du Feu et Archipel des Malouines*. Paris: Firmin Didot Frères, 1840.

Lafayette, Marquis de. "Letter from Lafayette to Jefferson, December 10, 1817." *The Papers of Thomas Jefferson: Retirement Series*, edited by J. Jefferson Looney et al., vol. 12, September 1, 1817, to April 21, 1818, 245–48. Princeton University Press, 2004–17.

Lafinur, Juan Crisóstomo. "Fragmentos ineditos del curso de filosofia dictado en Buenos Aires por el dr. D. J. C. Lafinur." In *Noticias historicas sobre el orijen y desarrollo de la enseñanza publica superior en Buenos Aires desde la época de la estincion de la Compañia de Jesus En El Año 1767 hasta poco despues de fundada la universidad en 1821*, edited by Juan María Gutiérrez, 87–120. Buenos Aires: Imprenta del Siglo, 1868.

Lafinur, Juan Crisóstomo. "Ideolojia." In *Noticias historicas sobre el orijen y desarrollo de la enseñanza publica superior en Buenos Aires desde la época de la estincion de la Compañia de Jesus en el año 1767 hasta poco despues de fundada la universidad en 1821*, edited by Juan María Gutiérrez, 7–82. Buenos Aires: Imprenta del Siglo, 1868.

Lajer-Burcharth, Ewa. *Necklines: The Art of Jacques-Louis David after the Terror.* Yale University Press, 1999.

Lamarck, Jean-Baptiste. *Philosophie zoologique ou exposition des considérations relatives à l'histoire naturelle des animaux*. Paris: Musée Histoire Naturelle, 1809.

Lamas, Andrés. *Apuntes históricos sobre las agresiones del dictador argentino D. Juan Manuel de Rosas: Artículos escritos en 1845 para el Nacional de Montevideo*. Montevideo: n.p., 1849.

Lamas, Andrés. *Escritos politicos y literarios de D. Andres Lamas durante la guerra contra la tirania de D. Juan Manuel Rosás: Acompañados de documentos, en gran parte inéditos, y de noticias importantes para la historia de la época (1836 á 1852) y para la vida política del autor.* Buenos Aires: Casa editora calle de Cangallo 1227, 1877.

"La más horrenda." *Museo de Ambas Américas*. Valparaíso: Imprenta de M. Rivadeneyra, 1842.

Lamennais, Félicité Robert de. *Le livre du people*. In *Oeuvres completes*, 2:627–57. Brussels: Hauman & Company, 1839.

Lamennais, Félicité Robert de. *Palabras de un creyente*. Paris: Libreria de Rosa, 1834.

Lamennais, Félicité Robert de. *Words of a Believer*. Unknown translator. New York: C. de Behr, 1834.

Lampillas, Francisco Javier. *Ensayo historico-apologético de la literatura española contra las opiniones preocupadas de algunos escritores modernos italianos*. Translated by Josefa Amar y Borbón. Zaragoza: Blas Miedes, 1782.

Lampillas, Francisco Javier. *Saggio storico-apologetico della letteratura spagnuola contro le pregiudicate opinioni di alcuni moderni scrittori italiani*. Genoa: Felice Repetto, 1778.

"Larrañaga." In *Anales del Museo Nacional de Montevideo*, vol. 1, 1–4. Montevideo: Imprenta Artística, 1894.

La Serna, Miguel. *The Corner of the Living: Ayacucho on the Eve of the Shining Path Insurgency*. University of North Carolina Press, 2012.

Lasserre, Pierre. *Le romantisme français: Essai sur la revolution dans les sentiments et dans les idées au XIX siècle*. 11th ed. Société du mercure de France, 1907.

"Latest from Buenos Ayres." *National Intelligencer*, May 15, 1838.

Lauro, Claudia Rosas. *Del trono a la guillotina: El impacto de la Revolución Francesa en el Perú 1789–1808*. Institut Français d'Études Andines, 2006.

"La virtud del pueblo sea la primera ley." *Telégrafo Mercantil, Rural, Politico-Económico, e Historiográfico de Río de la Plata*, no. 8 (May 2, 1801): 83–88.

Lawrence, William. *Lectures on Comparative Anatomy, Physiology, Zoology, and the Natural History of Man*. 9th ed. London: H. G. Bohn, 1848.

Learned, M. D. "Ferdinand Freiligrath in America." In *Americana Germanica*, vol. 1, 54–74. New York: Macmillan Company, 1897.

Lefebvre, Henri. *The Sociology of Marx*. Columbia University Press, 1982.

León, María Teresa Berruezo. *La lucha de hispanoamerica por su independencia en Inglaterra 1800–1830*. Ediciones de Cultura Hispánica, 1989.

Lerminier, Eugène. *Introduction générale a l'histoire du droit*. Paris: Alexandre Mesnier, 1829.

Leroux, Pierre. *De l'humanité, de son principe et de son avenir: où se trouve exposée la vraie définition de la religion et où l'on explique le sens, la suite et l'enchaînement du mosaïsme et du christianisme*. 2 vols. Paris: Perrotin, 1840.

Leroux, Pierre. "De l'individualisme et du socialisme." *Revue Encyclopédique*, 1834.

Leroux, Pierre. *Réfutation de l'éclectisme: Où se trouve exposée la vraie définition de la philosophie, et où l'on explique le sens, la suite, et l'enchainment des divres philosophes depuis Descartes*. Paris: C. Gosselin, 1839.

Leroux, Pierre. "Trois discours sur la situation actuelle de la société et de l'esprit humain." In *Oeuvres de Pierre Leroux*, vol. 1, 365–81. Paris: Société Typographique, 1850.

Lesch, John E. "Systematics and the Geometrical Spirit." In *The Quantifying Spirit in the Eighteenth Century*, edited by Tore Frängsmyr, J. L. Heilbron, and Robin E. Rider, 73–112. University of California Press, 1990. http://ark.cdlib.org/ark:/13030/ft6d5nb455/.

Levasseur, Auguste. *Lafayette in America, in 1824 and 1825: Or, Journal of Travels, in the United States*. New York: White, Gallaher & White, 1829.

Lewis, Thomas A. *Religion, Modernity, and Politics in Hegel*. Oxford University Press, 2011.

Lida, Raimundo. "Sarmiento y Herder." *Memoria del segundo congreso internacional de catedráticos de literatura iberoamericana, Agosto 1940*, edited by Instituto Internacional de Literatura Iberoamericana. University of California Press, 1941.

Lightman, Bernard. "Introduction." In *Global Spencerism: The Communication and Appropriation of a British Evolutionist*, edited by Bernard Lightman, 1–13. Brill, 2015.

Linberg, Hening Gottfried. "Preface." In Victor Cousin, *Introduction to the History of Philosophy*, i–viii. Boston: Hilliard, Gray, Little, and Wilkins, 1832.

Lloyd, Henry Martin. "Reason and Rationality Within the 'Enlightenment of Sensibility,' or Étienne Bonnot de Condillac and French Philosophy's First 'Linguistic Turn.'" In *Rethinking the Enlightenment: Between History, Philosophy, and Politics*, edited by Geoff Boucher and Henry Martin Lloyd, 151–76. Lexington Books, 2018.

Lluch, Ernest, and Lluís Argemí i d'Abadal. "Physiocracy in Spain." *History of Political Economy*, no. 26 (1994): 613–27.

Locke, John. *Some Thoughts Concerning Education*. Vol. 9: *The Works of John Locke, in Ten Volumes*. 11th ed. London: W. Otridge and Son, 1812.

Locke, John et al. *Familiar Letters Between Mr. John Locke, and Several of His Friends: In which are Explained, His Notions in His Essay Concerning Human Understanding, and in Some of His Other Works*. London: F. Noble, 1742.

"London Lyrics: Stage Wedlock; Doctor Gall." *New Monthly Magazine and Literary Journal* 7 (1823): 427–30.

López, Enrique Giménez. *La Compañía de Jesús, del exilio a la restauración: Diez estudios*. Universidad de Alicante, 2017.

López, M. Fernández, and D. R. del V.Orellana. "Manuel Belgrano y las máximas de Quesnay." *Revista de Economía y Estadística* 25, no. 1 (1984): 83–126. https://revistas.unc.edu.ar/index.php/REyE/article/view/3950116.

López, Vicente Fidel. *Evocaciones históricas: Autobiografía; La cran [sic] semana de 1810; El conflicto y la entrevista de Guayaquil*. El Ateneo, 1929.

"Los editores." *El Tribuno* 2, no. 1 (April 18, 1827). In *Dorrego: Tribuno y periodista*, edited by Alberto del Solar. Imprenta de Coni Hermanos, 1907.

Löwy, Michael, and Robert Sayre. *Romanticism Against the Tide of Modernity*. Duke University Press, 2002.

Lucerna, "Physiognomy and Craniology." *Fayetteville Observer*, August 26, 1824: 1.

Luzán, Ignacio. *Real Academia de buenas letras de la ciudad de Barcelona: Origen, progressos, y primera Junta general . . . con los papeles que en ella se acordaron*. Madrid: Franc. Suria, 1756.

Lynch, John, ed. *Andrés Bello: The London Years*. Richmond, 1982.

Lynch, John. *Argentine Caudillo: Juan Manuel de Rosas*. SR Books, 2006.

Lynch, John. *Argentine Dictator: Juan Manuel de Rosas, 1829–1852*. Oxford University Press, 1981.

Mably, Abbé Gabriel Bonnot de. *De la législation: Ou principes des loix*. Amsterdam, 1776.

Mably, Abbé Gabriel Bonnot de. *Des droits et des devoirs du citoyen*. A. Kell, 1790.

Mably, Abbé Gabriel Bonnot de. *Observations sur le gouvernement et les loix des États-Unis d'Amerique*. Dublin, 1785.

Mably, Abbé Gabriel Bonnot de. *Principes de morale*. Paris, 1784.

Macdonell, Lady Anne Lumb. *Reminiscences of Diplomatic Life*. A. & C. Black, 1913.

MacGregor, John. *The Progress of America*. Vol. 1. London: Whittaker and Company, 1847.

Macintyre, Iona. "Corinne in the Andes." In *Connections After Colonialism: Europe and Latin America in the 1820s*, edited by Matthew Brown and Gabriel Paquette, 179–90. University of Alabama Press, 2013.

Mackau–Arana Treaty. In *Tratados de los estados del Rio de la Plata y constituciones de las repúblicas sud-americanas: Colección formada por las publicaciones oficiales hechas en los estados respectivos, con los textos en ingles, frances, italiano y portugues, en frente del texto espanõl, en los tratados concluidos con potencias estranjeras*, 155–59. Montevideo: Comercio del Plata, 1848.

Maine de Biran, Pierre. *Influence de l'habitude sur la faculté de penser*. Paris: Henrichs, 1803.

Maine de Biran, Pierre, and Victor Cousin. *Leçons de philosophie de M. Laromiguière, jugées par M. Victor Cousin et M. Maine de Biran*. Paris: Johanneau, 1829.

Majluf, Natalia. "The Creation of the Image of the Indian in 19th-century Peru: The Paintings of Francisco Laso (1823–1869)." PhD diss., University of Texas at Austin, 1995.

Majluf, Natalia. "De la rebelión al museo: Genealogías y retratos de los incas 1781–1900." In *Los incas, reyes del Perú*, edited by Thomas Cummins, 253–319. Banco de Crédito, 2005.

Malta Romeiras, Francisco. *Jesuits and the Book of Nature: Science and Education in Modern Portugal*. Brill, 2019.

Malte-Brun, Conrad. *Universal Geography, Or, A Description of All the Parts of the World, on a New Plan, According to the Great Natural Divisions of the Globe. . . .* Philadelphia: A. Finley, 1827.

Mancini, Jules. *Bolívar y la emancipación de las colonias españolas*. Translated by Carlos Docteur. La Vda de C. Bouret, 1914.

Mandair, Arvind. "The Repetition of Past Imperialisms: Hegel, Historical Difference, and the Theorization of Indic Religions." *History of Religions* 44, no. 4 (2005): 277–99.

"Manifesto of Independence." *Morning Chronicle* [1801], August 5, 1818. British Library Newspapers. link.gale.com/apps/doc/BC3207280104/GDCS?u=tamp44898&sid=bookmark-GDCS&pg=2&xid=563bc680.

Manning, Patrick. *Navigating World History: Historians Create a Global Past*. Palgrave Macmillan, 2003.

Mansilla, Lucio. *Rozas: Ensayo histórico-psicológico*. Paris: Garnier Hermanos, 1899.

Marichal, Juan. *Cuatro fases de la historia intelectual latinoamericana (1810–1970)*. Fundación Juan March and Ediciones Cátedra, 1978.

Masiello, Francine. *Between Civilization and Barbarism: Women, Nation, and Literary Culture in Modern Argentina*. University of Nebraska Press, 1992.

Mathon de la Cour, Charles-Joseph. *Discours sur les meilleurs moyens de faire naître et d'encourager le patriotisme dans une monarchie; Qui a remporté le prix dans l'académie de Châlons-sur-Marne, le 25 août 1787. Par M. Mathon de La Cour* . . . Paris: Chez Cuchet, 1788.

Mayans y Siscar, Gregorio. "Aviso Primero." In *Avisos de Parnaso*, edited by J. B. Corachán. Valencia: Viuda de Antonio Bordazar, 1747.

Mayer, Roland. "Sleeping with the Enemy: Satire and Philosophy." In *The Cambridge Companion to Roman Satire, Cambridge*, edited by Kirk Freudenburg, 146–59. Cambridge University Press, 2005.

Mayhew, R. "William Guthrie's *Geographical Grammar*, the Scottish Enlightenment and the Politics of British Geography." *Scottish Geographical Journal* 115, no. 1 (1999): 19–34.

Mazade, Charles de. "De l'américanisme et des républiques du sud- La Société argentine, Quiroga et Rosas (*Civilization i Barbarie*, de M. Domingo Sarmiento)." *Revue des Deux Mondes* 16 (1846): 625–58.

Mazlish, Bruce. *The New Global History*. Routledge, 2006,

Mazzini, Giuseppe. "General Instructions for the Members of the Young Italy." In *Life & Writings of Joseph Mazzini*, 96–113. London: Smith, Elder, & Company, 1864.

Mazzini, Giuseppe. *Joseph Mazzini: His Life, Writings, and Political Principles*. Introduction by William Lloyd Garrison. New York: Hurd and Houghton, 1872.

McGrath, Larry Sommer. *Making Spirit Matter: Neurology, Psychology, and Selfhood in Modern France*. University of Chicago Press, 2021.

"Medicina." *La Argentina* 1, no. 2 (May 15, 1822): 71.

"Medicina: Origen y estado de esta ciencia en Buenos Aires." *La Abeja Argentina*, no. 1 (April 15, 1822): 461–68.

Medina, Marcela Guerrero. "Stepping Back, Unmasking Tango: (Re)Covering the African Elements in Tango." *Delaware Review of Latin American Studies* 9, no. 1 (2008). http://www.udel.edu/LAS/Vol9–1GuerreroMedina.html.

Meek, Ronald L. *Economics of Physiocracy*. Routledge, 2013.

Melman, Billie. "Claiming the Nation's Past: The Invention of an Anglo-Saxon Tradition." *Journal of Contemporary History* 26, no. 3/4 (1991): 575–95.

"Memoria sobre esta cuestion." *Argos de Buenos Aires* 2, no. 18 (March 1, 1823): 1.

Mensch, Jennifer. "Kant and the Skull Collectors: German Anthropology from Blumenbach to Kant." In *Kant and His German Contemporaries*, edited by Corey W. Dyck, 192–210. Cambridge University Press, 2018.

Meranze, Michael. *Laboratories of Virtue*. University of North Carolina Press, 1996.

Mercado, Juan Carlos. *Building a Nation: The Case of Echeverria*. University Press of America, 1996.

Michelena, Carmen. "Reformas y rebeliones en la crisis del Imperio borbónico: Dos intentos revolucionarios ilustrados: De san blas Madrid, 1795 a La Guaira 1797." PhD diss., University of Sevilla, 2007. https://www.academia.edu/5338633/Reformas_y_rebeliones_en_la_crisis_del_Imperio_borb%C3%B3nico.

Michelet, Jules. *Histoire de France*. Vol. 1. Paris: Hachette, 1833.

Michelet, Jules. *History of France.* Vol. 1. Translated by Walter K. Kelly. London: Chapman and Hall, 1844.

Michelet, Jules. *History of France.* Vol. 1. Translated by G. H. Smith. New York: D. Appleton and Company, 1845.

Michelet, Jules. "Letter to Charles Darwin, 15 November 1872." *Darwin Correspondence Project,* "Letter no. 8685." https://www.darwinproject.ac.uk/letter/?docId=letters/DCP-LETT-8685.xml.

Michelet, Jules. *Principes de la philosophie de l'histoire: Traduits de la Scienza Nova de Vico.* In *Oeuvres de Michelet.* Vol. 1. Brussels: Meline, Cans et Comp., 1840.

Mignet, François. "Notice historique sur la vie et les travaux de M. Victor Cousin." *Edinburgh Review* 351 (July 1890): 454–91.

Mill, John Stuart. "Michelet's History of France." *Edinburgh Review* 159, no. 325 (January 1844): 1–39.

Miller, John. *Memoirs of General Miller.* Vol. 1. London: Longman, 1828.

Miller, Samuel. *A Discourse, Delivered April 12, 1797: At the Request of and Before the New- York Society for Promoting the Manumission of Slaves, and Protecting Such of Them as Have Been or May Be Liberated.* New York: T. & J. Swords, 1797.

Miller, Samuel. *The Medical Works of Edward Miller, M.D.* New York: Collins and Co., 1814.

Mirabeau, Honoré Gabriel Riqueti, Count of. *Opinion du Comte de Mirabeau sur la noblesse ancienne et modern.* Paris: Chez Chaignieu jeune, 1815.

Miranda, Francisco Antonio Gabriel. *Correspondance du général Miranda.* Paris: Barrois, 1794.

Miranda, Francisco de. "Carta de Francisco de Miranda a Manuel Gual 1799." In *Miranda: Aventurero de la Libertad.* http://www.franciscodemiranda.info/es/documentos/cartagual.htm.

Miranda, Francisco de. "Exhaustivo y documentado alegato por la emancipación de Colombia: Proceso histórico de la iniciativa." In Francisco de Miranda, *América espera,* 365–70. Fundacion Biblioteca Ayacucho, 1982.

Mirón, Virginia Amelia Cruz. "Un orizabeño distinguido: José María Tornel y Mendívil." In *Veracruzanos en la independencia y la revolución,* edited by Abel Juárez Martínez, 233–57. Universidad Veracruzana, 2010.

"Mister Southern." *Revista de Derecho, Historia y Letras* 2 (1898): 6–12.

Mitre, Bartolomé. "Estudio Biográfico." In *Poesías de José Rivera Indarte, con biografía del autor escrita por el coronel de artillería D. Bartolome Mitre,* iii–lxxxv. Buenos Aires: Imprenta de Mayo, 1853.

Mitre, Bartolomé. *Historia de San Martín y de la emancipación Sudamericana según nuevos documentos.* Buenos Aires: Imprenta de La Nación, 1887.

Molina, Abad. *Histoire naturelle du Chile.* Translated by M. Gruvel. Paris: Née de la Rochelle, 1789.

Molina, Abad. *Saggio sulla storia naturale del Chili del signor Abate Giovanni Ignazio Molina.* Bologna: Stamperia di S. Tommaso d'Aquino, 1782.

Montújar, Carlos. "Biaje de Quito a la Lima de [. . .] con el Barón von Humboldt y don Alejandro Bonpland." *Boletín de la Sociedad Geográfica de Madrid* 25 (1889): 1–19.

Moore, John. *A View of Society and Manners in France, Switzerland, and Germany: With Anecdotes Relating to Some Eminent Characters, in Two Volumes.* Vol. 1. London: W. Strahan and T. Cadell, 1779.

Moreau de la Sarthe, Jacques-Louis. *Tratado histórico y práctico de la vacuna: Que contiene en com-*

pendio el orígen y los resultados de las observaciones y experimentos sobre la vacuna [. . .]. Madrid: Imprenta Real, 1803.

Moreen, Édouard. "Éloge de Jean-Théodore Lacordaire." In *Mémoires de la Société Royale des Sciences de Liége.* 2nd series, vol. 3, i–xl. Liége: Desoer, 1873.

Morell, J. R. "Socialistic Mystics." In Wilh. Gottlieb Tennemann, *A Manual of the History of Philosophy: Translated from the German of Wilh. Gottlieb Tennemann by the Rev. A. Johnson.* Revised, enlarged, and continued by J. R. Morell. London: G. Bohn, 1852.

Moreno, Mariano. "Prologo á la traducción del Contrato Social." In *Escritos*, edited by Mariano Moreno and Norberto Piñero, 375–83. Buenos Aires: Imprenta de Pablo E. Coni, 1896.

"The Morning Chronicle." *Morning Chronicle*, April 1, 1839.

Morse, Jedidiah. *The American Geography: Or, a View of the Present Situation of the United States of America* [. . .]. 2nd ed. London: Printed for John Stockdale, Piccadilly, 1789.

Morton, Samuel George. "Introductory Essay." In *Crania Americana; Or, A Comparative View of the Skulls of Various Aboriginal Nations of North and South America: To Which Is Prefixed an Essay on the Varieties of the Human Species*, 1–95. London: J. Dobson, 1839.

Morton, Samuel George. *Types of Mankind: Or Ethnological Researches* [. . .]. Edited by Josiah Clark Nott and George Robins Gliddon. Philadelphia: Lippincott, Grambo & Company, 1854.

Mosse, George L. *The Image of Man: The Creation of Modern Masculinity.* Oxford University Press, 1996.

Movsichoff, Paulina. *Juan Crisóstomo Lafinur: La sensualidad de la filosofía; novela.* Ediciones Fundación Victoria Ocampo, 2006.

"Mr. Clay's Speech." *Scioto Gazette* (Chillicothe, Ohio) 6, no. 52 (February 8, 1827): 2.

"Mr. Fenimore Cooper and the Times on Blackguardism." *Morning Chronicle*, April 21, 1838.

Mukerji, Chandra. "Dominion, Demonstration, and Domination: Religious Doctrine, Territorial Politics, and French Plant Collection." In *Colonial Botany: Science, Commerce, and Politics in the Early Modern World*, edited by Londa Schiebinger and Claudia Swan, 21–33. University of Pennsylvania Press, 2007.

"Multiple Classified Advertisements." *National Intelligencer* 10, no. 2920 (May 23, 1822): 1.

Muñoz, Juan Ramón. "La Sociedad de la Mazorca: Rasgos notables de la historia de la Revolución Arjentina." In *Revista del Pacífico: Publicación Literaria y Científica.* Vol. 3. Valparaíso: Imprenta del Mercurio, 1860.

Muratore, Ludovico Antonio. *La filosofía moral declarada y propuesta a la juventud.* Vol. 1. Translated by Fray Antonio Moreno Morales. Madrid: por Benito Cano, 1787. [Original Italian publication: Lodovico Antonio Muratori, *La filosofia morale esposta e proposta a i giovani da Lodovico Antonio Muratori* . . . Milan: Nella regio-ducal Corte, 1736.]

Muratori, Lodovico Antonio. *Il cristianesimo felice nelle missioni de' padri della Compagnia di Gesù nel Paraguai.* Vol. 2. Venice: Giambatista Pasquali, 1752.

Muratori, Ludovico Antonio. *La filosofia moral declarada, y propuesta a la juventud.* 2nd ed. Vol. 1. Madrid: Joachin Ibarra, 1791.

Myers, Jorge. "Giuseppe Mazzini and the Emergence of Liberal Nationalism in the River Plate and Chile, 1835–60." In Bayly and Biagini, *Giuseppe Mazzini*, 323–46.

Myers, Jorge. "Identidades porteñas: El discurso ilustrado en torno a la nación y el rol de la prensa: *El Argos de Buenos Aires*, 1821–1825." In *Construcciones impresas: Panfletos, diarios y revistas en la formación de los estados nacionales en América Latina, 1820–1920*, edited by Paula Alonso, 39–63. FCE, 2004.

Myers, Jorge. *Orden y virtud: El discurso republican del regimen de Rosas*. Universidad Nacional de Quilmes, 1995.

Myers, Jorge. "Philosophical Clio: The Birth of River Plate Historical Discourse (1830–1852)." *Varia Historia* 31, no. 56 (2015): 331–64.

Nafría, Juan Carlos Domínguez. "La América española y Napoleón en el Estatuto de Bayona." *Revista Internacional de Estudios Vascos*. Special Issue, "Les origines du constitutionnalisme et la Constitution de Bayonne du 7 juillet 1808." *Cuadernos* 4 (2009): 315–46.

Navia-Osorio y Vigil, Álvaro. *Reflexiones militares del mariscal de campo don Alvaro Navia Ossorio, vizconde de Puerto . . . Que tratan de sorpresas de plazas y quarteles, y de tropas en campaña . . . Sigue à los Indices del tomo un projecto del Autor para un diccionario universal*. Turin: Juan Francisco Mairesse, 1727.

Nieto Olarte, Mauricio. *Remedios para el imperio: Historia natural y la apropiación del Nuevo Mundo*. Instituto Colombiano de Antropología e Historia, 2000.

"Noticias." *Telégrafo Mercantil, Rural, Político-Económico, e Historiógrafo del Río de la Plata* 5, no. 1 (April 15, 1801): 70.

"Noticias estadisticas." *La Abeja Argentina* 1, no. 9 (December 15, 1822): 354.

Nott, Josiah Clark. "Introduction to *Types of Mankind*." In Morton, *Types of Mankind*, 49–62.

Novoa, Adriana. "The Dilemmas of Male Consumption in Nineteenth-Century Argentina: Fashion, Consumerism, and Darwinism in Domingo Sarmiento and Juan B. Alberdi." *Journal of Latin American Studies* 39 (2007): 771–95.

Novoa, Adriana. "From Virile to Sterile: Feminization, Masculinity and Darwinism in Late Nineteenth-Century Argentina." *Ometeca* (Corrales, NM) 12 (2008): 152–71.

Novoa, Adriana. "The Reception of Evolutionary Ideas in Argentina, 1860–1910." PhD diss., University of California, San Diego, 2000. ProQuest (9981234).

Novoa, Adriana. "Science, Sensibility and Gender in Argentina, 1820–1852." *Perspectives on Science* 28, no. 2 (2020): 318–40.

"Nuebo renombre de Apolo." In *Telégrafo Mercantil, Rural, Político-Económico, e Historiógrafo del Río de la Plata* 33, no. 22 (December 6, 1801): 262–64.

Nuñez, Ignacio. *An Account, Historical, Political, and Statistical, of the United Provinces of Rio de la Plata, with an Appendix Concerning the Usurpation of Monte Video, by the Portuguese and Brazilian Governments*. Translated from the Spanish. London: R. Ackerman, 1825.

Núñez, Ignacio. *Noticias históricas, políticas, y estadísticas de las Provincias Unidas del Río de la Plata: Con un apéndice sobre la usurpación de Montevideo por los gobiernos portugués y brasilero*. Appendix. London: R. Ackermann, 1825.

O'Brien, General. *Monte Video, Buenos Ayres, and the River Plate: Correspondence with the British Government Relative to the War between Buenos Ayres and Monte Video, and the Free Navigation of the River Plate, with an Appendix, Detailing Some of the Acts Committed by Rosas, Governor of Buenos Aires*. London: Reynell and Weight, 1845.

"The Observer, No. VII: Friday, January 1–4, 1785." *Massachusetts Centinel* 34, no. 2 (January 15, 1785): 2.

O'Neal, John C. *Changing Minds: The Shifting Perception of Culture in Eighteenth-Century France*. University of Delaware Press, 2002.

O'Neal, John C. *The Authority of Experience: Sensationist Theory in the French Enlightenment*. Pennsylvania State University Press, 1996.

Olson, Richard. *Science and Scientism in Nineteenth-Century Europe*. University of Illinois Press, 2007.

"On the Reality of the Rise of the Coast of Chile, in 1822, as Stated by Mrs. Graham: Extract from Mr. President Greenough's Address to the Geological Society, Delivered on the 4th of June, 1834." *American Journal of Science and Arts* 28, no. 2 (July 1835): 248–52.

One of a Number. "For the Centinel." *Massachusetts Centinel* (Boston) 35, no. 2 (January 17, 1785): 2.

Orain Arnaud. "The Moral Theory of Condillac: A Path Toward Utilitarianism." *Revue de Philosophie Économique* 13, no. 2 (2012): 93–117. https://doi.org/10.3917/rpec.132.0093.

Orden, Cipriano [Pedro Cerviño]. "Carta de D. Cipriano Orden Vetoño en orden al modo de hacer utiles los terrenos que nos rodean." *Semanario de Agricultura, Industria y Comercio* 1, no. 26 (March 16, 1803).

Orden, Cipriano [Pedro Cerviño]. "Concluye la carta de D. Cipriano Orden." *Semanario de Agricultura, Industria y Comercio* 1, no. 29 (April 6, 1803).

Orden, Cipriano [Pedro Cerviño]. "Concluye la carta de D. Cipriano Orden Vetoño." *Semanario de Agricultura, Industria y Comercio* 1, no. 85 (May 2, 1804): 274–75.

Orden, Cipriano [Pedro Cerviño]. "Vetoño en orden al modo de hacer utiles los terrenos que nos rodean." *Semanario de Agricultura, Industria y Comercio* 1, no. 27 (March 23, 1803).

Orecchia, José M. Olivero. "La Junta de Montevideo en 1808, una situación interna con repercusiones internacionales: Algunos aspectos de los intereses y acciones portuguesas." *Revista Digital Estudios Historicos* 3 (December 2010). http://estudioshistoricos.org/edicion_3/jose-olivero.pdf.

"Origen y estado de la medicina en Buenos Aires." *La Abeja Argentina*, no. 1 (1822): 25.

Orrje, Jacob. "The Logistics of the Republic of Letters: Mercantile Undercurrents of Early Modern Scholarly Knowledge Circulation." *British Journal for the History of Science* 53, no. 3 (2020): 351–69.

Ovid. *Publius Ovidius Naso, A New Translation of Ovid's Metamorphoses into English Prose.* 2nd ed. Translated by N. Bailey. London: Joseph Davidson, 1753.

Pacheco, José Ramon. *Una revolucion en la républica Argentina: Artículo de la Revista de los Dos Mundos.* Translated and edited by J. R. Pacheco. Mexico City: Ignacio Cumplido, 1835. [Original text: Lacordaire, "Un voyageur, 'Une révolution dans la république Argentine.'" *Revue des Deux Mondes* 1, no. 1 (1835): 23–40.]

Pagden, Anthony. *The Burdens of Empire: 1539 to the Present.* Cambridge University Press, 2015.

Page, Anthony. "'A Species of Slavery': Richard Price's Rational Dissent and Antislavery." *Slavery & Abolition* 32, no. 1 (2011): 53–73.

Parise, Agustin. "Slave Law and Labor Activities During the Spanish Colonial Period: A Study of the South American Region of Rio de la Plata." *Rutgers Law Record* 32, no. 1 (Spring 2008): 2–30.

Parra París, Lisímaco. "La recepción de Bentham en la Nueva Granada." In *Philosophy in and from Colombia*, edited by María del Rosario Acosta López and Miguel Gualdrón Ramírez, 183–93. Special issue, *Philosophical Readings* 11, no. 3 (2019).

Parra-León, Caracciolo. *Filosofía universitaria venezolana.* In *Obras.* Editorial J. B, 1954.

Patterson, Henry. "Notice of the Life and Scientific Labors of the Late Samuel Geo. Morton, M.D." In Morton, *Types of Mankind*, xvii–lvii.

Pauw, Cornelius. *Recherches philosophiques sur les Américains: Ou, Mémoires intéressants pour servir à l'histoire de l'espèce humaine.* Vol. 1. Berlin: George Jacques Decker, 1768.

Paz, José María. *Memorias postumas del brigadier general D. Jose M. Paz* [. . .]. Vol. 3. Buenos Aires: La Revista, 1855.

Paz, Octavio. *Children of the Mire: Modern Poetry from Romanticism to the Avantgarde*. Harvard University Press, 1991.

Paz, Octavio. *The Labyrinth of Solitude: And the Other Mexico; Return to the Labyrinth of Solitude; Mexico and the United States*. Translated by L. Kemp, Y. Milos, and R. Phillips Belash. Grove Press, 1985.

Pazos Kanki, Vicente. "Continúa el proyecto politico literario." *El Grito del Sud* 1, no. 6 (August 18, 1812): 41–48.

Pazos Kanki, Vicente. *Memorias historico-politicos de don Vicente Pazos*. Vol. 1. London: n.p., 1834.

Peaden, Catherine Bobbs. "Condillac and the History of Rhetoric." *Rhetorica: A Journal of the History of Rhetoric* 11, no. 2 (1993): 135–56.

Pelliza, Mariano A. *Monteagudo, su vida y sus escritos*. Vol. 1. Buenos Aires: Imprenta de Mayo, 1880.

Penrose, Mehl Allan. *Masculinity and Queer Desire in Spanish Enlightenment Literature*. Routledge, 2016.

Pettegrew, John. *Brutes in Suits: Male Sensibility in America, 1890–1920*. Johns Hopkins University Press, 2007.

Pfau, Thomas. *Romantic Moods: Paranoia, Trauma, and Melancholy, 1790–1840*. Johns Hopkins University Press, 2005.

"Phrenology." *Daily National Journal* 1, no. 224 (April 29, 1825): 1.

"Physical Evidences of the Characteristics of Ancient Races Among the Moderns." *Fraser's Magazine for Town and Country* 6 (August to December 1832): 673–79. London: James Fraser, 1832.

Piazza, Marco. "Maine de Biran and Gall's Phrenology: The Origins of a Debate About the Localization of Mental Faculties." *British Journal for the History of Philosophy* 28 (May 2020): 5. https://doi.org/10.1080/09608788.2020.1760785.

Picornell y Gomila, Juan. "A las reales sociedades económicas del reyno." In Charles-Joseph Mathon de la Cour, *Discurso sobre los mejores medios de excitar y fomentar el patriotismo en una monarquía . . . : Premiado por la Academia de Ciencias, Artes, Agricultura y Bellas Letras de Chalons de Marne el 25 de agosto de 1787 / por Mr. Mathon de la Cour*. Translated by Don Juan Picornell y Gomila. Madrid: Oficina de Aznar, 1790.

Pinel, Philippe. *Nosographie philosophique, ou La méthode de l'analyse appliquée à la médecine*. Paris: Maradan, 1797.

Planta y methodo, que, por determinación de la Academia Española: Deben observar los academicos, en la composición del nuevo diccionario de la lengua castellana. Madrid: Imprenta Real, 1713.

Pluche, Noël-Antoine. *Espectaculo de la naturaleza, o Conversaciones a cerca de las particularidades de la historia natural*. Vol. 7. Translated by P. Estevan de Terreros y Pando. Madrid: Ibarra, 1757.

P. M. "Algo sobre la agricultura." *Los Principios Politicos-Religiosos* 1, no. 8 (1871): 124–25. https://www.google.com/books/edition/Los_Principios_politico_religiosos/mUItAAAAYAAJ?hl=en&gbpv=1.

Poldervaart, Saskia. "Theories About Sex and Sexuality in Utopian Socialism." *Journal of Homosexuality* 29, no. 2/3 (1995): 41–68, 57–58.

Pomian, Krzysztof. "Franks and Gauls." In *Realms of Memory: Conflicts and Divisions*, edited by Pierre Nora and Lawrence D. Kritzman, 27–79. Columbia University Press, 1996.

Portocarrero y Guzmán, Pedro. *Theatro monarchico de España* [. . .]. Madrid: Juan Garcia Infançon, 1700.

Portugués, Joseph Antonio. *Coleccion general de las ordenanzas militares, sus innovaciones, y aditamentos, dispuesta . . . con separacion de clases, por . . . Joseph Antonio Portugues, etc.* Madrid: Imprenta de Antonio Marín, 1764.

Posadas, Gervasio Antonio. “Logias masónicas en Buenos Aires, 1806.” Archivo General de la Nación Argentina: Ensayo de Gervasio Antonio Posadas a, transcripto por A. L. Documentos Escritos, Colección Andrés Lamas, Legajo, 52.

Poskett, James. *Materials of the Mind: Phrenology, Race, and the Global History of Science, 1815–1920.* University of Chicago Press, 2019.

Poskett, James. “Phrenology, Correspondence, and the Global Politics of Reform, 1815–1848.” *Historical Journal* 60, no. 2 (2017): 409–42.

Potofsky, Allan. “The Political Economy of the French-American Debt Debate: The Ideological Uses of Atlantic Commerce, 1787 to 1800.” *William and Mary Quarterly*, Third Series 63, no. 3 (2006): 489–516.

Potts, Alex. *Flesh and the Ideal: Winckelmann and the Origins of Art History.* Yale University Press, 2000.

Pratt, Mary Louise. “Women, Literature, and National Brotherhood.” In *Women, Culture, and Politics in Latin America*, edited by Emilie L. Bergmann, 48–73. University of California Press, 1992.

Price, Richard. *Observations on the Importance of the American Revolution, and the Means of Making It a Benefit to the World* [. . .]. 2nd ed. Dublin: L. White, W. Whitestone, P. Bryne, P. Wogan, J. Cash and R. Marchbank, 1785.

Price, Richard. *Richard Price and the Ethical Foundations of the American Revolution: Selections from His Pamphlets, with Appendices.* Duke University Press, 1979.

Prieto, Aldolfo. *Los viajeros ingleses y la emergencia de la literatura argentina, 1820–1850.* Sudamericana, 1996.

“Progresos del papel periódico que se publica en Santa Fe de Bogotá.” *Mercurio peruano de historia, literatura, y noticias públicas que da á luz la Sociedad academica de amantes de Lima, y en su nombre J. Calero y Moreira*, 164–71. Lima: Imprenta Real, 1791.

“The Progress of America.” *Athenaeum: Journal of Literature, Science, the Fine Arts, Music and the Drama*, no. 1023 (June 5, 1847): 591–92.

“Prospecto del periódico que se intenta publicar con el título de Correo de Comercio.” Buenos Aires: Imprenta de Niños Expósitos, 1810, 1.

Protocolo de la Negociacion de Paz, promovida por los SS. Ministros Plenipotenciarios de los gobiernos interventorcs, iniciado el 21 de Marzo y terminado el 8 de Junto de 1818. Publicacion Official [*sic*]. Montevideo, 1848.

“Proyecto del Dr. Moreno.” *Revista del Rio de la Plata* 4, no. 15 (1872).

Pufendorf, Samuel Freiherr von. *Les devoirs de l’homme et du citoien tels qu’ils lui sont prescrits par la loi naturelle.* Amsterdam: P. De Coup & G. Kuyper, 1735.

Pufendorf, Samuel Freiherr von. *Of the Law of Nature and Nations.* London: J. and J. Knapton, 1728.

Pufendorf, Samuel Freiherr von. *The Whole Duty of Man According to the Law of Nature.* 2nd ed. London: Printed by B. Motte for C. Harper, 1698.

Quesada, Ernesto. “El movimiento intelectual argentino revistas y periódicos.” In *Reseñas y críticas*, 119–43. Buenos Aires: F. Lajouane, 1893.

Quesnay, François. “Maximes générales du government d’un royaume Agricole.” *L’Esprit des Journaux, François et Étrangers* 4 (April): 178–89. France: Valade, 1775.

Quinet, Edgar. "La Alemania. (Extracts from Quinet)." *La Moda, Gacetín Semanal*, no. 8, December 25, 1837, 1–4.

Rabin, Sheila J. "Jesuit Science Before 1773: A Historiographical Essay." *Jesuit Historiography Online*. http://dx.doi.org/10.1163/2468-7723_jho_COM_196375.

"The Races of Men." *Vermont Chronicle* 22, no. 29 (July 21, 1847): 114.

Raillard, Matthieu P. "Petimetres, pseudoeruditos and eruditos a la violeta: Anti-Models and the Eighteenth-Century Spanish Republic of Letters." *Revista de Estudios Hispanicos* 42, no. 1 (2008): 1–19.

Raj, Kapil. "Beyond Postcolonialism . . . And Postpositivism: Circulation and the Global History of Science." *Isis* 104, no. 2 (2013): 337–47.

Rapport, Mike. *1848: Year of Revolution*. Basic Books, 2009.

Raynal, Abbé Gustave. *Philosophical and Political History of the Settlements and Trade of the Europeans in the East and West Indies*. Revised, Augmented, and Published, in Ten Volumes, by the Abbé Raynal. Translated from the French by J. O. Justamond, F. R. S. London: A. Strahan; and T. Cadell, 1788.

Raynal, Guillaume-Thomas. *Histoire philosophique et politique, des établissemens et du commerce des Européens dans les deux Indes par Raynal*. Amsterdam: J. E. Dufour et Philipp, 1770.

Raynal, Guillaume-Thomas. *Historia política de los establecimientos ultamarinos de las naciones europeas por Eduardo Malo de Luque*. 5 vols. Madrid, 1784–1790.

Real Academia Española. *Diccionario de la lengua castellana, en que se explica el verdadero de las voces . . . con las phrases o modos de hablar, los proverbios o refranes . . .* Vol. 5. Madrid: Imprenta de Francisco del Hierro, 1737.

"Real cedula de S. M. y Señores del Consejo, por la qual se declara, que no solo el Oficio de Curtidor, sino tambien los demas Artes y Oficios de Herrero, Sastre . . . son honestos y honrados; y que el uso de ellos no envilece la familia . . . ni la inhabilit." https://bibliotecavirtual.defensa.gob.es/BVMDefensa/es/consulta/registro.do?id=44255.

Real Sociedad Bascongada de los Amigos del Pais. *Extractos de las Juntas Generales celebradas por la Real Sociedad Bascongada de los Amigos del País en la villa de Bilbao por julio de 1787*. Vitoria: Baltasar de Manteli, 1788.

Rebok, Sandra. *Humboldt and Jefferson: A Transatlantic Friendship of the Enlightenment*. University of Virginia Press, 2014.

Recopilación de las leyes y decretos promulgados en Buenos Aires desde el 25 de mayo de 1810 hasta fin de diciembre de [1840] con un indice general de materias, vol. 2. Buenos Aires: Imprenta del Estado, 1836.

Recopilación de las leyes y decretos promulgados en Buenos Aires desde enero de 1841 hasta la fecha. Buenos Aires: Imprenta de Mayo, 1858.

"Reglando los estudios preparatorios." In *Registro Nacional, Provincias Unidas del Río de la Plata: libro segundo*, 129–30. Buenos Aires: Imprenta de los Expositos, 1826.

Rengger, Johann Rudolph, and Marcelin Longchamps. *The Reign of Doctor Joseph Gaspard Roderick de Francia in Paraguay: Being an Account of Six Years' Residence in That Republic, from July, 1819 to May,* 1825. Translated by J. R. Rengger. London: T. Hurst, E. Chance, 1827.

Richards, Robert J. "Ideology and the History of Science." *Biology and Philosophy* 8, no. 1 (1993): 103–8.

Richards, Robert J. *The Romantic Conception of Life: Science and Philosophy in the Age of Goethe*. University of Chicago Press, 2010.

Ricketts, Mónica. *Who Should Rule? Men of Arms, the Republic of Letters, and the Fall of the Spanish Empire*. Oxford University Press, 2017.

Riskin, Jessica. 2002. *Science in the Age of Sensibility: The Sentimental Empiricists of the French Enlightenment.* University of Chicago Press.

Rivadavia, Bernardino. *Mensaje del Gobierno a la Sala de Representantes de la Provincia de Buenos Aires.* Buenos Aires: Imprenta de la Independencia, 1824.

Rivera Indarte, José. *El voto de America: O sea breve examen de esta cuestion.* Madrid: Imprenta Real, 1835.

Rivera Indarte, José. *La intervención en la guerra actual del Río de la Plata.* Río: Mercantil de Lopes, 1845.

Rivera Indarte, José. *Las Tablas de Sangre: Es acción santa matar a Rosas.* Montevideo: Imprenta del Nacional, 1843.

Rivera Indarte, José. *Poesías.* Mayo, 1943.

Rivera Indarte, José. *Rosas y sus opositores.* 2nd ed. Buenos Aires: Imprenta de Mayo, 1853. https://www.loc.gov/item/06010626/.

Rivera Indarte, José. "Un opúsculo de Rivera Indarte." *La Revista de Buenos Aires: Historia Americana, Literatura y Derecho. Periódico Destinado á la República Argentina, la Oriental del Uruguay y del Paraguay* 23, no. 92 (December): 402–21. Buenos Aires: Imprenta de Mayo, 1870.

Rock, David. "British Communities and Foreign Intervention in Nineteenth-Century South America: The Rio de la Plata in the 1840s." In *The British Abroad Since the Eighteenth Century*, vol. 2, 154–75. Palgrave Macmillan, 2013.

Rodríguez, Martín. *Diario de la expedición al desierto.* Notes by Andrés Carretero. Editorial Sudestada, 1969.

Rodríguez Calderón, Juan Jacinto. *Don Líquido ó el Currutaco vistiéndose: Escena unipersonal; Para representarse en casa particular.* Valencia: José Ferrer de Orga, 1816.

Rodríguez Mohedano, Rafael, and Pedro Rodríguez Mohedano. *Historia literaria de España, desde su primera población hasta nuestros días.* Vol. 1. Madrid: Imprenta Francisco Xavier Garcia, 1769.

Rodríguez Molas, Ricardo. *Historia de la tortura el orden represivo en la Argentina.* Eudeba, 1984.

Roger, Philippe. *The American Enemy: The History of French Anti-Americanism.* University of Chicago Press, 2006.

Romay, Tomás. *Memoria sobre la introducción y progresos de la vacuna en la Isla de Cuba: Leida en Juntas Generales celebradas por la Sociedad Económica de la Havana el 12 de diciembre de 1804.* Havana: Imprenta de la Capitania General, 1805.

Romero, José Luis. *A History of Argentine Political Thought.* Stanford University Press, 1963.

Root, Regina A. *Couture and Consensus: Fashion and Politics in Postcolonial Argentina.* University of Minnesota Press, 2010.

Rosas, and Some of the Atrocities of His Dictatorship in the River Plate: In a Letter to the Right Honourable the Earl of Aberdeen, &c. London: Simmonds and Clowes, 1844.

Rosas, Juan Manuel de. *Manifiesto de las razones que legitiman la declaración de guerra contra el gobierno del General D. Andres Santa Cruz, titulado presidente de la Confederación Perú-Boliviana.* Buenos Aires: Imprenta del Estado, 1837.

Rosell, Antonio Gregorio. *Instituciones matemáticas: tomo I: Contiene la aritmética propia y los principios de algebra.* Madrid: Imprenta Real, 1785.

Rotger, Susana. *Captive Women: Oblivion and Memory in Argentina.* University of Minnesota Press, 2002.

Rousseau, George Sebastian. *Enlightenment Crossings: Pre- and Post-Modern Discourses; Anthropological.* Manchester University Press, 1991.

Rousseau, Jean-Jacques. "A Discourse on a Subject Proposed by the Academy of Dijon: What Is the Origin of Inequality Among Men, and Is It Authorised by Natural Law?" 1754. The Constitution Society, Jean-Jacques Rousseau. https://www.constitution.org/jjr/ineq.htm.

Rousseau, Jean-Jacques. *Emile*. In *Œuvres complétes de J. J. Rousseau, citoyen de Genève: Émile*. Paris: Bélin, 1793.

Royal Spanish Academy. *Diccionario de la lengua castellana: En que se explica el verdadero sentido de las voces, su naturaleza y calidad, con las phrases o modos de hablar, los proverbios o refranes, y otras cosas convenientes al uso de la lengua. . . .* Madrid: F. Del Hierro, 1734.

Royal Spanish Academy. *Diccionario de la lengua castellana compuesto por la real academia*. Española, Reduced to one volume. Madrid: Imprenta de J. Ibarra, 1780.

Russell, Joseph, and Others. "To James Madison from Joseph Russell and Others, 6 June 1802." *Founders Online*, National Archives. https://founders.archives.gov/documents/Madison/02-03-02-0344.

Russo, Elena. "Sociability, Cartesianism, and Nostalgia in Libertine Discourse." *Eighteenth-Century Studies* 30, no. 4 (1997): 383–400.

Safier, Neil. *Measuring the New World: Enlightenment Science and South America*. University of Chicago Press, 2008.

Saint-Robert, Chevalier de. *Le Général Rosas et la question de la Plata, par le chevalier de Saint-Robert*. Paris: Gerdes, 1848.

Sala-Valldaura, Josep Maria. "Gurruminos, petimetres, abates y currutacos en el teatro breve del siglo XVIII." *Revista de Literatura* 71, no. 142 (2009): 429–60.

Salessi, Jorge. *Médicos maleantes y maricas: Higiene, criminología y homosexualidad en la construcción de la nación Argentina (Buenos Aires, 1871–1914)*. B. Viterbo Editora, 1995.

Sampay, Arturo Enrique. *Las ideas políticas de Juan Manuel de Rosas*. Juarez, 1972.

Sampson, Ezra. *The Youth's Companion: Or An Historical Dictionary* [. . .]. Albany: Websters and Skinners, 1813.

Sans Souci Alias Free and Easy or an Evening's Peep into a Polite Circle an Entire Entertainment. In Three Acts. Boston: Warden and Russell, 1785. https://quod.lib.umich.edu/e/evans/N15151.0001.001?rgn=main;view=fulltext.

Santa Cruz, Andrés. *Contra-Manifesto to that published by the government of Buenos-Ayres, stating the grounds on which it pretends to justify its declaration of war against the Peru-Bolivian Confederacy*. Lima: Eusebio Aranda, 1837.

Sarmiento, Domingo F. "Argirópolis." In *Obras de D. F. Sarmiento: Argirópolis, capital de los Estados Confederados*, vol. 23, 1–108. Buenos Aires: Imprenta Mariano Moreno, 1896.

Sarmiento, Domingo F. "Avíos y monturas." *Mercurio*, July 25, 1842. In *Obras de D. F. Sarmiento*, vol. 1. Buenos Aires: Félix Lajouane, 1885.

Sarmiento, Domingo F. "Colonización inglesa en el Río de la Plata." In *Obras de D. F. Sarmiento: Argirópolis, capital de los Estados Confederados*, vol. 23, 324–41. Buenos Aires: Imprenta Mariano Moreno, 1896.

Sarmiento, Domingo F. "Concluye el análisis del artículo Romanticismo." In *Obras de D. F. Sarmiento*, vol. 1, 308–14. Santiago: Imprenta Gutenberg, 1885.

Sarmiento, Domingo F. "Continúa el exámen del artículo Romanticismo." In *Obras de D. F. Sarmiento*, vol. 1, 302–8. Santiago: Imprenta Gutenberg, 1885.

Sarmiento, Domingo F. "Cuadros de Monvoisin." In *Obras de D. F. Sarmiento*, vol. 2, 123–28. Buenos Aires: Félix Lajouane, 1887.

Sarmiento, Domingo F. "Cuestión de Magallanes convertida en reclamo." *La Crónica*, December 9, 1849. In *Obras de D. F. Sarmiento: Cuestiones americanas; Límites con Chile*, vol. 35, 37–46. Imprenta Mariano Moreno, 1900.

Sarmiento, Domingo F. "Decreto del gobernador de Salta alzándose con el poder." In *Obras de D. F. Sarmiento: Argirópolis, capital de los Estados Confederados*, vol. 23, 222–38. Buenos Aires: Imprenta Mariano Moreno, 1896.

Sarmiento, Domingo F. "El comunicado del otro Quindam." In *Obras de D. F. Sarmiento*, vol. 1, 224–31. Santiago: Imprenta Gutenberg, 1887.

Sarmiento, Domingo F. *El gran Sarmiento: Las cartas que develan al hombre de acción y su intimidad*. Editorial El Ateneo, 2001.

Sarmiento, Domingo F. "El romanticism segun *El Semanario*." *Mercurio*, July 25, 1842. In *Obras de D. F. Sarmiento*, vol. 1, 287–92. Santiago: Imprenta Gutenberg, 1885.

Sarmiento, Domingo F. *Facundo: Civilización y barbarie*. Santiago: Imprenta del Progreso, 1845. In *Obras de D. F. Sarmiento*, vol. 7. Santiago De Chile: Imprenta Gutenberg, 1889.

Sarmiento, Domingo F. "Fisiolojia del Paquete." In *Obras de D. F. Sarmiento*, vol. 2, 9–18. Santiago: Imprenta Gutenberg, 1885.

Sarmiento, Domingo F. *Juan Facundo Quiroga*. In *Obras de D. F. Sarmiento*, vol. 7. Santiago: Imprenta Gutenberg, 1887. https://www.gutenberg.org/ebooks/33267.

Sarmiento, Domingo F. "La Cuestión del Plata." In *Obras completas de D. F. Sarmiento*, vol. 2, 60–89. Santiago: Imprenta Gutenberg, 1885.

Sarmiento, Domingo F. "Las filípicas de los Andes." In *Obras de D. F. Sarmiento: Argirópolis, capital de los Estados Confederados*, vol. 23, 197–222. Buenos Aires: Imprenta Mariano Moreno, 1896.

Sarmiento, Domingo F. "Los estudios históricos en Francia." In *Obras de D. F. Sarmiento*, vol. 6, 199–201. Santiago: Imprenta Gutenberg, 1887.

Sarmiento, Domingo F. *Travels in the United States in 1847*. Translated and with an introduction by Michael Aaron Rockland. Princeton University Press, 1970.

Sarmiento, Domingo Faustino. "Al señor H. Southern, encargado de negocios de S. M. B. cerca del gobierno de Buenos Aires." *Cronica* (January 1850). In *Obras de D. F. Sarmiento*, vol. 6, 260–80. Santiago: Imprenta Gutenberg, 1886–1914.

Sarmiento, Domingo Faustino. "Investigaciones sobre el sistema colonial de los españoles." *Progreso*, September 27, 1844. In *Obras de D. F. Sarmiento*, vol. 2, 211–18. Buenos Aires: Félix Lajouane, 1887.

Sarmiento, Domingo Faustino. "Las procesiones de Semana Santa." In *Obras de D. F. Sarmiento*, vol. 2, 140–45. Buenos Aires: Félix Lajouane, 1887.

Sarmiento, Domingo Faustino. *Obras de D. F. Sarmiento: Costumbres, progresos*. Vol. 42. Imprenta Mariano Moreno, 1900.

Sarmiento, Domingo Faustino. *Viajes por Europa, África y América 1845–1847*. In *Obras completas de Domingo Faustino Sarmiento*. Vol. 5. Buenos Aires: Félix Lajouane, 1887.

"Saturday, July 5, 1845." *Newcastle Journal*, July 5, 1845: 2.

Schell, Patience A. *The Sociable Sciences: Darwin and His Contemporaries in Chile*. Palgrave Macmillan, 2013.

Schiebinger, Londa. "Feminine Icons: The Face of Early Modern Science." *Critical Inquiry* 14, no. 4 (1988): 661–91.

Schiebinger, Londa. *The Mind Has No Sex? Women in the Origins of Modern Science*. Harvard University Press, 1989.

Schiebinger, Londa. *Nature's Body: Gender in the Making of Modern Science*. Rutgers University Press, 2004.

Schwartzman, Simon. *Space for Science: The Development of the Scientific Community in Brazil.* Pennsylvania State University Press, 2010.

Scipio. "News." *Public Advertiser,* October 28, 1774. gale.com/apps/doc/Z2001151298/GDCS?u=tamp44898&sid=bookmarkGDCS&xid=c740d3ff.

Scott, Joan W. "Against Eclecticism." *differences* 16, no. 3 (2005): 114–37.

Scotto, José Arturo. *Las diabluras del tirano Juan Manuel de Rosas.* Buenos Aires: Biblioteca Histórica, 1896.

Secord, James A. "Edinburgh Lamarckians: Robert Jameson and Robert E. Grant." *Journal of the History of Biology* 24, no. 1 (1991): 1–18.

Secord, James A. "Global Darwin." In *Darwin,* edited by William Brown and Andrew Fabian, 31–57. Cambridge University Press, 2010.

Secord, James A. "Knowledge in Transit." *Isis* 95, no. 4 (2004): 654–72.

Senillosa, Felipe. *Gramática española o principios de la gramática general aplicados a lalengua castellana.* Buenos Aires: Imprenta de los Expósitos, 1817.

"Señor Editor." *Los Amigos de la Patria y de la Juventud,* Suplemento al, no. 1 (November 1815): 2.

Sera-Shriar, Efram. *The Making of British Anthropology, 1813–1871.* Routledge, 2015.

Shapin, Steven. "The Image of the Man of Science." In *The Cambridge History of Science.* Vol. 4: *Eighteenth-Century Science,* edited by Roy Porter, 159–83. Cambridge University Press, 2003.

Shapin, Steven. 1975. "Phrenological Knowledge and the Social Structure of Early Nineteenth- Century Edinburgh." *Annals of Science* 32, no. 3 (1975): 219–43.

Shepperson, George. "The Intellectual Background of Charles Darwin's Student Years at Edinburgh." In *Darwinism and the Study of Society: A Centenary Symposium,* edited by Michael Banton, 17–37. Quadrangle Books, 1961.

Sheriff, Mary D. *The Exceptional Woman: Elisabeth Vigée-Lebrun and the Cultural Politics of Art.* University of Chicago Press, 1997.

Shovlin, John. *The Political Economy of Virtue: Luxury, Patriotism, and the Origins of the French Revolution.* Cornell University Press, 2007.

Siebers, Tobin. "Kant and the Politics of Beauty." *Philosophy and Literature* 22, no. 1 (1998): 31–50.

"The Siege of Montevideo." *Dublin Review* 26 (March–June 1849): 44.

Simon, Jules. *Thiers, Guizot, Rémusat.* Paris: Lévy, 1885.

Simon, Walter M. *European Positivism in the Nineteenth Century: An Essay in Intellectual History.* Cornell University Press, 1963.

Smith, Bonnie G. "The Rise and Fall of Eugène Lerminier." *French Historical Studies* 12, no. 3 (1982): 377–400.

Smith, George. *The Use and Abuse of Free-Masonry: A Work of the Greatest Utility to the Brethren of the Society, to Mankind in General, and to the Ladies in Particular.* London: G. Kearsley, 1783.

Smith, Robert Sidney. "The Wealth of Nations in Spain and Hispanic America, 1780–1830." *Journal of Political Economy* 65, no. 2 (1957): 104–25.

Smith, Theresa Ann. *The Emerging Female Citizen: Gender and Enlightenment in Spain.* University of California Press, 2006.

Smollet, Tobias George. *Travels Through France and Italy: Containing Observations on Character, Customs, Religion, . . . with a Particular Description of the Town, Territory, and Climate of Nice . . .* Vol. 1. Dublin: J. Hoey, Sen. A. Leathly, P. Wilson, 1766.

"Sobre la libertad de escribir." *Gazeta de Buenos-Ayres,* June 21, 1810: 29–31.

"Sobre mujeres." *Los Amigos de la Patria y de la Juventud,* no. 5 (April 1816): 51.

Sociedad de Beneficiencia de la Capital (Buenos Aires, Argentina). "Acta de instalación de la Sociedad." In *Origen y desenvolvimiento*, 31–36.

Sociedad de Beneficiencia de la Capital (Buenos Aires, Argentina). "Comienzos de la acción de la Sociedad." In *Origen y desenvolvimiento*, 37–62.

Sociedad de Beneficiencia de la Capital (Buenos Aires, Argentina). *Origen y desenvolvimiento de la Sociedad de Beneficencia de la capital 1823–1912*. Estab. tip. M. R. Giles, 1913.

Sociedad de Beneficiencia de la Capital (Buenos Aires, Argentina). "Origen y fundación de la Sociedad." In *Origen y desenvolvimiento*, 15–18.

Société Ethnologique. "Liste des membres de la Societé." *Mémories de la Société Ethnologique*, xvi–xxi. Paris: Mme. Ve Dondey Dupré, 1841.

"Sofismas politicos." *La Abeja Argentina* 1, no. 6 (August 15, 1822): 236.

"Soliloquio." *Los Amigos de la Patria y de la Juventud*, no. 5 (April 1816): 35–37.

Solomianski, Alejandro. "Gabino Ezeiza y su recuperación dentro del imaginario de la identidad nacional Argentina." *Cincinnati Romance Review* 30 (2011): 53–69.

Sommer, Doris. *Foundational Fictions: The National Romances of Latin America*. University of California Press, 1991.

Son of Candor. "For the Centinel." *Massachusetts Centinel* (Boston) 37, no. 2 (January 26, 1785): 1. Readex: America's Historical Newspapers.

Spurzheim, Johann Gaspar. *Phrenology: Or, the Doctrine of the Mental Phenomena*. 5th American ed., greatly improved by the author, from the 3rd London ed. Boston: Marsh, Capen and Lyon, 1838.

S. S. El Regañon, "Diálogo sobre alguna cosa importante." In *La Moda, Gacetín Semanal*, no. 19 (Buenos Aires), March 24, 1838: 1–2.

Sterne, Laurence. *A Sentimental Journey Through France and Italy*. London: 1768.

Stewart, Ian B. "William Frédéric Edwards and the Study of Human Races in France, from the Restoration to the July Monarchy." *History of Science* 8, no. 3 (2020): 275–300.

Strang, Cameron B. *Frontiers of Science: Imperialism and Natural Knowledge in the Gulf South Borderlands, 1500–1850*. University of North Carolina Press, 2018.

Strangford, Lord Clinton. "Exmo. Señor Lord Visconde Strangford. Buenos Aires, December 28, 1813." In Núñez, *Noticias históricas, políticas, y estadísticas*, 293–94.

Strangford, Lord Clinton. "Letter of Strangford to the Government of the United Provinces, Rio de Janeiro, November 27, 1813." In Núñez, *Noticias históricas, políticas, y estadísticas*, 292.

Strasser, Ulrike. *Missionary Men in the Early Modern World: German Jesuits and Pacific Journeys*. Amsterdam University Press, 2020.

Strube, Julian. "Socialism and Esotericism in July Monarchy France." *History of Religions* 57, no. 2 (2017): 197–221.

Strube, Julian. "Socialist Religion and the Emergence of Occultism: A Genealogical Approach to Socialism and Secularization in 19th-Century France." *Religion* 46, no. 3 (2016): 359–88.

Szuchman, Mark. *Order, Family, and Community in Buenos Aires, 1810–1860*. Stanford University Press, 1988.

Tacitus, Cornelius. *Las historias*. Vol. 3. Translated by Cayetano Sixto and Joaquín Ezquerra. Madrid: Imprenta Real, 1794.

Taillandier, Saint-René. "La littérature politique en Allegmane: Le movement constitutionnel en Prusse." *Saint-René Taillandier* 9, no. 5 (March 1, 1845): 859–99.

Taillandier, Saint-René. "La littérature politique en Allegmane: Poésis nouvelles de M. Henri Heine." *La Revue des Deux Mondes* 9, no. 2 (January 15, 1845): 297–332.

Tanghe, Koen B., and Mike Kestemont. "Edinburgh and the Birth of British Evolutionism: A Peek Behind a Veil of Anonymity." *BioScience* 68, no. 8 (2018): 585–92.

Tavárez, Simó, and Fidel José. "La invención de un imperio commercial Hispano, 1740–1765." *Magallánica, Revista de Historia Moderna* 2, no. 3 (July–December 2015): 54–73.

Taylor, Keith. "Saint-Simon and the Conquest of the Future." *Futures* 9, no. 1 (1977): 58–64.

Tesler, Mario. *Agresión militar de los EE.UU. a las Islas Malvinas y el gaucho Antonio Rivero*. Editorial Dunken, 2013.

Teulon, Fabrice. "Fiasco théorique de l'Idéologie chez Destutt de Tracy." *Studies on Voltaire and the Eighteenth Century* 12 (2001): 121–31.

Thierry, Augustin. *Histoire de la conquête de l'Angleterre par les Normands: De ses causes et de ses suites jusqu'à nos jours, en Angleterre, en Ecosse, en Irlande et sur le continent*. Paris: F. Didot, 1825.

Thierry, Jacques Nicolas Augustin. *History of the Conquest of England by the Normans*. . . . Translated from the French by C. C. Hamilton. Vol. 1. London: B. Whittaker, 1825.

Thierry, M. Amédee. *Histoire des Gaulois depuis les temps les plus reculés jusqu'à l'entière soumission de la Gaule à la domination romaine*. Paris: Didier, (1828) 1881. https://www.gutenberg.org/ebooks/36058.

Thiry, Paul-Henri, Baron d'Holbach. *La moral universal o los deberes del hombre fundados en su naturaleza*. Barcelona: Impreso por Soler y Gaspar, 1835.

Thiry, Paul-Henri, Baron d'Holbach. *The System of Nature or, the Laws of the Moral and Physical World*. 2 vols. Translated from the French (n.p., 1770). Vol. 1: https://www.gutenberg.org/files/8909/8909-h/8909-h.htm. Vol. 2: https://www.gutenberg.org/files/8910/8910-h/8910-h.htm.

Thomson, Sinclair. "Cuando sólo reinasen los indios: recuperando la variedad de proyectos anticoloniales entre los comuneros andinos (La Paz, 1740–1781)." *Argumentos Estudios Críticos de la Sociedad*, no. 50 (September): 15–47.

Tiedemann, Frederick. "On the brain of the Negro, compared with that of the European and the Ourang-Outang." *Abstracts of the Papers Printed in the Philosophical Transactions of the Royal Society of London* 3 (June 9, 1836): 398–99.

Tirado y Rojas, Mariano. *La masonería en España: Ensayo histórico*. Madrid: Ulan Press, 1983.

"To the Observer." *Massachusetts Centinel* 34, no. 2 (January 15, 1785): 2.

Tornel y Mendívil, José María. *Tejas y los Estados Unidos de América en sus relaciones con la República Mexicana*. Mexico: Ignacio Cumplido, 1837.

Torres, Fernando Villegas. "El costumbrismo americano ilustrado: El caso peruano. Imágenes originales en la era de la reproducción técnica." *Anales del Museo de América*, no. 19: 7–67.

Torrubia, Fray Joseph. *Centinela contra Francs-Masones: discurso sobre su origen, instituto, secreto y juramento, descúbrese la cifra con que se escriben y las acciones, señales y palabras con que se conocen*. . . . Madrid: Imprenta de Ramon Ruiz, 1793.

Tosh, John. "What Should Historians Do with Masculinity? Reflections on Nineteenth-Century Britain." *History Workshop* 38 (1994): 179–202. http://www.jstor.org/stable/4289324.

Tresch, John. *The Romantic Machine: Utopian Science and Technology After Napoleon*. University of Chicago Press, 2012.

Turner, Sharon. *The History of the Anglo-Saxons*. Vol. 1. London: Longman, 1807.

Turner, Sharon. *The Sacred History of the World, as Displayed in the Creation and Subsequent Events to the Deluge: Attempted to Be Philosophically Considered, in a Series of Letters to a Son*. Vol. 1. London: Longman, 1832. US ed.: New York: J. & J. Harper, 1832.

Urbas, Joseph. "In Praise of Second-Rate French Philosophy: Reassessing Victor Cousin's Contribution to Transcendentalism." *Revue Française d'Études Américaines* 140, no. 3 (2014): 37–51. https://doi-org.ezproxy.lib.usf.edu/10.3917/rfea.140.0037.

Urbinati, Nadia. "The Legacy of Kant: Giuseppe Mazzini's Cosmopolitanism of Nations." In Bayly and Biagini, *Giuseppe Mazzini*, 11–36.

Valdés, Manuel Antonio. *Gazeta de Mexico: Compendio de noticias de Nueva España desde principios de los años de 1796 y 1797.* Vol. 8. Mexico: Imprenta de D. Felipe de Zuñiga y Ontiveros, 1797.

Valverde, Antonio Sánchez. *La America vindicada de la calumnia de haber sido madre del mal venereo.* Madrid: Imprenta de Dn. Pedro Marín, 1785.

Van Evra, James. "Richard Whately and the Rise of Modern Logic." *History and Philosophy of Logic* 5, no. 1 (1984): 1–18.

Van Miert, Dirk. "What Was the Republic of Letters? A Brief Introduction to a Long History." *Groniek: Historisch Tijdschrift*, no. 204/5 (2014): 269–88.

Varaigne. "Précis historique sur l'état actuel de la République Argentine (Buenos Ayres)." *Revue Encyclopédique, ou Analyse Raisonnée* 25 (July 1827): 5–17.

Varela, Florencio. *Auto-biografía de D. Florencio Varela, natural de Buenos-Ayres, redactor del "Comercio del Plata."* Montevideo: Imprenta del Comercio del Plata, 1848.

Varela, Florencio. *Escritos políticos, económicos y literarios.* Buenos Aires: Orden, 1859.

Varela, Florencio. *Observations on Occurrences in the River Plate as Connected with Their Foreign Agents and the Anglo-French Intervention.* Montevideo, 1843.

Vartanian, Aram. "Man-Machine from the Greeks to the Computer." *Dictionary of the History of Ideas: Studies of Selected Pivotal Ideas.* Vol. 3, edited by Philip P. Wiener, 131–46. Scribner's, 1973.

Velarde, Juan Anselmo de [Manuel José de Labardén]. "Cartas de F. Juan Anselmo de Velarde al redactor del Semanario." *Semanario de Agricultura, Industria y Comercio* 6, no. 1 (October 27, 1802): 41–48.

Vélez, Rafael. *Preservativo contra la irreligión o planes de la filosofía contra la Religión y el Estado.* Manresa: Brusi, 1813.

Vermeir, Koen, and Michael Funk Deckard. "Philosophical Enquiries into the Science of Sensibility." In *The Science of Sensibility: Reading Burke's Philosophical Enquiry*, edited by Koen Vermeir and Michael Funk Deckard, 3–56. Springer Netherlands, 2011.

Verney, Luís António. *Essay sur les moyens de rétablir les sciences & les lettres in Portugal.* Paris: Chez P. Al. Le Prieur, 1762.

Verney, Luís António. *Verdadeiro metodo de estudar para ser util à republica, e à Igreja* Naples: n.p., 1746.

Verney, Luis António. *Verdadero metodo de estudiar para ser util a la Republica y a la Iglesia, proporcionando al estilo y necessidad de Portugal: Expuesto en varias cartas.* Madrid: Joachín Ibarra, 1760.

Vickers, Brian. "Introduction." In *The Man of Feeling*, edited by Brian Vickers, Stephen Bending, and Stephen Bygrave, vii–xxiv. Oxford University Press, 2001.

Vico, Giambattista. *Cinque libri di Giambattista Vico de' principj d' una scienza nuova d'intorno alla comune natura della Nazioni.* Naples: Felice Mosca, (1725) 1730.

Vico, Giambattista. *Principi di una scienza nuova.* Naples: Nella stamperia Muziana, 1744.

Vidal, Emeric. *Picturesque Illustrations of Buenos Ayres and Monte Video: Consisting of Twenty-four Views: Accompanied with Descriptions of the Scenery, and of the Costumes, Manners, &c. of the Inhabitants of Those Cities and Their Environs.* London: R. Ackermann, 1820.

Vidal, Fernando. *The Sciences of the Soul: The Early Modern Origins of Psychology.* Translated by Saskia Brown. University of Chicago Press, 2011.

Vieytes, Hipólito. "Motivos por que se hace dificultosa la subsistencia de este periódico." *Semanario de Agricultura, Industria y Comercio* 1, no. 37 (June 1, 1803): 289–96.

Vieytes, Juan Hipólito. "Prospecto." *Semanario de Agricultura, Industria y Comercio*, no. 1 (1802): i–viii.

Vila, Anne C. *Enlightenment and Pathology: Sensibility in the Literature and Medicine of Eighteenth-Century France.* Johns Hopkins University Press, 1998.

Vilain, Robert. "'An Excess of Savage Force?' Faust in French: Stapfer, Delacroix, and Goethe." *Princeton University Library Chronicle* 73, no. 3 (2012): 313–71.

Viñuales, Alvaro Perpere. "Felicidad pública y civilidad en el análisis de Juan Hipólito Vieytes sobre el desarrollo económico del Virreinato del río de la Plata." *Cultura Económica* 32, no. 87 (2018): 66–73.

Viola, Miguel Navarro. "La abolición de la esclavitud en Portugal: Mirada restrospectiva sobre el Río de la Plata." *La Revista de Buenos Aires* 18, no. 72 (1869): 554–606.

Voltaire. *Letters Concerning the English Nation.* London: C. Davis, 1733. French publication: Voltaire, *Lettres écrites de Londres sur les Anglais et autres sujets: Suivant la copie imprimée à Londres.* Amsterdam: J. Des Bordes, 1735.

Walker, Charles F. *The Tupac Amaru Rebellion.* Harvard University Press, 2014.

Weinberg, Félix. "Estudio preliminar." In Marcos Sastre, Juan B. Alberdi, Juan María Gutiérrez, and Esteban Echeverría, *El Salón Literario*, edited by Félix Weinberg, 9–101. Hachette, 1958.

Weiner, Dora B. "Mind and Body in the Clinic: Philippe Pinel, Alexander Crichton, Dominique Esquirol, and the Birth of Psychiatry." In *The Languages of Psyche: Mind and Body in Enlightenment Thought*, edited by George Sebastian Rousseau, 331–404. University of California Press, 1990.

Wellman, Kathleen Anne. *La Mettrie: Medicine, Philosophy, and Enlightenment.* Duke University Press, 1992

Whately, Richard. *Elements of Logic.* 2nd ed. London: J. Mawman, (1826) 1827.

Whitington, G. T. *The Falkland Islands.* London: Smith, Elder, & Co., 1840.

Whittembury, Guillermo, Klaus Jaffé, Cesia Hirshbein, and David Yudilevich. "Charles Darwin, Robert Fitzroy and Simón Rodríguez Met in Concepción, Chile, After the Earthquake of February 20, 1835." *Interciencia* 28, no. 9: 549–53.

Williams, Alberto. "Estética músical y conciertos sinfónicos." *La Biblioteca* 3 (February 1897): 260–64.

Williams, Tony. *The Pox and the Covenant: Mather, Franklin and the Epidemic That Changed America's Destiny.* Sourcebooks, 2010.

Williamson, Gillian. *British Masculinity in the "Gentleman's Magazine," 1731 to 1815.* Springer, 2016.

Wilson, Edward O., and José M. Gómez Durán. *Kingdom of Ants: José Celestino Mutis and the Dawn of Natural History in the New World.* Johns Hopkins University Press, 2010.

Wilson, Elizabeth. "The Invisible Flâneur." *New Left Review* 1, no. 191 (January–February 1992): 90–110.

Wilson, Elizabeth. *The Sphinx in the City: Urban Life, the Control of Disorder, and Women.* University of California Press, 1992.

Wright, T. R. *The Religion of Humanity: The Impact of Comtean Positivism on Victorian Britain.* Cambridge University Press, 1986.

Yates, Frances Amelia. *Rosicrucian Enlightenment.* Routledge, 2013.

Zaccaria, Francesco Antonio. *Saggio critico della corrente letteratura straniera . . .* Modena: Remondini, 1756.

Zamudio Varela, Graciela. "Los pintores de la Real Expedición Botánica a Nueva España (1787–1803)." In *Colecionismos, práticas de campo e representações*, edited by Maria Margaret Lopes and Alda Heizer, 29–41. Editora da Universidade Estadual da Paraíba, 2011.

Zarate, Gil de, and D. Antonio. *De la instrucción pública en España*. Vol. 3. Madrid: Imprenta del Colegio de Sordo-Mudos, 1855.

Zeballos, Estanislao Severo, ed. *Cancionero popular de la Revista de derecho, historia y letras*. Vol. 1. J. Peuser, 1905.

Zinny, Antonio. "Bibliografia periodistica de Buenos Aires, hasta la caida del gobierno de Rosas." *La Revista de Buenos Aires: Historia Americana, Literatura y Derecho* 4, no. 41 (September 1866): 449–80.

Zinny, Antonio, and Gregorio Funes. *Efemeridografia argirometropolitana hasta la caida del gobierno de Rosas*. Buenos Aires: Imprenta del Plata, 1869.

Zúñiga, Antonio R. *La Logia "Lautaro" y la independencia de América*. Est. gráfico J. Estrach, 1922.

INDEX

Note: Page numbers in *italics* indicate figures.